Cultural Heritage Science

The preservation and interpretation of our cultural heritage is one of the major challenges of today's society. Cultural Heritage Science is a highly interdisciplinary book series covering all aspects of conservation, analysis and interpretation of artworks, objects and materials from our collective cultural heritage. The series focuses on science and conservation in three main fields

- Art technology
- Active conservation and restoration
- Preventive conservation and risk management

The series addresses conservators and conservation scientists at museums, institutes, universities and heritage organizations. It also provides valuable information for curators and decision makers at museums and heritage organizations. Cultural Heritage Science comprises two subseries, one focusing on advanced methods and technology for conservation experts, the second presenting the latest developments in conservation science. All titles in the book series will be peer reviewed. Titles will be published as printed books and as eBooks, opening up the opportunity to include electronic supplementary material (videos, high-resolution figures, special data formats, and access to databases).

More information about this series at http://www.springer.com/series/13104

Shigeo Aoki · Yoko Taniguchi · Stephen Rickerby ·
Michiyo Mori · Takayasu Kijima · Su Bomin ·
Fumiyoshi Kirino
Editors

Conservation and Painting Techniques of Wall Paintings on the Ancient Silk Road

Springer

Editors
Shigeo Aoki
Tokyo University of the Arts
Tokyo, Japan

Yoko Taniguchi
University of Tsukuba
Tsukuba, Ibaraki, Japan

Stephen Rickerby
Private Wall Painting Conservator
Herefordshire, UK

Michiyo Mori
Tokyo University of the Arts
Tokyo, Japan

Takayasu Kijima
Tokyo University of the Arts
Tokyo, Japan

Su Bomin
Dunhuang Academy
Dunhuang, Gansu, China

Fumiyoshi Kirino
Tokyo University of the Arts
Tokyo, Japan

ISSN 2366-6226 ISSN 2366-6234 (electronic)
Cultural Heritage Science
ISBN 978-981-33-4160-9 ISBN 978-981-33-4161-6 (eBook)
https://doi.org/10.1007/978-981-33-4161-6

Cover illustration: Frying deity, western side of West Giant Buddha niche, Bamiyan, Afghanistan(1970s). It was demolished during internal conflict. Photo courtesy of the Committee of the Kyoto University Scientific Mission to Central Asia (Kyoto University, Institute for Research in Humanities)

This Springer imprint is published by the registered company Springer Nature Singapore Pte Ltd.
The registered company address is: 152 Beach Road, #21-01/04 Gateway East, Singapore 189721, Singapore

Preface

In March 2019, the Public Collaboration Center at the Tokyo University of the Arts organized 'The International Colloquium on the Conservation of Asian Wall Paintings and the Exchange of Painting Technologies' and a subsequent symposium, both commissioned by Japan's Agency for Cultural Affairs. Motivated by concern for the study and conservation of wall paintings along the Silk Road, experts from across Eurasia representing the fields of art history, archaeology, conservation, science and cultural heritage attended both events. Participants presented the latest findings from recent conservation projects and ongoing research. The results published in this book provide us with an opportunity to consider the variety and diversity of culture and religion that flourished on the Silk Road over its long history and vast geographical expanse. We would like to thank the Agency of Cultural Affairs, the Japanese Government and the Tokyo University of the Arts for facilitating this.

The Silk Road derives its name from the lucrative trade in Chinese silk, starting in around the second century BC. Over the next millennia, the network of trade routes that developed between the East and the West, over both land and sea, connected East Asia with Southeast Asia, South Asia, Central Asia, Persia the Arabian Peninsula and Southern Europe. This encouraged unprecedented economic, cultural and religious exchange among the civilizations of the known world.

The contributions in this volume open a window on this world through the rich medium of wall paintings. As discussed by Taniguchi and Mori in their chapter, wall paintings are an 'immoveable heritage' that embody much more than the sum of their parts. Through their creation, they express artistic and technical ingenuity, the exchange across cultures not just of materials but also of ideas. The human desire to decorate the places we inhabit, in both life and death, has left glimpses of how people like us lived hundreds of years ago. Wall paintings have been used by rulers to impose their authority, by the wealthy to display their opulence, and by the less wealthy to demonstrate their aspirations. Cults and religions have developed imagery for decorating the walls of tombs, churches and temples to convey spiritual as well as temporal preoccupations. Most notably, wall paintings were integral to the development and spread of Buddhism along the Silk Road, as attested by the Buddhist cave sites of Dunhuang and Kizil in China, Ajanta in India, and Bamiyan, Foladi and Mes Aynak in Afghanistan, all sites that are considered by contributors here.

In recent years, we have been able to cull a wider range of information from original painting materials owing to advances in the accuracy of non-invasive analytical and imaging techniques, and better readings of data from minute samples, particularly of organic materials, which have in the past proved difficult to detect and identify. This book testifies to collaborative research projects and information exchanges between specialists working with wall paintings in Turkey, India, Afghanistan, Uzbekistan, Tajikistan and China, among other locations. With the expansion of these studies, important findings have been made, encouraging discussion of the relationships between painting technologies in the Eurasian continent. Perhaps the most exciting discovery, reported by Taniguchi, is of drying oils used as binding media in the mid-seventh century Buddhist cave paintings at Bamiyan, some 400 years before their first known use in Europe. As several authors acknowledge, the transmission of materials and techniques along the Silk Road was evidently so vibrant and fluid that it is difficult to pinpoint with certainty where innovations occurred. Based on analysis of wall paintings at Kizil dating from the fifth–seventh century AD, for example, Zhibo Zhou, Ling Shen and Hui Zhang suggest that the technique of applying glazes to tin in imitation of gilding may have originated in Central Asia before spreading westward to sites such as Bamiyan; but they also recognize that technological transfer was multidirectional, with the possibility that certain organic colorants that were not native to Kizil may have been sourced from as far away as South Asia or Persia.

Inevitably some wall paintings attract greater analytical attention than others. Su Bomin describes investigative advances that have been made in studying the Buddhist paintings at Mogao, China, by the Dunhuang Academy over a period of almost 40 years. Despite this and the commitment of future resources, the task of studying a site that preserves 492 painted caves spanning a 1000-year history is overwhelming. The sheer number of paintings, the complexity of their original materials and techniques, and the difficulty of comprehending alterations that have occurred with their aging and exposure, mean that technical knowledge will remain partial and imperfect. Similar obstacles are discussed by other contributors. In Shimadzu's review of the original technology of the Ajanta wall paintings, it emerges that past and current investigations have only scratched the surface of understanding: the presence and use of organic components continue to challenge analytical detection, for instance. Effects of physical history at painted sites and monuments further complicate the task of unraveling original painting techniques: Blaensdorf's account of the polychromy of the Giant Buddha statues at Bamiyan demonstrates how difficult it is to be certain about original painting intentions when historic renewal and modern destruction have also occurred.

Although the dissemination of ideas along ancient trade routes undoubtedly led to the sharing of painting materials and techniques within certain regions and even beyond them, it is also true that every wall painting is different. Barbu's study of azurite and smalt in the medieval churches of Romania illustrates a distinctly regional approach to the use of these blue pigments, but also highlights significant differences in the way they were applied in individual churches. The comprehensive survey of the rock-cut churches of Cappadocia carried out over a 13-year period by Andaloro and Pogliani similarly illustrates aspects of continuity and diversity in wall painting techniques in another Byzantine context. Hidaka's account of the technical characteristics of painted churches in the Balkans reveals yet more variants in this interconnected region at more or less the same time. Such differences should not be a surprise. Varying circumstances of patronage and expenditure, disparities in levels of expertise and availability of materials, constraints and opportunities imposed in different contexts, as well as local resistance to or acceptance of innovation, are only a few of the many factors that influence wall painting outcomes. Nevertheless, we all too often make assumptions that paintings of a certain type, such as those belonging to the Byzantine tradition, are all alike.

We do well to remember historic and technical idiosyncrasies when considering our approaches to preserving wall paintings. An example of the risks of basing treatment decisions on assumptions instead of collected data is found in Rickerby's chapter on the conservation of a Nabataean wall painting in Petra, Jordan. Here, analysis showed that sophisticated materials and techniques were used in the painting's creation, including organic colorants and media, and extensive gold leaf. Previously its technique was considered to be *fresco*, and cleaning proposals were formulated on this basis: had they been implemented, the damage to this unique painting would have been irrevocable. All the contributions to this book are evidence that analytical procedures are now incorporated into conservation projects as perhaps never before. While this is a welcome development, it is also true that much analysis is carried out without clear purpose. As Hidaka reminds us, it is essential that scientific information on painting materials and techniques 'not only improves technical and art-historical understanding, but also constitutes an effective base for future conservation efforts'. Effective conservation has to be the main outcome of technical inquiry.

But what do we mean by effective conservation, and how achievable is it? Rickerby opens this book with a critique of approaches to wall painting conservation that have developed since the second half of the twentieth century. Most current practice now prioritizes preserving wall paintings *in situ*. This is easier said than done, of course. Contributors relate a familiar range of natural and human threats that continually challenge and undermine our technocratic efforts at monumental preservation. Hayakawa describes the constant threat of natural disasters that overshadow the conservation of Japan's ancient tumuli. For the rock-cut churches of Cappadocia, environmental exposure and inherent geological vulnerability pose perennial risks. As Ertugrul Gulyaz observes, the region's rock-cut sites are widely distributed over an area of approximately 25,000 km^2, and currently there are not enough specialists to carry our adequate conservation measures. Similar circumstances of scattered or remote distribution, environmental deterioration and poor protection are emphasized or referred to in the chapters on Uzebekistan, Bamiyan and Petra (Aripdjanov; Taniguchi; Blaensdorf; Rickerby). Even for sites that benefit from the dedicated protection of national or regional agencies—such as at the Mogao and Kizil grottoes—conservation is a limitless challenge (Su Bomin; Zhibo Zhou, Ling Shen and Hui Zhang).

Alongside natural hazards, human actions also cause immense harm. The destruction of the Great Buddha statues in Bamiyan by the Taliban in 2001 is an especially egregious example. Kijima's chapter on the conservation of wall painting fragments detached from Bamiyan, Foladi and Mes Aynak for illicit sale on international art markets further highlights the wider depredations that befell Afghanistan's Buddhist heritage following this event. The research of Blaensdorf on the original technology of the statues, and of Taniguchi on their associated paintings, is a poignant reminder that precious little material evidence is left to decipher at Bamiyan. While acts of cultural violence provoke understandable shock, they also raise difficult questions in relation to other human actions: a sudden instance of religious iconoclasm at a single site grabs our attention, but how does this weigh against the cumulative harm that mass tourism brings to many painted sites, even as much modern conservation practice is also tied to tourism agendas? Ertugrul Gulyaz's simple but explicit statement that 'mass tourism is not appropriate for the nature of Cappadocia's rock-cut churches' could equally be said of other sites described here, including Mogao, Petra and Ajanta. Threats from tourism also converge with those from climate change. At Mogao, where salt-related deterioration is a key concern, environmental conditions inside the painted caves are highly sensitive to combined events of rainfall and tourist traffic, both of which are increasing.

Confronted with these proliferating circumstances of deterioration, destruction and loss, it is not surprising that approaches to conserving wall paintings are divergent. A number of responses are covered in this book. Aripdjanov describes continuing reliance on detachment of archaeological wall paintings in Uzebekistan, a legacy of Soviet-era interventions, but he closes with a plea for 'the strengthening of site management and conservation practices, to ensure the promotion and transmission of this cultural heritage to the next generation'. Japan's history of conservation has also grappled with issues of site preservation at the painted tumuli of Takamatsuzuka and Kitora. Both Hayakawa and Udagawa relate the lengthy efforts to conserve these paintings *in situ*—in the case of the Takamatsuzuka tumulus, for a period of 40 years after their discovery—and the eventual decisions to transfer them to museum environments, while still holding open the option that they might one day be returned to their original locations. In contrast, models of *in situ* conservation are discussed for wall paintings sites at Cappadocia and Petra (Andaloro and Pogliani; Ertugrul Gulyaz; Rickerby).

Despite their differences of approach, a number of shared conclusions can be drawn from these examples. Firstly, whether attempts are made to preserve wall paintings *in situ* or *ex situ*, conservation is not a finite undertaking. The paintings from the Takamatsuzuka tumulus have been under intense conservation scrutiny for almost 50 years (Hayakawa; Udagawa). Lee Hwa Soo, Song You Na, Han Gyu Seong and Han Kyeong Soon describe the range of new scientific investigations required before retreating the Buddhist wall paintings in the Josadang Shrine of the Buseoka Temple in Korea, which were first removed from their original built context in 1916. The experiences of preserving wall paintings *in situ* attest to similar long-term efforts that have involved returning to the same site repeatedly. None of our endeavors are guaranteed certainty. This raises other issues relating to the resources that are expended on the paintings of a single site when our concerns for painted heritage are so much larger.

There is much then to question about the future direction of conservation practice, a theme that is explored in Rickerby's opening chapter, with a warning that many of our current approaches demand radical rethinking. This was true when the International Colloquium was held in March 2019, but is more urgent now. The COVID-19 pandemic has since changed our world. The impacts of this new reality on conservation can only be guessed at, as the story is still unfolding. But some repercussions are becoming clear. For now, social distancing measures and limitations on travel have curtailed or stopped the types of research and conservation projects that are reported in this book; even laboratory-based analysis cannot be easily done. Even if these restrictions prove to be temporary, they may last long enough to alter how wall paintings are studied and conserved in the future. Meanwhile, tourism has plunged around the world, and religious activities in painted churches and temples are restricted. Consequences are emerging that pull in different directions: absent tourism will mean lost incomes for conservation activities (and for local communities associated with cultural sites), but will also reduce many of the risks to wall paintings described by a number of contributors. Stark choices lie ahead, but also new opportunities.

It has quickly become a truism that our experiences in relation to Covid-19 are unprecedented. Yet the painted sites described in this book tells us otherwise. The development of wall paintings along the Silk Road testifies to human creativity when infectious diseases, most notoriously the Plague (the 'Black Death'), were also transmitted along the same trade routes. Temples and churches were places of refuge and sanctity from such afflictions, reflected in wall paintings dedicated to plague saints in the Christian tradition and to those depicting Bhaiṣajyaguruvaiḍūryaprabha in Buddhism. That ancient paintings have endured the vicissitudes of their times and remain with us today is a source of hope in our current period of uncertainty. We may come to different realizations of their cultural identity and meaning. We may recognize anew the indispensible values they represent in their fragile physical form, and how easily these can be lost. Perceptions of risk may change and we may develop more nuanced responses to their preservation. Doing less may, some day, make more sense.

Yoko Taniguchi
Conservation Science and Archaeological Science, Faculty of Humanities and Social Sciences, University of Tsukuba
Tsukuba, Japan

Stephen Rickerby
Consultant
Conservation of Wall Painting Department
Courtauld Institute of Art
University of London, London, UK

Contents

1 **Rethinking Our Models of Wall Painting Conservation: Are We Doing the Right Thing?** ... 1
Stephen Rickerby

2 **On Painting Materials, Techniques, and the Conservation of Wall Paintings Across Eurasia** ... 17
Yoko Taniguchi and Michiyo Mori

3 **Deterioration and Conservation of Cappadocian Wall Paintings** ... 23
Murat Ertugrul Gülyaz

4 **Materials and Techniques of Cappadocian Wall Paintings** ... 43
Maria Andaloro and Paola Pogliani

5 **At the Western Edge of the Silk Road: Challenges of Conserving a Unique Nabataean Wall Painting in Petra, Jordan** ... 89
Stephen Rickerby

6 **pXRF and FTIR Spectrometry Applied to the Study of Azurite and Smalt in Romanian Medieval Wall Painting** ... 105
Olimpia Hinamatsuri Barbu

7 **Technical Characteristics of Church Wall Paintings in the Balkans** ... 119
Midori Hidaka

8 **Painting Materials and Techniques of the Ajanta Wall Paintings** ... 137
Yoshiko Shimadzu

9 **Materials and Techniques of the Polychromy of the Giant Buddha Statues in Bāmiyān** ... 157
Catharina Blaensdorf

10 Materials and Technologies of the Bamiyan Wall Paintings 177
Yoko Taniguchi

11 Conservation of Detached Wall Paintings from the Bamiyan, Foladi and Mes Aynak Sites 197
Takayasu Kijima

12 Deterioration and Conservation of Wall Paintings in Uzbekistan 223
Otabek Aripdjanov

13 The Wall Painting Techniques and Materials of Kizil Grottoes 235
Zhibo Zhou, Ling Shen, and Hui Zhang

14 Wall Painting Materials and Techniques of the Mogao Grottoes ... 253
Su Bomin

15 Analysis and Diagnosis of the Buddhist Wall Paintings in the Josadang Shrine, Buseoksa Temple, Korea 265
Hwa Soo Lee, You Na Song, Gyu Seong Han, and Kyeong Soon Han

16 Conservation of Wall Paintings on Plaster in the Tumuli of Japan 287
Noriko Hayakawa

17 The Protection and Utilization of Wall Paintings of Ancient Tumuli in Japan 305
Shigemasa Udagawa

Chapter 1
Rethinking Our Models of Wall Painting Conservation: Are We Doing the Right Thing?

Stephen Rickerby

Abstract Modern principles of conservation prioritise preservation of significance and authenticity. For wall paintings, this means stabilising them in situ in their present condition and focusing on measures that slow deterioration. Understanding the condition and deterioration of wall paintings is both complex and challenging; their conservation treatment is equally difficult. Since resources are always limited, and the needs of our painted cultural heritage are limitless, science-based decision making maximises conservation efforts. However, these approaches are undermined in practice by short-term and reactive agendas, the distribution of conservation across diverse sectors and personnel, and the growth of a conservation industry that is dominated by other considerations. Consequently, many current models of practice are failing to address key concerns. A radical rethinking of priorities is urgently required. Proactive models of conservation that broaden concepts of context and reemphasise its importance offer ways to address our problems before they reach crisis levels.

Keywords Wall paintings · Reactive and proactive conservation · Principles · Significance · Context

1.1 Introduction

It was with the dramatic words 'Conservation has changed' that Sharon Cather began her article on the state of wall painting conservation in 2006, which she presented at the colloquium and symposium on the theme of 'Mural Paintings of the Silk Road: Cultural Exchanges between East and West', organised by the National Research Institute for Cultural Properties (NRICP), Tokyo, in that year [1, p. 173]. Describing advances in conservation knowledge and understanding that had occurred since the middle of the twentieth century, Cather highlighted the complex and challenging nature of modern conservation practice, and defined ways of addressing

S. Rickerby (✉)
Consultant, Conservation of Wall Painting Department, Courtauld Institute of Art, University of London, Vernon Square, Penton Rise, King's Cross, London WC1X9EW, UK
e-mail: stephen.rickerby1@btinternet.com

S. Aoki et al. (eds.), *Conservation and Painting Techniques of Wall Paintings on the Ancient Silk Road*, Cultural Heritage Science,
https://doi.org/10.1007/978-981-33-4161-6_1

these demands through rigorous application of the scientific method and a hierarchical approach of effectiveness in choosing appropriate conservation measures. This model of conservation provides the gold standard against which our efforts in the preservation of immovable painted heritage are still assessed.

More than a decade later, however, it is time to revisit the state of wall painting conservation and ask how circumstances have changed in the intervening period. Threats such as natural disasters and human conflicts are not new. Others, such as urbanisation, competing economic demands and impacts of mass tourism are also familiar, but have increased alarmingly. Newer threats—such as effects of climate change—are presenting additional problems, the full consequences of which remain to be seen. The sense of urgency is heightened, and wall paintings and our efforts to preserve them are facing a crisis.

1.2 Conservation Defined

The aims and objectives of site conservation as defined by existing charters and guidelines are specific and incontrovertible, and can be summarised succinctly.[1]

They are to preserve the site and its historic features in their existing condition and to slow deterioration. New interventions should be minimal and only carried out when absolutely necessary, namely when there is established risk of serious damage or loss. New materials and techniques are to be selected on the basis of recognised conservation requirements, and should be appropriately evaluated and tested, to be certain that they do not contribute to future damage and deterioration. Respect for the varied and numerous significance values of the site and all of its features must guide conservation decisions.

How do we translate these principles into the practice of conserving wall paintings? The empirical means of knowledge acquisition known as the scientific method underpins pure science, and science-based disciplines ranging from medicine to nature conservation. Wall painting conservation, although a relatively new discipline, is also science-based and follows the same model. Cather has described how the nature of wall paintings—their physical dependency on supporting structures; their exposure to the surrounding environment as part of an open system; their vast number, large size and discontinuous state; their wide dispersal, often in remote or inaccessible places; and their varying ages and physical histories—influence the condition and deterioration of their diverse original (and added) materials (Figs. 1.1, 1.2 and 1.3) [1, 2]. We therefore approach this complicated and dynamic set of problems as a diagnostic undertaking.

In the fields of pure science, steps in the scientific method broadly proceed as follows:

- make observations of phenomena;
- ask questions about the phenomena and begin information gathering;
- form a hypothesis about the status of the observed phenomena;

Fig. 1.1 Amheida, Dakhleh Oasis, Egypt, excavated Roman house, first century AD. The room shown here was excavated to reveal its decorative wall paintings for the first time in centuries. The exposure of the original materials (including the mud-brick support) to new environmental conditions led to sudden deterioration, which was addressed by back-filling the site. This example demonstrates how wall paintings are a vulnerable physical interface for environmental deterioration (© *Stephen Rickerby*)

Fig. 1.2 Ateni Monastery, Georgia, eleventh century wall paintings. The scheme here is vast and survives in a discontinuous state. Consequently, it is extremely difficult to make sense of the diverse conditions that are present and to diagnose causes of deterioration. These circumstances are typical of many wall paintings (© *Stephen Rickerby*)

Fig. 1.3 Tomb of Nebattauy (QV60), Valley of the Queens, Luxor, Egypt, 19th Dynasty. Periodic flash floods have had a devastating effect on the expansive clay-rich limestone geology of the tombs in the Valley of the Queens, leading to catastrophic rock collapses and loss of painting, as shown here. Inherent geological issues at wall painting sites such as this often present problems that cannot be eradicated (© *Stephen Rickerby*)

- test the hypothesis scientifically;
- analyse the data, and accept or reject the hypothesis;
- establish a theory based on sufficiently reproducible outcomes.

When acquired information (both empirical and measureable) leads to the rejection of a hypothesis, the model demands that we step back and repeat steps at any point during the investigation (the 'iterative process'). The rationale is to avoid biased interpretation of observations and results. Another basic expectation is that data and conclusions can be objectively scrutinised.

For diagnosis of wall paintings, the same model is used for the same reasons. Painting conditions are the phenomenological starting point of our observations. Given the many complex and imperfectly understood ways in which wall paintings change, and the difficulties involved in even recognising and naming their conditions, let alone in diagnosing deterioration, a scrupulous iterative process is essential (Fig. 1.4). This is driven by other imperatives, too. Identifying objective conservation priorities is important because our professional resources are finite, while the demands that wall paintings place on us are growing. This is reflected in the painted sites from along the Silk Road represented at this colloquium, and the extraordinary conservation challenges they face, which indicate only a small fraction of the greater problem.

In discussing resources, we must also recognise the finite nature of wall paintings: if mistakes are made in their treatment, we risk causing permanent damage

Fig. 1.4 Mogao Grottoes, Dunhuang, China: Cave 260, Northern Wei, c. 525. In this small area of the scheme, the diverse materials of the painting have undergone a wide range of deterioration processes, reflecting aspects of their original technology, the passage of time and circumstances of physical history. Diagnosing these complex changes requires a rigorous scientific approach (© *Courtauld Institute of Art*)

or irreversible loss. Modern conservation practice continues to fail in these areas. Wall paintings undergo cycles of treatment and retreatment, which ignore underlying processes of deterioration. Wall paintings very quickly reach states of complete condition breakdown and treatment deadlock (Fig. 1.5). Original materials, contaminants (such as salts), and added treatment materials become inextricably combined. The material nature of the wall painting is irretrievably compromised and weakened. Cycles of treatment failure of approximately 50–100 years' duration, or even less, can be enough to render the condition of a wall painting effectively untreatable. In the timescales of many wall paintings, which can be hundreds if not thousands of years'

Fig. 1.5 Church of St Mary the Virgin, Cropredy, England, fifteenth century wall painting. This medieval wall painting was coated with an organic varnish, probably in the late-nineteenth or early-twentieth century, in an effort to preserve it. This has badly discoloured and is now promoting severe paint flaking and loss, as this detail shows. Wall paintings very quickly reach states of complete condition breakdown as a result of poor remedial treatment choices (© *Stephen Rickerby*)

old, this is not a long time. This emphasises our responsibility to define, choose and implement the *right* conservation measures.

Assuming that we have correctly diagnosed our wall painting problems—and this is a big assumption—how then do we implement appropriate conservation measures? Cather articulated a hierarchical approach based on order of effectiveness, which can be reiterated here [2, p. 64, 1, pp. 177–178]:

- preventive: measures that address *causes* of detrimental change;

- passive: measures that attempt to control or limit the factors that *activate* the problem;
- remedial: physical intervention.

Some examples can be used to illustrate these measures. Damage to wall paintings from direct effects of tourism, such as site attrition and vandalism, can be *prevented* by installing barriers, rerouting visitors and controlling their numbers. Deterioration such as paint flaking and loss may be *activated* by exchanges of moisture between interior and exterior environments. In this case, we cannot alter the nature of the painting, but we can intervene *passively* by trying to stabilise environmental conditions. Lastly, we can directly intervene on the paintings themselves, resorting to a range of *remedial* treatments.

Our objective is to achieve maximum *effective* impact, without causing new harm to paintings. To do this, we base conservation decisions on a scientific assessment of constraints and opportunities, in which the linkage between approaches and principles, resources and ethics, then extends into practice. To accomplish greatest success with the limited resources available for conservation, we need to identify the highest conservation return, using a systematic and objective approach in advance of significant expenditure of resources.

1.3 Conservation Undermined

There is broad agreement that this is what we should be doing, but less unity in our responses to conserving wall paintings. There are many reasons for this, which are related to the nature of the conservation options themselves, and to the difficulties and biases we have in implementing them. Even though preventive options are our best choice, they usually address only single-issue problems (e.g., direct damage from tourism), whereas damage and deterioration typically affect wall painting sites in complex combinations. The obstacles to achieving passive measures successfully are numerous. Wall paintings and their problems need first to be thoroughly understood, demanding access to multiple skills and other resources, including time. Patience is required on the part of conservation managers and administrators, who often feel impelled towards more immediate and demonstrable forms of intervention. At many sites, passive measures may be very difficult or impossible to achieve. Where they are possible, unintended consequences can occur: some deterioration may be controlled, but different deterioration may also be created.

For these and other reasons, remedial treatment remains the prevalent option in conservation practice. Despite the evidence of failed treatment cycles, faith in the permanency of material interventions is undiminished. Remedial treatments offer ‘quick fixes’ that appeal to preferences for immediate action. Although *good* conservation is both less expensive and more effective when it is guided by accurate and reliable information acquired through long-term planning, few are willing to make this investment. Instead, we make reactionary decisions and interventions, which give the impression of solving the problem with fewer resources. This inclination is partly related to the organisation and economics of conservation in much of the world, whereby remedial treatment is relegated to a menial activity done by unskilled and lower paid workers. But even on well-resourced conservation projects, remedial treatment is usually the principal component. Achieving conspicuous results is a powerful motivating factor. This specifically translates into interventions that renew and complete. Cleaning and reintegration are therefore favoured. But other interventions that address the *discontinuities* in wall paintings, making them appear whole again, or at least substantially visually improved, also fall into this category: examples include fixing and consolidating flaking and powdering paintings. Despite our stated aims to preserve wall paintings in the aged condition in which we find them, these tendencies to improvement testify to other human biases.

1.4 Conservation Repackaged and Repurposed

Among the many ways in which conservation has changed in the last few decades is in its organisation and implementation, and in the structures and bodies that generate conservation endeavours. The latter have proliferated and diversified vastly, taking the discipline in unpredicted directions. These developments have embedded many of the biases mentioned above.

The classic model of the conservation project emerged in the 1980s. The major wall painting conservation programmes that stand out in the last two decades of the twentieth century include: Leonardo's *Last Supper* in Milan [3] the Sistine Chapel in Rome [4] the Brancacci Chapel in Florence [5] Piero della Francesca's frescoes in Arezzo [6] the baroque church of Die Wies in Germany [7] and the New Kingdom Tomb of Nefertari in Egypt [8]. As projects of international importance with exceptional levels of funding, they benefited from planning, inclusive investigations and gradual implementation. Their pioneering approaches to diagnosis, scientific analysis and environmental studies have since become expected components of other conservation endeavours. The level of professional implementation that distinguished these now historic projects also raised the status of the conservation scientist and practicing conservator.

The concept of the flagship project has continued, and has expanded to include regional clusters of sites. But prestigious single- and multi-site projects have become alarmingly expensive and unwieldy, and although data are difficult to pin down, these trends do not appear to be abating. The ever-increasing availability of new technical resources feeds this escalation, as the minimum requirements that are considered to define projects expand exponentially. We incorporate many more types of expertise and project partners than ever before, and this involves greater organisation and costs more. The increased financial, technical and administrative volume is predominantly inwardly directed and circular, benefiting the agencies, institutions and experts who are already in the professional loop. On mega-projects, conservation has become a self-generating enterprise.

The idea of the 'model' project is still promoted, whose outcomes are intended to have wider benefits. Prestigious projects may fulfil this function in ways that are not just dependent on their access to better resources, for example by demonstrating effective diagnostic approaches and ethical practice. But in the wider context, the over-resourced single site, or even the multi-site project, cannot be a realistic model for the majority that are under-resourced or without resources. These discrepancies often mean that influences, if they do occur, are misunderstood and misapplied in their broader application. There is a high risk of unregulated detrimental outcomes. The tendency of projects to include a component of local training—which is usually of a short-term nature—is a vehicle that encourages these unwanted consequences.

This project-type is not the only model. Conservation has devolved to many structures and organisations at supranational, international, national and regional levels. These are divided—and often contested—between religious institutions, public and private bodies (or partnerships of both), governmental and non-governmental agencies, and commercial and non-profit organisations. Insufficient or overabundant funding, and varying levels of conservation oversight and expertise—including none at all—influence outcomes. Different agendas are exerted on conservation by these different arrangements. In tourist-dependent economies, agencies tasked with the care of cultural heritage are often combined with those of tourism, which drives conspicuous interventions such as cleaning and restoration of wall paintings above other measures. At many Buddhist sites in Asia, conservation has become a side activity of Western-sponsored charities. Well-meant but misguided projects of this type have caused irreparable damage to important wall paintings. At major archaeological sites, conservation is often repurposed in the service of large-scale reconstruction and restoration; that this often boosts local employment is considered an added incentive. On large architectural development projects—such as the adaptive reuse of historic buildings—the conservation of wall paintings preserved in them is often subsumed within a predetermined design-led agenda. Heritage management companies that oversee wall paintings as one item of a large site portfolio may be motivated by profit-led considerations.

These are just some of the many ways in which wall painting conservation has changed since the advent of the classic model project in the 1980s. Although conservation has long been organised and defined in various ways, a crucial difference now is that major endeavours are no longer the main preserve of a few dedicated professional bodies and institutions. In the dispersal of conservation across many sectors and personnel, each with different objectives, direct knowledge of the discipline in no longer a fundamental requirement. One consequence of these developments is that key components of conservation planning become separated from their core diagnostic purposes, and are carried out mainly to fulfil organisational expectations. Conservation steps such as condition mapping, technical investigations and environmental studies are rolled out routinely, stipulated and implemented without clear understanding of their need, function and sequence. The growing number of experts now involved in these activities exacerbates fragmentation of what is meant to be an interdependent process. This wastes resources and undermines the rationale of the iterative approach. The concepts and principles pioneered by a developing profession three decades ago have been uncritically adopted and are only paid lip service by the wider community now involved in conservation.

1.5 Current State

Where have all these changes and developments brought us? Many of our approaches and models are on the verge of being unjustifiable, owing to the sheer amount of resources they both require and absorb; their disengagement from defining precepts; their implementation by a wider group of professionals (and non-professionals) with little or no direct knowledge of conservation; and their questionable wider impacts. The case is emphasised by the growing number of cultural sites and places that are at risk globally, and the changing and urgent threats they face. Broadening definitions of heritage, from historic urban environments to cultural landscapes, stretch our concerns endlessly. The regular risk lists compiled by organisations such as UNESCO, ICOMOS and the World Monuments Fund can only highlight the random few.[2] Most endangered heritage does not make it on to any list, and sites with wall paintings are only a proportion of this increasing patrimony. The competition for resources is tightly rationed: wall paintings in need, even highly deserving ones, cannot always be the recipient of qualified attention, which has to be shared with many other types of cultural heritage. Meanwhile, the profusion of threatened wall paintings at the margins of and beyond protective oversight has encouraged a rise in unregulated conservation endeavours.

Even for wall paintings that benefit from effective conservation, their needs do not end. Conservation is a process of constant adjustment to changing events and circumstances. Almost 30 years after the completion of conservation in the Tomb of Nefertari in Egypt, changes in both climate and visitor patterns mean that its current and future care requires continual calibration of environmental controls. A 9-year project focused on the conservation and management of the Tomb of Tutankhamen recently concluded by the Getty Conservation Institute and the Egyptian authorities has successfully stabilised the tomb's interior environment against a current range of identified threats, but it is recognised that these will change over time, requiring further adjustment [9]. One of our greatest challenges is the sheer number of wall paintings in need, and the fact that for each and every painted site, requirements will continue to shift and expand, often unpredictably, irrespective of the conservation attention they may or may not have received.

Among the many factors changing wall painting conservation, one stands out. The alliance of global development and tourism is driving and shaping models of conservation to an unprecedented extent. Latest figures indicate that in 2018 there were 1.4 billion international tourist arrivals worldwide, a 56-fold increase since 1950.[3] Global tourism is booming and shows no signs of reversing any time soon. Promoting more tourism is seen as a sure means of underpinning development. Conservation of key cultural sites, including those with wall paintings, is locked into major infrastructure projects, under the title of 'sustainable cultural tourism'.

We should be under no illusions about development projects that use 'cultural assets' as their driver, and the premise that large-scale tourism is predictable and controllable. The model of constant growth that usually underlies these initiatives is ultimately incompatible with the finite nature of painted cultural heritage. At sites where tourism is a dominant feature already, the evidence is manifest. Some 25 years ago, the rise of domestic tourism in China was cautiously seen as a positive development that could be harnessed to conservation's benefit [10]. Today, sites such as the Mogao Grottoes, at Dunhuang, are overwhelmed by tourism, and more frequent rain events at times of high tourism compound threats to its unique Buddhist cave paintings (Su Bomin, this volume). With further robust growth in tourism expected, it is short-sighted to promote development models that could push the same problems onto more 'cultural destinations'.

On the other hand, it is also unwise to imagine that tourism is a constant, given economic uncertainties and geopolitical instability (Fig. 1.6). Also, not all projects that select cultural sites for their development focus can be successful: there is not an automatic link between tourism and economic growth and poverty reduction. These contradictions expose the shaky foundations of our confidence in such models. But for now, tourism is a major determinant of conservation outcomes. Wall painting conservation projects incorporate more components related to this than ever before, including tourism infrastructure, environmental mitigation measures and lighting initiatives. While some of these activities may be seen as preventive and passive measures, in the broader context they do not tackle the actual *cause* of our problems, which is tourism itself.

Fig. 1.6 Tomb of Tutankhamen, Valley of the Kings, Egypt, 18th Dynasty. The pressures and expectations associated with mass tourism are a major determinant of conservation outcomes at important cultural heritage sites. But tourism is not a constant. In 2009, a year before Egypt's 'Arab Spring', tourism to Tutankhamen's tomb in the Valley of the Kings was at record levels (left image). But between 2010 and 2017, this dropped dramatically and only recovered very slowly (right image) (© *Stephen Rickerby*)

An overriding factor characterises the state we are in. Wall painting conservation is dominated by reactive rather than proactive choices. We react to an immediate risk, or one in the near future, such as a site that is threatened by an imminent development (e.g., the building of a road or airport, the construction of a dam). Worse still, we react when a disaster has already occurred, despite warning indicators. We react to a chance opportunity (e.g., the provision of funding for a specific site, stemming from political, religious or personal motives). We react to imperatives that are imposed upon us, which may not be directly associated with conservation requirements (e.g., the treatment of a painted site to enhance its tourism potential). We react to wider political and economic developments, such as the rush to invest in emerging economies (where business goes, conservation follows). None of these approaches appropriately addresses our key conservation concerns.

1.6 Rethinking Conservation

To summarise, modern principles of conservation make clear distinctions and preferences for stabilising wall paintings in situ in their present condition; they also privilege preservation of significance and authenticity over other considerations. This obliges us to understand wall paintings in their context. Recognising the constraints and difficulties of doing this, we should approach our problems scientifically, and by prioritising options that achieve long-term effectiveness. When human and financial resources are scarce, and the needs of our painted cultural heritage are limitless, science-based decision making makes all the difference in maximising the success of conservation efforts. However, this approach is undermined in practice by competing and short-term agendas, by the distribution of conservation across diverse sectors and personnel, and by the growth of a conservation industry that is dominated by other considerations. Lastly, biases towards renewal skew conservation endeavours.

But there are indications of shifting attitudes and values. Although recent reports of cultural sites facing a crisis of 'overtourism' (or even 'peak tourism') and 'tourism-phobia' are overstated—and are only the latest spin on problems that have been present for decades—they hint at both a new mood and different expectations.[4] Communities in popular tourism destinations have taken to the streets to protest at the absence of 'trickle-down' benefits, and assert that tourism is now causing more harm than good. At the same time, tourists report disappointing experiences at over-crowded and commercialised sites, and seek out alternative destinations for a better appreciation of cultural authenticity. An example of this is the growing popularity of the rock-cut churches of Tigray, Ethiopia, where visitors are independently motivated to see wall paintings in their remote spiritual context, and appreciate them in their existing un-restored states (Fig. 1.7). On the other side of this equation, local sensitivities relating to religious use incorporate a protective responsibility for the inherited condition of wall paintings: wear and tear accumulated through acts of faith and worship have importance. In combination, these circumstances are shaping

Fig. 1.7 Maryam Bahera, Tigray, Ethiopia, fifteenth century. The rock-cut painted churches of Tigray, Ethiopia, survive in remote mountainous locations, offering intrepid visitors an experience of an ancient spiritual world. Tourists want to see the wall paintings in an authentic state, while local religious practice values wear and tear accumulated through acts of faith. These combined circumstances are shaping different conservation approaches, which emphasise restraint and respect for existing conditions (© *Stephen Rickerby*)

different conservation approaches, which emphasise restraint and respect for existing conditions [11].

These are reminders that appreciation of cultural heritage has not always been driven by mass demands, or by expectations that it should be presented in ways that reverse or mask temporal processes of change. Indeed, the appreciation of impermanence is a counter-trend that runs through many cultures. The Japanese aesthetic of wabi-sabi (侘寂) rejects artifice and recognises beauty in imperfection and the natural processes of decay. In Kintsugi (金継ぎ), the art of repairing broken pottery, this philosophy is translated into a celebration of the flawed condition of the historic object. Similar ideas are found in China and Korea. In China, ritual bronze artefacts dating from the Shang and Zhou dynasties were collected in the Song dynasty, when they were valued for their aged appearance [12]. These same bronzes, and their antiquarian imitations made in the Song and Yuan dynasties, were transported to the Korean peninsula, where they were revered during the Koryŏ dynasty as ritual embodiments of the passage of time. In the west, the Romantic movement, which originated in the late-nineteenth century as a reaction against rationalism and industrialisation, adopted a reverence for the position of creativity within the natural world, celebrating the material past in its ruined state.

As similar attitudes re-emerge in reaction to a painted heritage that is being overcommoditised, many project-based models of conservation begin to look profligate and outmoded. Their tendency to concentrate on prestigious sites reinforces the impression that wall paintings have become products of tourism development aims. But if instead we are relearning to value wall paintings in the state in which we find them—which, crucially, also includes a respect for their meaningful place within specific contexts and environments—this signals a need to shift the emphasis of our efforts and to find better models of conservation.

Cather highlighted how in the field of ecology, the preservation of natural habitats provides a model for cultural heritage and its recognition of context as integral [1, p. 173]. We have been slow in following this lead. Nature conservation provides many examples of proactive approaches to habitat preservation, which start before endangered species are threatened with extinction.[5] In cultural heritage, we continue to do the opposite: we intervene reactively when our 'species populations'—our painted heritage sites—are already in a critical state. Unlike in ecology, we do not have the possibility to hope for species regeneration, which makes the case for adopting proactive models more compelling. They encourage us to initiate conservation actions at an early stage when they can be most effective. They point us to the need for better assessment systems that provide reliable and accurate information about the quality and quantity of wall paintings, and the threats they face. They promote the efficient use of scant resources to protect environments that hold the highest cultural values, rather than wasting those resources on ineffective reactionary endeavours.

A few examples of heritage conservation are showing the way. At the archaeological World Heritage site at Cantabria, Spain, efforts to preserve its Palaeolithic painted caves have shifted from a conventional approach—use of biocides to stop growth of microorganisms, installation of barriers to prevent visitor damage, periodic closing and monitoring of caves—to a proactive management plan [13]. This has strengthened legislation to prevent potential damage from construction, mining, quarrying and forestry activities; it has reinforced buffer zones and environmental impact policies; it has initiated full documentation of the caves with new technologies; it is managing their climates through intensive environmental monitoring and modelling; and it is minutely tracking the evolution of microbial colonies. In a subterranean ecosystem, the response of microorganisms to human presence and actions is a critical concern. What is notable about these efforts is that they focus on the role of people—including in relation to conservation activities—as the greatest threat to the site, with the implication that exclusion of humans is the best option.

This far-reaching approach is not possible at all sites. But this example's emphasis on the entire context—and on the anthropogenic factors that undermine its integrity—offers a different viewpoint on our problems, more akin to preserving an ecological habitat. Conservation has changed sufficiently to provide us with the technical and diagnostic tools to address even the most challenging problems, but in a heritage environment dominated by tourism development and other commercial imperatives, such an approach is still generally regarded as too radical. However, responsible conservation demands both radical rethinking and a different commitment to the future. If we fail to change, the diminishment of our painted heritage will only become greater.

Notes

1. Charters and guidelines that define approaches to the conservation of cultural sites have proliferated both internationally and nationally since the early twentieth century. The most up-to-date and comprehensive example is the *Principles for the Conservation of Heritage Sites in China*, first issued in 2010 and revised in 2015, which can be downloaded at: http://www.getty.edu/conservation/publications_resources/pdf_publications/china_principles_revised_2015.html.
2. For recent lists of cultural sites at risk, see https://whc.unesco.org/en/danger/; https://www.icomos.org/en/get-involved/inform-us/world-heritage-monitoring-report; https://www.wmf.org/2018Watch.
3. See http://www2.unwto.org/press-release/2019-01-21/international-tourist-arrivals-reach-14-billion-two-years-ahead-forecasts, in which the UNWTO reports the 1.4 billion tourist arrivals as a target that has been achieved two years ahead of its long-term forecast.
4. https://moneyweek.com/472323/are-we-reaching-peak-tourism/; https://www.responsibletravel.com/copy/overtourism-solutions.
5. For examples see https://conservationcorridor.org/2014/06/proactive-approaches-freshwater-conservation/; https://link.springer.com/article/10.1007/s10531-011-0013-4; https://www.sciencebase.gov/catalog/item/505a8c78e4b0c8380cd7e6e8.

Acknowledgements I am entirely indebted to Professor Sharon Cather, who sadly passed away in June 2019, as the inspiration for this paper. Its contents are also greatly informed by discussions with Lori Wong, Project Specialist, the Getty Conservation Institute. Lastly, I thank my wife and partner conservator, Lisa Shekede, for her valuable insights and support.

References

1. Cather, S.: Issues in the conservation of mural paintings: past and present. In: Yamauchi, K., Tanaguchi, Y., Uno, T. (eds.) Mural Paintings of the Silk Road Cultural Exchanges Between East and West (Proceedings of the 29th Annual International Symposium on the Conservation and Restoration of Cultural Property, National Research Institute for Cultural Properties, Tokyo, January 2006), London, pp. 173–179 (2007)
2. Cather, S.: Assessing causes and mechanisms of detrimental change to wall paintings, In Gowing, R., Heritage, A. (eds.) Conserving the Painted Past: Developing Approaches to Wall Painting Conservation (Postprints of an English Heritage Conference, 1999), London, pp. 64–74 (2003)
3. Barcilon, P.B., Marani, P.C.: Leonardo, The Last Supper, translated by Harlow Tighe (University of Chicago Press, Chicago, 2001)
4. Mancinelli, F. ed.: La Cappela Sistina: Documentazione e Interpretazioni Vol. 2: Rapporto sul Restauro degli Affreschi della Volta (Monumenti, Musei e Gallerie Pontificie), Istituto Geografico de Agostino, Novara (1994)

5. Baldini, U., Casazza, O.: Brancacci Chapel, New York (1992)
6. Centauro, G.: Dipinti Murali di Piero della Francesca: La Basilica di S. Francesco ad Arezzo: Indigini su Sette Secoli, Milan (1990)
7. Die Wies: Geschichte und Restaurierung, Arbeitsheft, Bayerisches Landesamt für Denkmalpflege, 1992
8. Corzo, M.A., Afshar, M. (eds.): Art and Eternity: The Nefertari Wall Paintings Conservation Project 1986–1992. The Getty Conservation Institute, Singapore (1993)
9. Wong, L., Beltran, V., Yang, C., Maekawa, S., Lardinois, S., Hussein, H., Agnew, N.: Improving environmental conditions in the Tomb of Tutankhamen. Stud. Conserv. **63**(sup1), 307–315 (2018)
10. Collins, R.E.: Management of tourism at Buddhist sites. In Agnew, N. (ed.) Conservation of Ancient Sites on the Silk Road (Proceedings of the International Conference on the Conservation of Grotto Sites, Mogao Grottoes at Dunhuang, October 1993), The Getty Conservation Institute, Los Angeles, pp. 41–45 (1997)
11. Rickerby, S., Shekede, L., Priday, B., Amare, K., Hagos Desta, M.: Exercising restraint: conserving wall paintings within a living religious tradition in Tigray, Ethiopia. In: ICOM-CC 18th Triennial Conference Preprints, Copenhagen, 4–8 September 2017, Bridgland, J. (ed), art. 1005, Paris, International Council of Museums (2017)
12. Won Lee, J.: Between ritual and elegance: the perception of Chinese ritual bronzes in Korea. Int. J. Arts Humanit. Manage. Stud. **4**(5), 64–71 (2018)
13. Ontañón, R., Vicente Bayam, V., Herrera, J., Rodríguez, R.: The conservation of prehistoric caves in Cantabria, Spain (2014). https://www.researchgate.net/publication/271211021_The_conservation_of_prehistoric_caves_in_Cantabria_Spain

Chapter 2
On Painting Materials, Techniques, and the Conservation of Wall Paintings Across Eurasia

Yoko Taniguchi and Michiyo Mori

Abstract Wall paintings are an immovable heritage and are therefore rarely subject to controversies regarding their authenticity. Each painting conveys an abundance of information through its materials, which may be both local and distant in origin. In the study of wall painting techniques, it is essential to take account of many aspects, such as the relationship between paint layers and supporting layers; layering techniques and optical effects; the nature and combination of pigment particles; painting application techniques and the tools used by ancient artists; and the state of conservation of paintings. Over the last decade, the scope of investigative techniques has increased greatly. These include the development of new analytical procedures that offer the retrieval of more accurate information from minute samples. The growth of international and transnational research bodies has also promoted more and better multi-disciplinary studies. Nevertheless, establishing a full understanding of wall paintings is difficult, as this may require travelling great distances to view them in rock cut-caves, temples and churches. In situ conservation is also extremely difficult, especially if sites are very remote and their paintings survive in poor condition. In the past, wall paintings were detached for display in museums outside their countries of origin, or they underwent treatments that have since caused further harm. Today, at sites such as China's Mogao and Kizil grottoes, Yugoslavia's built churches, and Cappadocia's rock-cut churches, our single greatest challenge remains how best to preserve paintings in situ, while also keeping their contextual authenticity.

Keywords Wall paintings · Immovable heritage · Palette · Materials · *In situ* · Technology · Artisans · Colourants

---

Y. Taniguchi (✉)
Conservation Science and Archaeological Science, Faculty of Humanities and Social Sciences, University of Tsukuba, 1-1-1 Tennoudai, Ibaraki 305-8577, Japan
e-mail: Taniguchi.yoko.fu@u.tsukuba.ac.jp

M. Mori
Eurasian Cultural Exchange Center, Public Collaboration Center, Tokyo University of the Arts, 12-8, Ueno-Park, Taito, Tokyo 110-8714, Japan

S. Aoki et al. (eds.), *Conservation and Painting Techniques of Wall Paintings on the Ancient Silk Road*, Cultural Heritage Science,
https://doi.org/10.1007/978-981-33-4161-6_2

2.1 The Nature and Characteristics of Wall Paintings

Wall paintings are an immovable heritage and, unlike portable objects, controversies rarely emerge regarding their authenticity or origin. Rather, wall paintings embody myriad regional contexts, as conveyed by the indigenous and imported materials and techniques used to create them, and by the artisans who often traveled from distant locations to produce them. Since wall paintings can reflect both local and distant origins, they are repositories of abundant information.

Wall paintings represent the dynamic relationships that link people, materials and technologies. In decorating the architectural surfaces of many and varied monuments—including palaces, sacred spaces, and tombs—wall paintings offer a unique view of world history. The materials embedded within them may have travelled across Asia, Central Asia, and the far Roman Empire via trade. In their material form, they embody the transmission of ideas from one cultural region to another, or from a regional center to its periphery.

The study of wall paintings must therefore be multidisciplinary to appreciate fully their cultural and technical complexity within the varied historical and geographic contexts spanning the Eurasian continent. In this regard, scientific research of constituent materials and painting techniques has been an effective means of uncovering similarities and differences across this far-reaching terrain.

Specialists from East, South and Central Asia, and from Eastern and Western Europe, have contributed to this book. Their combined contributions illustrate that when viewed trans-regionally, wall paintings offer a unique window into the layers of culture embedded within art across multiple regions. The insights gained provide a broader perspective than that often offered by individual museum exhibits, which are generally viewed isolated from their original contexts.

On rare occasions, historical sources provide direct information about the patronage and authorship of wall paintings. More often, however, such information is lacking. The study of original materials and painting techniques offers an alternative means of elucidation. Where documentary sources are incomplete, wall paintings themselves contain information that may reveal the highly organized workshops and patrons behind their making.

The process of producing full-scale wall paintings is itself a multidisciplinary task, incorporating numerous stages that usually involve multiple personnel from patron to craftsman. For this reason, there are various approaches to researching wall paintings. Historical and technical approaches seek to identify how wall paintings were imagined and designed. For instance, they may explore the use of pattern books, or how compositions were transferred and executed. Stylistic analysis involves considering distinguishing aspects such as brushwork, form and palette in the creation of wall paintings, while technical analysis determines material composition. This book combines insights from these approaches to identify processes and divisions of labor pertaining to the creation of wall paintings.

2.2 Research of Colouring Techniques and Their Potential

With the notable exception of *fresco*, wall painters used paints that were produced by mixing pigment particles or colouring agents with a binding medium. Colours are sourced or manufactured from naturally occurring inorganic minerals, or they may be organic substances derived from plant or animal matter such as madder, lac, indigo, or gamboge. Inorganic pigments include natural minerals such as azurite or cinnabar, and artificially made materials such as lead white, vermillion, or Han blue. In ancient China, artisans seemed to have produced lead white by combining lead with vinegar; lead white could then be converted to red by way of a firing process. Vermillion was produced by heating sulphur and mercury, while Han blue is a man-made barium copper silicate. Organic pigments and colourants were produced by different modes of extraction and refinement and, in wall painting, they were generally applied onto a mordant. Although organic colours were used extensively in ancient paintings, textiles and craft materials, their presence has not, until recently, been widely recognized. The fading and deterioration of organic colours makes their detection difficult, requiring specialist scientific research, which has been limited to date.

The binding media of wall paintings include plant gums (such as gum arabic), animal glues, egg, casein and drying oils. Used singly or in varying combinations, their interaction with pigments and colourants results in a rich variety of paint effects. Paint materials may be adjusted to produce different tones and hues, for example, by adding white additives, varying the reflective index of their combined materials. Additionally, varying the thickness of paint layers or layering them produces other diverse effects. For example, the hue and tone of reds may be adjusted through different layering techniques, mixtures of pigments, or other modifications to colouring agents. These examples illustrate just a few of the many ways in which wall paintings exploit the physical and chemical possibilities presented by the combination of their constituent materials. This highlights the need to consider these and other components—such as paint application processes—as a composite whole in the study of original materials and techniques.

In Asia, the binding medium of ancient wall paintings was once believed to consist solely of animal glue, an assumption that curtailed further scientific investigation. However, over the last 10 years, analytical studies have increased considerably, for several reasons. These include the development of new analytical techniques, which have allowed greater opportunity to apply scientific methodologies to minute samples. The growth in the number of international and transnational research bodies, and the span of multi-disciplinary studies fusing biology, chemistry, physics, art history and archaeology, have also promoted greater scientific inquiry. New analytical research stemming from these developments is reflected throughout this book.

In addition, non-invasive methods of analysis and multi-spectral imaging techniques have made it possible to acquire more analytical information in a shorter timeframe. It is critically important, however, that these non-invasive procedures are grounded in knowledge of ancient painting technologies. This includes in-depth study of ancient and medieval texts and treatises providing insight into how paintings were made and the original materials used. Techniques that enable the procurement of information from paint samples should not be overlooked, despite their invasive nature. While non-invasive techniques can be highly informative, they provide incomplete data owing to methodological limitations and difficulties. In contrast, examination of minute paint samples provides a tremendous amount of technical information, particularly with respect to layering techniques and particle adjustments that cannot be analysed from the surface alone. Therefore, a combination of these examination approaches is required. In doing this, European institutions have accumulated substantial knowledge and experience over many years, whereas Asian institutions have been slower to make progress. However, in recent years, the gap has been closing, as evidenced by the findings presented in this book.

An example of the technical complexities associated with assessing wall paintings is found at the Kizil and Bezeklik grottoes, both located in Xinjiang Province, China. Here, only the colours blue, green, brown and white are readily visible. The intense blue originates from lapis lazuli and the pale green from atacamite, but over time the yellow pigment orpiment has altered to white, while lead-based red pigments have turned dark brown. In other words, only the chemically stable pigments have survived visibly unchanged. However, using modern analytical techniques, colours not otherwise readily visible—or those altered from their original appearance—may be detected and identified. Consequently, painting schemes that have undergone these radical changes can be better appreciated for the mastery of their ancient artisans.

Similar processes of change affect Bamiyan's Cave N(a) in Afghanistan. At first sight, only reds appear to be present on the Thousand Buddha paintings; however, investigation revealed that yellow orpiment had once been present, too, but had altered to an almost invisible form, as in Kizil. In other words, the colour palettes of the Kizil and Bamiyan paintings may have been closely comparable in the past. It has also been observed that wall paintings at both sites share iconographic and stylistic similarities. Given that both regions came under the influence of the Turkic Khaganate during their heyday, the incidence of commonalities despite geographical distance comes as no surprise. If the paintings were intended to achieve similar effects in these two regions, an analysis of similarities and differences in their application would provide an interesting window into the history of cultural exchange. Similar outcomes are visible in China's Mogao grottoes as well. Today, with the wider availability of modern analytical techniques, there is greater potential for the rediscovery of 'lost' colours.

Previous art historical studies of wall paintings in the region were based primarily on detailed observations with the naked eye. Such findings should be reconsidered using modern scientific techniques and methodologies, in particular to clarify the differences between how paintings appear now in their aged and altered states, and how they would have appeared when they were first created. With this objective in mind, the consideration of non-visible trace elements left by degraded paint materials is especially important. For example, trace residues of binding media may help in determining the presence of paint materials that have otherwise disappeared. Such information may then be directly associated with specific painters or artisans, as is the case at Bamiyan.

Over the first three phases of Bamiyan's decoration, from the fifth to the eighth century, water-soluble binding media were used in the application of the paintings. Then, as the 3rd phase took root in the mid-seventh century, drying oils first appeared. This could imply that different workshops took over painting practice at this time. But few assumptions can be made about the nature of this transition. Painting techniques using oil are described in the ancient Buddhist painting treatise, the *Chitra Sutra*, which may therefore indicate a development originating in northern India. However, in terms of pictorial style, Bamiyan's oil-based paintings do not differ much from those of previous phases, and also share characteristics with Central Asian paintings. Moreover, most caves using oil painting technology also have *lanterndecke*-style ceilings, which appeared during the 3rd phase. Because this *lanterndecke*-style seems to have originated in wooden architecture, we can surmise that oil painting technology was originally used on wooden temples and was later applied to cave temples during Bamiyan's 3rd phase.

It has also been noted that the oil-based paintings at Bamiyan incorporate a new style of frontally seated Buddha. A workshop practice that utilised oil painting technology must therefore have introduced this new concept to Bamiyan. At the same time, more usual iconography (such as *Maitreya* and *Parinirvana* scenes) persisted in tempera paintings, as well as being found in oil-based paintings. It can be safely assumed that Buddhist monks and donors decided the iconography of these works, so the circumstances mentioned above indicate that two separate workshops painting in different techniques were simultaneously carrying out commissions that were directed by the same philosophical branch of Buddhism.

The influences and processes of development underlying these painting styles in Bamiyan will certainly continue to be debated. Since only a few historical texts refer to painting techniques in Bamiyan, further scientific study has the potential to build a clearer picture of how workshops and patrons interacted at this time.

2.3 Conclusion

The expansion of scientific study of Asian wall paintings—particularly regarding their use of inorganic pigments, organic colourants, and organic binding media—has led to new interpretations and better understanding of their technologies. Recent studies, such as those at Bamiyan, demonstrate that Asian paintings were produced using materials and processes far more complex and diverse than that of merely crushing powdered minerals and combining them with animal glue. This book provides instead numerous other examples of wall paintings that were created using sophisticated materials and colouring agents. Ongoing research continues to open new channels to understanding and expanding the field of technological study and cultural exchange in Eurasia.

In the past, wall paintings at many archaeological sites in the region—such as those in caves, temples and churches—were detached and relocated for display in museums within and outside their countries of origin. In addition, many attempts at in situ preservation also resulted in the diminishment of their value. *In situ* conservation remains exceptionally difficult, especially for painted sites that are remote and survive in poor condition. As the contributions in this book demonstrate, wall paintings at China's Mogao and Kizil grottoes, in Yugoslavia's churches, and Cappadocia's rock-hewn churches are beset by multiple conservation problems. The biggest challenge for us today lies not just in preserving wall paintings in situ, but in also maintaining their contextual authenticity.

Chapter 3
Deterioration and Conservation of Cappadocian Wall Paintings

Murat Ertugrul Gülyaz

Abstract In 1985, the region of Cappadocia, Turkey, was included on the UNESCO World Heritage List as a combined cultural and natural site. Capadoccia's distinctive landscape is a product of volcanic activity, which resulted in the formation of tuff rock strata that were gradually eroded into dramatic peaks. The tuff is easily hollowed out, and for this reason the region has long been a place of rock-cut settlements. Carving techniques remained unchanged for centuries. In the fourth century A.D., Cappadocia became a religious center both for hermits and monastic communities. This led to the architectural development of rock-cut churches and monasteries. The Göreme Valley contains some of the most important rock-cut churches. These have been, and continue to be, a particular focus of conservation efforts regarding their rock structures and wall paintings. Environmental exposure leads to a wide range of problems, including rock erosion, rainwater infiltration through cracks and fissures, exfoliation and collapse of rock, and deterioration of wall paintings. Added to these natural problems are growing tourist numbers and inappropriate touristic developments. This paper describes efforts to address these various issues.

Keywords Cappadocia · Rock-cut church · Environmental deterioration · Wall paintings · Conservation and restoration

3.1 Introduction

Cappadocia, in Turkey, is one of the most impressive regions chosen as a place of settlement by various successive civilizations. The people who have lived and still live here have always been inseparable from nature and history: over hundreds of years, their lives have been shaped by both the climate and the natural environment.

Present-day Cappadocia is the area covered by the city provinces of Nevşehir (NYSSA), Aksaray (KOLONOEIA), Niğde (NAKIDA), Kayser (KAISAREIA) and Kırşehir (THERMAE). The smaller rocky region of Cappadocia is the area around

M. E. Gülyaz (✉)
Nevşehir (Cappadocia) Museums, 50100 Nevsehir, Turkey

S. Aoki et al. (eds.), *Conservation and Painting Techniques of Wall Paintings on the Ancient Silk Road*, Cultural Heritage Science,
https://doi.org/10.1007/978-981-33-4161-6_3

Fig. 3.1 Map of Cappdocia

Uçhisar, Göreme, Avanos, Ürgüp, Derinkuyu and Ihlara [1, pp. 48–49, 2, 3, pp. 8–9, 4, 5, pp. 1–4, 6, pp. 6–7] (Fig. 3.1).

Cappadocia is an UNESCO World Heritage site. It was listed in 1985 as 'Göreme National Park and Rock Sites of Cappadocia'. The Göreme valley was nominated as a mixed cultural/natural site. The eastern part of the Göreme valley was subsequently turned into an open-air museum by Turkey's Ministry of Culture and Tourism [7, pp. 5–27, 8, p. 164] (Fig. 3.2).

In its geological past, ancient Cappadocia covered a very large expanse of land, which was submerged beneath water. But after a volcanic eruption it became a great land mass. A great number of volcanoes were active in early geological periods, including those of Erciyes (3917 m), Hasandağ (3268 m) and Göllüdağ (2142 m), which were responsible for many volcanic eruptions in the Early Miocene Era (10 million years ago), and have continued until the present day. The lava produced by these volcanoes formed a stratigraphy of tufa, a variety of limestone, over 100 m thick across the region's plateaus. Other mineral substances in the layers include ignimbrite, soft tufa, ash, clay, sandstone, marn, basalt and other agglomerates [9, pp. 8–10, 10, pp. 6–11, 11, p. 45, 12, 13, p. 12, 14, p. 37, 15, 16, pp. 386–411].

The main reason people settled in Cappadocia was that the tufa formation could be hollowed out easily for habitation and use. Initially they preferred the steep hillsides of the valleys for excavating settlements, but later on, as available spaces became scarce, they hollowed out the insides of the rocks. The hollowing-out technique remained the same for centuries, and the tool-marks made on the walls and ceilings of these dwellings during their excavation have been preserved by the dry climate. The rock settlements were suitable for residence because they were warm in winter

Fig. 3.2 Göreme Open Air Museum

and cool in summer. Consequently, this troglodyte life-style continued for centuries, including up to the present day (Fig. 3.3).

In the fourth century A.D., Cappadocia became a religious center both for hermits and monastic communities. This resulted in the architectural development of rock-cut churches and monasteries. The Göreme Open Air Museum, one of the most important ancient settlements of Cappadocia, is not only remarkable as a place of natural wonder, but is also important for the history of architecture and painting (Fig. 3.4). It also has theological importance as the locus of early religious education and monasticism in the region. The same model was then introduced to the valleys of Soğanlı, Ihlara and Açıksaray [3, pp. 62–132; 17, 18, p. 13, 19, p. 3, 20, pp. 229–237].

Churches in Cappadocia were made by excavating structures into the tuff rock. Carving was made easy by the relative softness of the rock, which facilitated the creation of desired architectural plans. At the same time, great care had to be exercised, as mistakes could not be recovered: if a dome or pillar broke during carving, this could not be rectified (Fig. 3.5).

The plan type of a single nave with a barrel-vaulted ceiling is widespread in these structures. While this was the most appropriate architectural style for the religious communities living in the region and of monks who retired into seclusion, it also seems that this type of transverse rectangular plan was of Mesopotamian origin, and was brought to Cappadocia by incomers to the region. In general, the Göreme Open Air Museum churches were not very big, and it is even doubtful that they each accommodated as many as 20 monks. Churches usually contain a narthex, and internal architectural features include pedestals and columns (Fig. 3.6). In the apse area, the bench, bema, altar desk, and templon were all created by carving. The varied

Fig. 3.3 Subterranean rock-cut settlement, Nevşehir

Fig. 3.4 General view of Göreme Open Air Museum

Fig. 3.5 Karanlık Church, Göreme Open Air Museum

Fig. 3.6 Tokalı Church, Göreme Open Air Museum

rock-cut complexes include convents, monasteries, churches, chapels, prayer rooms and monk chambers, tomb spaces in columbarium style, and auxiliary spaces such as dining halls and dormitories, wineries and barns [21, p. 239, 22, pp. 126–127].

The rock-cut churches within the Göreme Open Air Museum usually date back to the ninth–thirteenth centuries, and those with wall paintings testify to the aesthetic values of the region's medieval painting traditions. As a consequence, the Göreme Open Air Museum has been and continues to be one of Turkey's most visited historical sites.

3.2 Conservation Context

The rock-cut churches of the Göreme Valley are exposed to various forms of deterioration, and for this reason it was chosen as the primary zone for structural conservation and rescue of the wall paintings (Fig. 3.7a, b). Climatic factors such as wind, rain and frost cause fracturing and cracking of the surface of the tuff rocks. Rainwater dispersed from the surrounding plains runs over the soft rock causing rapid erosion and formation of deep cracks. In winter, water that leaks through these cracks into

Fig. 3.7 **a** Virgin Mary Church in Göreme Valley. **b** Virgin Mary Church in Göreme Valley. Elevation mapping of the structural and macrostructural cracks that have occurred and still occur in the structure of rock churches and other surfaces remove

Fig. 3.7 (continued)

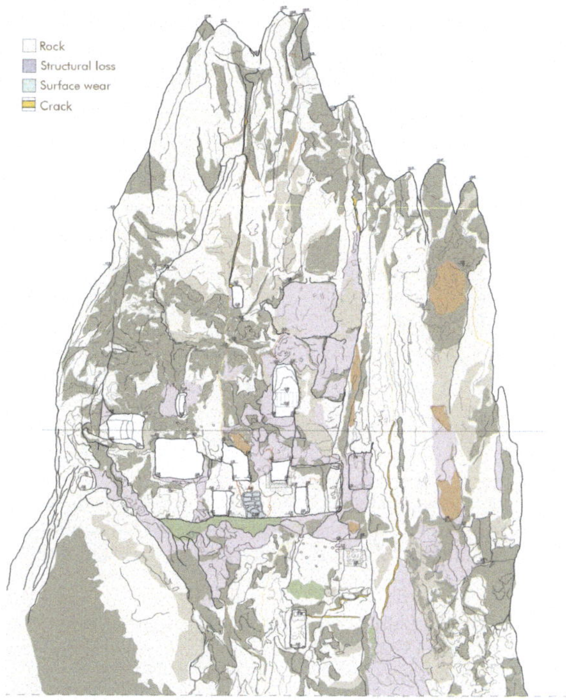

deeper grooves and crevices causes tension within the rock, as a result of freeze-thaw cycling (Fig. 3.8a, b).

Natural cracks are an inherent geological feature of the tuff rock, and in the process of carving, they were exposed and made worse. Cracks of this type are usually concentrated at the corners of architectural openings and windows, or in the apses (Fig. 3.9a–c). Consequently, when water reaches the structures through these access points, it accumulates between the painted plaster and the rock wall, which causes paintings to separate and fall off in large pieces. The leaking water also causes alteration of pigments, promotes the growth of microorganisms, and leads to the formation of salts and mineral layers on internal surfaces.

The earliest organized attempt to restore Cappadocian cave churches was the 'Göreme Historical National Park' project implemented by the National Park Department of the Ministry of Forestry in cooperation with the US National Park Service Planning Team, in 1967. This project covered a vast area, including Uçhisar, Göreme and Zelve valleys Ürgüp, Ortahisar and Çavuşin [23, p. 563]. Subsequent international exchange in 1969 on the conservation and restoration of the rock-hewn churches of Cappadocia led to the organization of an UNESCO mission in 1972. In that year, specialists from ICCROM (International Centre for the Study of Perseveration and Restoration of Cultural Property) prepared a report on the conservation needs of the region, drawing on investigations that included analysis of paint samples to understand the original technology of the paintings, and on measurements

Fig. 3.8 **a** Çarıklı Church, Göreme Open Air Museum. Climatic factors such as wind, rain and frost cause fracturing and cracking of the surface of the tuff rocks. **b** Çarıklı Church, Göreme Open Air Museum. Loose rock is a potential danger for visitors

Fig. 3.8 (continued)

of humidity levels in the painted walls and interiors of churches [24, p. 195]. As a result of the inspections and measurements made, three major problem areas were identified, and it was determined that the best way to eliminate these was to address them one by one.

The three major areas of concern were:

Fig. 3.9 **a** Environmental deterioration causes the tuff rock to exfoliate, leading to rock collapses from the vaults. **b** Natural cracks are an inherent geological feature of the tuff rock. In the process of carving, they were exposed and made worse. **c** Plan of the Virgin Mary Church, Göreme, showing crack formations, which can extend from the outer surface to the interior

Fig. 3.9 (continued)

Fig. 3.9 (continued)

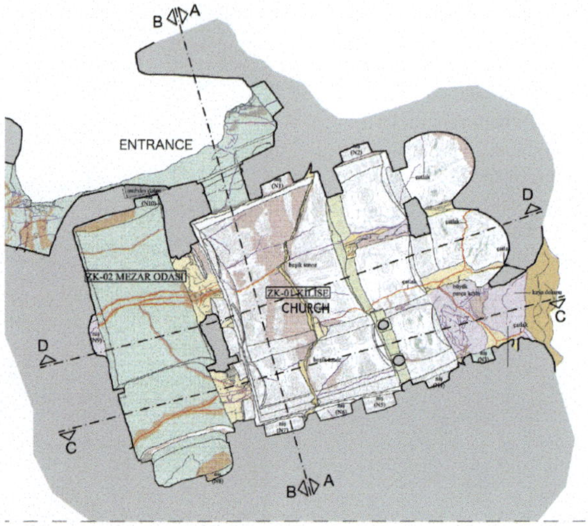

1. Structural Protection (protection of rock at risk of collapse)
2. Protection of Wall Paintings
3. Rock Consolidation.

For trial purposes, rock surfaces exposed to erosion were covered with a lime mortar prepared from local materials, to prevent further damage. While under normal circumstances one should exercise restraint in treating mural paintings, conservation of these was urgently required in many churches to prevent their complete loss.

3.3 Preservation of Churches

In 1973, the first international mission started preservation activities in the Göreme rock-cut churches. The project had a dual purpose: to carry out emergency conservation treatment of the mural paintings; and to give assistance in the training of a Turkish team of restorers. The mission was organized by ICCROM and comprised international experts together with specialists from the Middle East Technical University and the Ministry of Culture of the Republic of Turkey.

As already briefly described, the geological formation into which the churches were carved is a type of glassy tufa rock. Although this can be easily worked because of its soft and porous texture, these features also make it quite weak against external effects. The harsh seasonal climatic conditions and drastic temperature differences of the region especially result in erosion of rock surfaces. In addition, structural and macro-structural cracking is already present, and still occur, in the rock-cut churches. While some of these cracks are only present on interior surfaces, some formations extend from the outer surface to the interior. Water seeping through these fissures, as well as effects of interior condensation, cause the cracks to expand. In several churches, natural cracks exacerbated by rain and frost have contributed to partial rock collapses, resulting in loss of the architectural integrity of the structure. In the preservation of Cappadocian rock-cut churches it is therefore necessary to prioritise immediate/emergency treatment needs, and to establish the most appropriate methodological approaches. Periodic maintenance is a critical issue. Erosion is a natural characteristic of the region, which must be fought against.

As a priority, it is necessary to prevent water ingress and provide regular exterior maintenance. Hidden cracks and loose rocks around the Çarıklı Church complex were completely cleaned by expert mountaineers; other dangerous rocks in the Göreme Open Air Museum are also cleared periodically (Fig. 3.10). Exterior cracks must be filled with properly formulated mortar to prevent the ingress and accumulation of water, and the penetration of dust, soil, dirt and other contaminants into the microcracks on the interior of rock-cut structures, particularly those of great artistic value. The repair of such cracks is not a routine operation, but involves specialist treatment with conservation-grade materials (Fig. 3.11).

A critical issue is how best to protect exterior rock surfaces. The application of external renders to rock-cut churches has not always been successful. Renders of the wrong density that reduce permeability tend to fail, leading to their separation and fragmentation. Within the Göreme Open Air Museum site, several rock-cut structures with eroded ceilings have been protected in this manner (Fig. 3.12a, b). The approach of covering all rock surfaces with a mixture of hydraulic lime, tufa dust

Fig. 3.10 Hidden cracks and loose rocks around the Çarıklı Church complex were completely cleaned by expert mountaineers

Fig. 3.11 Göreme Open Air Museum

Fig. 3.12 **a** St. Basil Church, Göreme Open Air Museum. Before recent external treatment. The application of external renders to rock-cut churches in the past has not always been successful. **b** St. Basil Church, Göreme Open Air Museum, after external treatment

Fig. 3.12 (continued)

and riverbed sand has not been satisfactory in the long term. Over time, small cracks became larger and deeper, resulting in new deterioration. This occurred to the wall paintings in the Elmalı and Barbara complexes, as rainwater from above infiltrated through deep cracks and crevices in the rock. In subsequent preservation efforts, a different approach has been adopted. Eroded areas on the exterior are first repaired with suitable mortar. Then a wire-mesh is laid over the rock surface to provide a visible control of the depth of erosion. Finally, the top layer is again covered with mortar. This method has had good results in the Church of St. Basil (Fig. 3.13).

Acts of vandalism are another primary concern. In many cases, contemporary vandalism is encouraged by historical examples, such as the inscriptions carved by Christian pilgrims in the nineteenth century. Local shepherds were responsible for considerable destruction about fifty years ago: some damage was done unwittingly, but other damage was deliberate, such as that targeting the faces of figures. Besides the damage caused by people in the past, the high number of visitors to the churches today inevitably leads to the deterioration of tuff surfaces and wall paintings. All of these factors combined have resulted in irreparable damage to many painted churches (Fig. 3.14).

Fig. 3.13 In several churches, natural cracks caused by rain and frost eventually lead to separation and loss of important elements of the rock-cut surfaces. Specialist injection grouting is required, as shown here

Fig. 3.14 Vandalism is another primary concern. In many cases, contemporary vandalism is encouraged by the presence of historical examples

3.4 Original Materials and Techniques of the Wall Paintings

Following their excavation, newly cut church interiors were painted with simple decorative motifs, masonry lines and crosses in red paint directly on the rock. This kind of painting was quite usual in Cappadocian churches, functioning as an act of consecration. Subsequently, interiors were further embellished with paintings on plaster.

The themes on the wall paintings are generally taken from the Life of Christ, and from other Biblical subject matter. The paintings also include saints from the Christian world and important people of Cappadocia.

Two different technical approaches to wall painting can be broadly identified in Cappadocian churches:

(1) Painting applied directly onto the bare rock without an intervening layer of plaster. Basic mineral colours were used for painting simple decorations such as consecration crosses, and geometrical and floral designs. This type of painting is more usually found in the early Byzantine churches and chapels (Fig. 3.15). Examples include the church of St. Basil, Elmalı, and the Chapel of St. Barbara.
(2) Painting applied onto plaster. In this case, the technique was either that of *fresco*, whereby pigments suspended in water are applied onto fresh lime plaster; or *secco*, painting on dry plaster with the aid of a binding medium.

Wall paintings in the second category are particularly vulnerable to various types of deterioration. Over time the original paint binders gradually lose their adhesiveness, leading to paint flaking. Fibre additives in the plaster decay and disappear, resulting in decohesion and loss of paintings. The preservation endeavours initiated within the Göreme Open Air Museum aim to address these and other problems affecting the rock-cut churches and their paintings. These efforts have increased in pace over the last few years.

Fig. 3.15 St. Barbara Church. Painting applied directly onto the bare rock without an intervening layer of plaster

3.5 Conservation of the Wall Paintings

Conservation work has been carried out at the Tokalı and Karanlık churches in the Göreme valley [25, p. 303, 26, 27, pp. 61–72, 28, pp. 6–7, 29, p. 184, 30, 31, p. 80]. Here the earliest mural decoration comprises simple geometric designs applied directly to the rock surface in mineral colours of red ochre or green earth.

The components and interventions applied within the framework of the conservation programme included:

- Detailed documentation of the state of deterioration of the support, plaster layer(s), and paint layers; of surface accumulations and contaminants; and of the different processes and materials used in the remedial treatment of the mural paintings.
- Consolidation of the rock support (volcanic tuff rock).
- Re-attachment of the plaster layers to the rock support, and consolidation of the plasters.
- Consolidation and re-adhesion of the paint layers.
- Removal of dust and other types of surface accumulation from the paint layers by mechanical and chemical methods.
- Repairing deep cracks and losses in the support and plaster layers with suitable mortar.

From 1973 to 1980 successive missions were carried out, dealing principally with the conservation and restoration of the mural painting in Tokalı Church, the biggest church of the Göreme region, where the detachment of plaster layers was discovered to be extremely serious. In 1980 Tokalı Church was opened to the public. From 1981 to 1990 successive missions were carried out to conserve and restore the mural paintings in Karanlık Church.

Restoration of the Forty Martyrs Church near Ürgüp was accomplished between 2009 and 2013, under the leadership of the Nevşehir Museum Administration and the supervision of Prof. Maria Andaloro of Tuscia University [32, pp. 120–135, 33]. The wall paintings were found to be in a very good condition underneath a thick layer of soot from fires, which had provided them with protection. Cleaning was carried out during a comprehensive restoration project (Fig. 3.16a, b). Aqueous and reagent (ammonium bicarbonate) cleaning procedures were employed, applied either directly or through intervention layers, or in cellulose poultices, to address differences in condition and original technology. The timings of the applications were critical for obtaining homogeneous results.

Conservation efforts at the Tokalı Church began in 2012 and are still in progress. The project is also being carried out as a collaboration of Nevşehir Archaeological Museum and Tuscia University [32, pp. 133–140]. Another restoration and conservation project at Üzümlü Church in the Cappadocia region started in 2014 in collaboration with Prof. Yoko Taniguchi, Faculty of Humanities and Social Sciences, University of Tsukuba (Japan) [34, pp. 361–172] (Fig. 3.17).

Fig. 3.16 **a** Restoration of the Forty Martyrs Church near Ürgüp. Before cleaning. **b** After cleaning

Fig. 3.16 (continued)

3.6 Tourism and Its Impact

Since the 1990s, the number of visitors to Cappadocia has increased as the region has gained attention as a touristic attraction. It is clear that mass tourism is not appropriate for the nature of Cappadocia's rock-cut churches, and excessive numbers of visitors cause serious problems of erosion to the rock surfaces. There are also severe risks arising from the pressure of touristic developments, which are generally made without due regard for the integrity of the natural and cultural landscape. Control of tourist

Fig. 3.17 Üzümlü Church in Göreme Open Air Museum

activities has to be achieved, to avoid both unintentional and deliberate damage occurring to painted sites and their architectural contexts.

In order to prevent some of the more direct negative effects associated with the increasing number of visitors to the painted churches, certain measures have already been taken. The number of visitors allowed to enter at one time is restricted and the visiting time is limited to 3 min. Visitors are not allowed to take images inside churches. Wooden platforms with rails have been inserted and covered with synthetic carpets, to prevent erosion of the rock-cut floor surfaces and to protect the wall paintings from raised dust.

3.7 Final Remarks

Cappadocia covers an area of approximately 25,000 km^2, over which rock-cut settlements are widely distributed. Effective overall preservation is difficult. Isolated churches that lack protection are in danger of damage or destruction from human actions, such as vandalism, as well as from a variety of natural forces. Several rock-cut churches in the Göreme Valley are structurally unstable. In churches that are located near slopes, large cracks may appear as a result of disintegration of the rock mass; natural processes of erosion and cracking cause great damage to the churches and their unique rock formations.

It is important to take collaborative measures to prevent the loss of this incomparable cultural heritage, including the required actions for structural strengthening of churches at risk. At present, however, there are not enough full-time employed specialists in our region to carry out these tasks. There is a great need to establish reliable monitoring procedures that are then followed by local conservation efforts.

References

1. Giovannini, L.: Arts of Cappadocia. London (1971)
2. Jolivet Levy, C.: La Cappadoce Mémoire de Byzance. Paris (1997)
3. Hild, F., Restle, M.: Tabula Imprerii Byzantini II Kappadokien. Vienna (1981)
4. Ousterhout, R.: Byzantine Settlement in Cappadocia. Washington, DC (2005)
5. Strabon, Geography XI
6. Umar, B.: Karadeniz Kappadokiası (Pontus). İstanbul (2000)
7. Baggley, G.: Göreme Historical National Park Master Plan for Protection and Use. Ankara (1972)
8. Gülyaz, M.E.: World Heritage in Turkey. Ankana (2009)
9. Thierry, M.: Les paysages irréels d'une terre mystique. Histoire et Archeologie, pp. 8–10 (1982)
10. Coindoz, M.: La Planete Cappadoce, Dossier Histoire et Archeologie Nr.121, Novembre, pp. 6–11 (1987)
11. Sür, Ö.: Volkanoloji. Ankara (1989)
12. Gürel, A., Kadir, S.: Geology and mineralogy and origin of clay minerals of the Pliocene fluvial-lacustrine deposits in the Cappadocian Volcanic Province, Central Anatolia, Turkey. Clay and Clay Miner. **54**, 555–570 (2006)
13. Cooper, J.E., Decker, M.: Life and Society in Byzantine Cappadocia. Springer, New York (2012)
14. Eckert, S.: Fairy Chimneys of Cappadocia Evaluation of a Landscape. World Geology Group (2012)
15. Göz, E., Kadir, S., Gürel, A., Eren, M.: Geology, mineralogy, geochemistry, and depositional environment of a late Miocene/Pliocene fluviolacustrine succession, Cappadocian Volcanic Province Central Anatolia, Turkey. Turk. J. Earth Sci. **23**, 386–411 (2014)
16. Gülyaz, M.G.: Göreme National Park and Cappadocia. World Heritage in Turkey, Ankara (2012)
17. Rodley, L.: Cave Monasteries of Byzantine Cappadocia. Cambridge University Press, Cambridge (1985)
18. Ötüken, Y.: Göreme. Ankara (1987)
19. Jolivet-Levy, C.: Les Eglises Byzantines de Cappadoce. Paris (1991)
20. Akyürek, E.: Fourth to Eleventh Centuries: Byzantine Cappadocia. İstanbul (1998)
21. Akyürek, E.: Carved Churches, Carver Architects, Cappadocia, İstanbul (1988)
22. Kostof, S.: Caves of God: The Monastic Environment of Byzantine Cappadocia. MIT Press, Cambridge, MA (1972)
23. Özil, R.: The Conservation and Restoration of the Mural Paintings in the Rock Hewn Churches of Cappadocia, Cappadocia (1998)
24. Özil, R.: Nevşehir İli, Göreme Elmalı Kilise Duvar Resimlerinde Koruma Onarım-2003, 14, Müze Çalışmaları ve Kurtarma Kazıları Sempozyumu Nevşehir 30 Nisan-2 Mayıs 2014, Ankara, pp. 195–204 (2014)
25. Jerphanion, G.: Les Eglises rupestles de Cappadoce. Une nouvelle province de l'art byzantin, Paris (1925–1942)
26. Jolivet-Lévy, C.: Le riche décor peint de Tokalı Kilise à Göreme, Dossiers d'Archéologie (1982)
27. Restle, M.: Byzantine the Wall Painting in Asia Minor, I-III, Recklinghausen, Germany (1967)
28. Epstein, W.A.: Tokalı Kilise, Tenth-Century Metropolitan Art in Byzantine Cappadocia (Dumbarton Oaks Studies 22), Washington (1986)
29. Jolivet-Levy, C.: La Cappadoce Medievale. Paris (2002)

30. Andaloro, M.: The New Tokalı Kilise, or the Sistine Chapel of Cappadocia, I. Uluslararası Nevşehir Tarih ve Kültür Sempozyumu 2011, Ankara, pp. 5–25 (2012)
31. Andaloro, M.: The project on the rock paintings in Cappadocia. Research and Preservation in the Göreme Open Air Museum (Report 2011) 30. Araştırma Sonuçları Toplantısı, 28 Mayıs-1 Haziran 2012 Çorum, 2.Cilt, pp. 133–146 (2012)
32. Andaloro, M.: The project on the rock painting in Cappadocia (2012). Research and Preservation: 1. The Göreme Open Air Museum; 2. The Church of the Forty Martyrs in Şahinefendi. 31. Araştırma Sonuçları Toplantısı, 27–31 Mayıs 2013 Muğla, pp. 120–135 (2013)
33. Andaloro, M.: The Cappadocia wall paintings project: research and preservation: 1. The Rupestrian Habitat of the Göreme Area; 2. The Church of the Forty Martyrs in Şahinefendi, Restoration and Opening to the Public, 32. Araştırma Sonuçları Toplantısı, 2–6 Haziran Gaziantep 2014 2.Cilt, pp. 93–103 (2014)
34. Taniguchi, Y. et al.: Scientific research for conservation of the rock hewn Church of Üzümlü, Capadocia, 33. Araştırma Sonuçları Toplantısı, 11–15 Mayıs 2015 Erzurum, Ankara, pp. 361–378 (2016)

Chapter 4
Materials and Techniques of Cappadocian Wall Paintings

Maria Andaloro and Paola Pogliani

Abstract This paper presents the findings of a research mission and restoration project, *Rupestrian painting in Cappadocia: knowledge, conservation and enhancement*, which the University of Tuscia has been conducting in Cappadocia (Turkey) annually since 2006. Carried out in collaboration with our Turkish colleagues from the Archaeological Museum and the Regional Restoration Laboratory of Nevşehir, it is intended to promote knowledge and conservation of the extraordinary wall painting heritage of Cappadocia. Our paper focuses on the materials and techniques of the wall paintings of Cappadocia, based on research carried out over a 13-year period, derived firstly from a survey in the region and then from the experience acquired from conservation projects undertaken at two rock-cut churches, the Church of the Forty Martyrs in Şahinefendi and the New Tokalı church in the Göreme Open Air Museum. Findings indicate that the use of different materials and execution techniques can be related to four macro-phases into which we have divided painting practice in Cappadocia. In the overall development of painting up to the eleventh century, there was a general tendency to expand from simplified forms to more complex approaches, which is partly related to the development of more ambitious architectural spaces. In the tenth and eleventh centuries, painting procedures proliferated, which can be associated with the presence of different workshops coexisting in the extended territory between the Open Air Museum and Çavusin. This diversity encompasses a broad range, from the most modest wall painting schemes up to those of an exceptional nature, the latter exemplified by the painting in the New Tokalı church or the painting cycles of the so-called 'Column Churches'.

Keywords Wall paintings · Cappadocia · Byzantine art · Pigments · Technique

---

M. Andaloro · P. Pogliani (✉)
DIBAF - Blocco F, Tuscia University, Largo dell' Univerita' snc, 01100 Viterbo, Italy
e-mail: pogliani@unitus.it

S. Aoki et al. (eds.), *Conservation and Painting Techniques of Wall Paintings on the Ancient Silk Road*, Cultural Heritage Science,
https://doi.org/10.1007/978-981-33-4161-6_4

4.1 Introduction

This *International Colloquium on the Conservation of Asian Wall Paintings* provides a unique and ambitious forum to share knowledge and increase our understanding of the materials and execution techniques of wall paintings, which is reflected in the diverse conservation projects and studies presented in this publication. Spread over a vast territory, which ranges from Eastern Europe to the far regions of Asia, and spanning several centuries, this heritage has universal importance for the fields of history, art and religion. The principal wall paintings that are our concern belong to a number of important Buddhist cave sites in Asia, including the famous religious centres along the Silk Road—Bamiyan, Kızıl, Mogao—and a sample of Christian complexes situated in both Eastern Europe—Romania and the Balkan territory—and Asia, such as the Carpathians or the cave sites of Cappadocia in Turkey.

In this context, our contribution stems from a research mission and restoration project, *Rupestrian painting in Cappadocia: knowledge, conservation and enhancement*, which the University of Tuscia has been conducting in Cappadocia (Turkey) annually since 2006 (Fig. 4.1). It is a collaborative project with our Turkish colleagues from the Archaeological Museum of Nevşehir and the Regional Restoration Laboratory of Nevşehir, which is intended to promote knowledge and conservation of the extraordinary wall painting heritage of Cappadocia.

Our paper focuses on the materials and techniques of the wall paintings of Cappadocia, based on research carried out over a 13-year period, derived firstly from a survey in the region (2006–2018) (Figs. 4.2 and 4.3), and then from the experience acquired from conservation projects undertaken at two rock-cut churches, the Church of the Forty Martyrs in Şahinefendi and the New Tokalı church in the Göreme Open Air Museum [1–3]. Since the survey incorporated many sites, its terms of reference were broad, whereas the conservation projects enabled us to analyse the materials and

Fig. 4.1 Cappadocia, the Göreme valley landscape with the rupestrian evidence

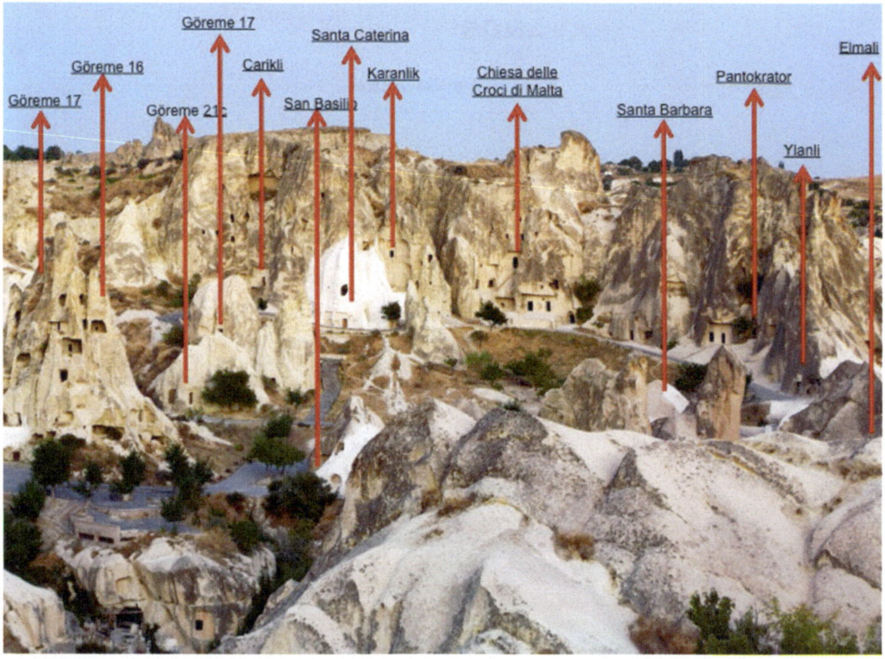

Fig. 4.2 The Göreme Open Air Museum. The arrows indicate the location of the rock-hewn churches

execution techniques of these wall paintings in a targeted, systematic and in-depth way. Combining all these findings, we were able to define certain general painting parameters, such as the persistent use of *secco* painting on *fresco* base layers, and also variations in technological approaches. In this latter regard, the churches of the Forty Martyrs in Şahinefendi (Figs. 4.4, 4.5, 4.6 and 4.7) and of New Tokalı in Göreme (Fig. 4.35) stand out.

A guiding principle of our research of the rock-cut churches of Cappadocia is that an intimate relationship exists between art-historical features of their wall paintings, and the materials and techniques used in their execution. This approach demands a collaboration of science on the one hand, and the historical and humanistic disciplines on the other. Research, to be valid, must be based on shared multidisciplinary findings. This is particularly the case for understanding wall paintings, which are characterized by their complexity, combining image and matter. Another key consideration is the historical vision involved in the creation of wall paintings, from ideation to execution. This requires placing them in their historical environment, contextualizing their use of materials and execution techniques in chronological and geographical frameworks. At the same time, our investigations in Cappadocia demonstrate that the use and evolution of painting materials and techniques are also rooted in specific localities, defined by unique circumstances of time and place. These trends sometimes run counter to or parallel with general historical trends.

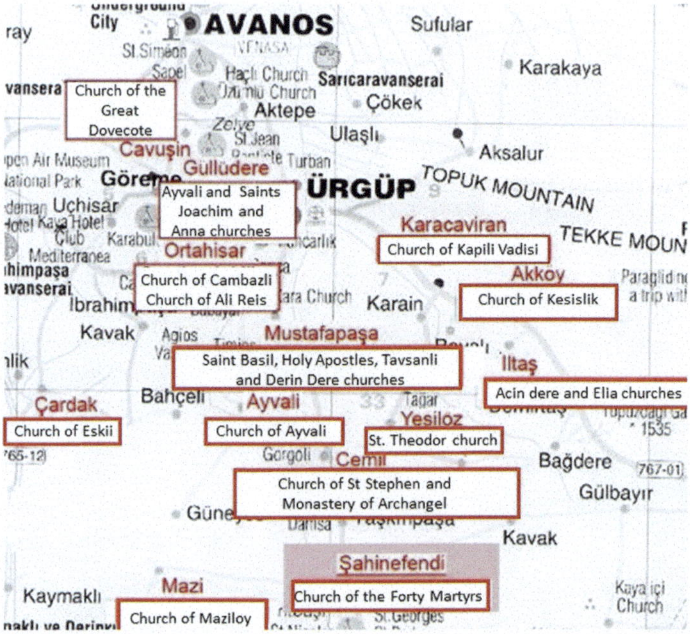

Fig. 4.3 The map shows the location of the survey sites of the University of Tuscia (2006–2018)

Fig. 4.4 Şahinefendi, church of the Forty Martyrs. View of the exterior

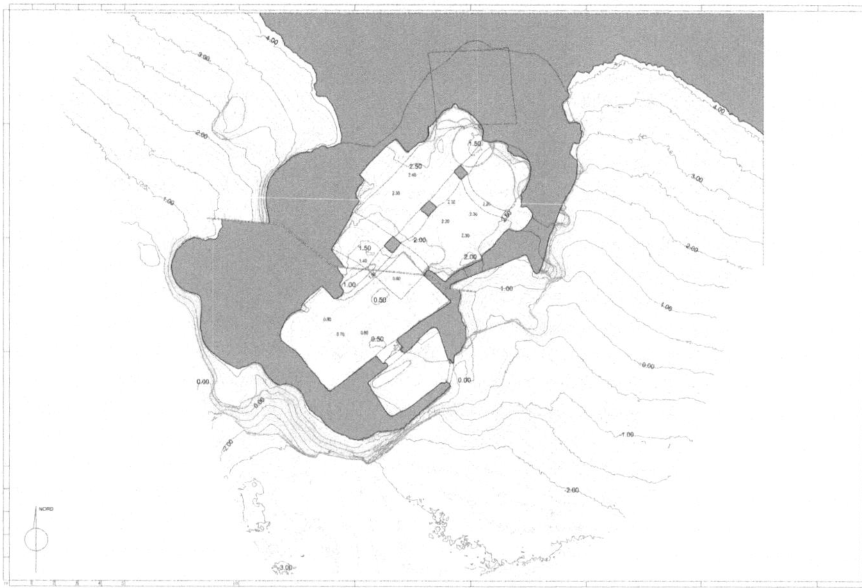

Fig. 4.5 Şahinefendi, church of the Forty Martyrs. Planimetry of the church of Şahinefendi (M. Carpiceci, Sapienza University)

4.2 Knowledge of Cappadocian Rock-Cut Wall Paintings Through Their Restoration: From Cesare Brandi's *Theory of Restoration* to the Restoration Worksites of the University of Tuscia

The rock-cut churches of Cappadocia were constructed over an extended period between the sixth and thirteenth centuries AD, producing a vast number of wall paintings that are characterized by their use of diverse materials and techniques. This abundance and diversity presents complex preservation issues. In an attempt to address these, we embarked on a large conservation project at two important painted sites. The first, which assumed the value of a pilot site from a methodological perspective, concerned the complex of paintings in the Church of the Forty Martyrs in Şahinefendi, which was started in 2007 and completed in 2013 (Figs. 4.7, 4.8, 4.9, 4.10 and 4.11) [1, 4]. The second, in the New Tokalı church, was begun in 2011 and is still ongoing (Fig. 4.35) [1, 5].

The remedial treatment rationale at both sites is that based on Cesare Brandi's *Theory of Restoration*, and the methodological approaches first applied under the Central Institute of Restoration in Rome, founded in 1939 [6–9]. Insights gained during the course of this work provided an extraordinary opportunity to increase our knowledge about wall painting practice, based on the integration of iconographic and stylistic studies with research into original techniques and materials.

Fig. 4.6 Şahinefendi, church of the forty Martyrs. View of the south aisle and south apse before restoration

Italian restoration approaches have had a significant recent history in Cappadocia, being first introduced in the churches of Tokalı and Karanlık in the Open Air Museum of Göreme in the 1970s and 1980s. In 1973, restoration work on the paintings of the church of Tokalı was started under the executive project proposed by Giorgio Torraca, a notable chemist, then deputy director of ICCROM (the international intergovernmental organization created by UNESCO for the conservation of cultural heritage) and Paolo Mora, chief restorer, Central Institute of Restoration, in Rome. Both were key members of the UNESCO international mission that laid the foundations for the recovery of the monuments of the Göreme valley, and which led to the listing of the rock-cut churches of Cappadocia as a World Heritage Site in 1985 [10]. In accordance with the theoretical and methodological approaches promoted at this time, the Tokali project involved joint activities to study the churches and their paintings in terms of their materials and techniques; photographic and graphic documentation of their state of conservation; and investigation of the methods and materials considered the most suitable for consolidating and cleaning the delicate paintings of Tokalı. In Cappadocia, this overall approach then became standard practice in subsequent

Fig. 4.7 Şahinefendi, church of the Forty Martyrs. View of the south aisle and south apse after restoration

restoration sites, for example in the churches of Karanlık, Elmalı, El Nazar and Saklı in the Open Air Museum of Göreme [11].

Our restoration activities in Cappadocia are carried out in accordance with the theoretical and methodological framework of Italian restoration, and are achieved thanks to an interdisciplinary team directed and coordinated by the University of Tuscia, made up of Italian and Turkish restorers, art-historians, archaeologists, architects, chemists, and photographers and documentation experts. Extensive long-term research in the region has allowed us to develop a profound understanding of Cappadocian painting from many points of view. Survey activities were focused within the borders of the Nevşehir province, the heart of Cappadocia, whose territory is characterized both by its unique natural landscape and by the density of its rock-cut sites, painted churches, and structures for monastic, civil and funeral use, all excavated by human hands (Fig. 4.1) [12–14]. Over two hundred painted churches are preserved in this area, created over the course of eight centuries—between the sixth and the thirteenth—in a succession of multiple chronological phases. Their wall paintings have been the subject of much previous research, which has enabled

Fig. 4.8 Şahinefendi, church of the Forty Martyrs. Paintings of the Forty Martyrs in the frozen lake on the vault of the north aisle before and after the restoration

us to acquire a great deal of data on the foundations and development of rock-cut churches in the region.

These development phases can be summarised, as follows: an original foundation phase dating from the sixth century to the early-ninth century, which precedes and in part coincides with a major period of dispute and iconoclasm, which witnessed the destruction of sacred images in the land of Byzantium [15–18]; a brief but significant post-iconoclastic phase between the ninth and beginning of the tenth century; the tenth century, the golden age of Cappadocia, characterized by the activities of many different workshops [19–21], and whose undisputed pinnacle of iconographic complexity, stylistic quality and the use of rare and precious original materials is achieved in the New Tokalı church [2]; a homogeneous nuclei of churches dating to the eleventh century; and a final expression of notable wall painting practice in the thirteenth century [22–24] [M. A.].

Fig. 4.9 Şahinefendi, church of the forty Martyrs. Detail from the painting of the Forty Martyrs in the frozen lake after restoration

4.3 Original Materials, Execution Techniques and Procedures: Integrating Knowledge of the Wall Paintings of Cappadocia

Until the Tuscia University project, relatively few studies of the original materials and techniques of Cappadocian rock-cut wall paintings had been made, particularly of the pigments and binders used. Among the most relevant is Marcel Restle's publication of 1967, which contains a pioneering study of 13 churches (from the sixth to the thirteenth centuries) in an area extending from Soğanlı to the Göreme valley [25]. Later, during the UNESCO and ICCROM restoration work at the Churches of Tokalı and Karanlık in the Göreme Open Air Museum, researches were conducted on the materials and techniques used [11, 26–28]. Art historians such as Nicole Thierry and Catherine Jolivet have illustrated the great variety of Cappadocia painting, sometimes including a discussion of technical features in an attempt to identify chronological periods [29, 30].

Fig. 4.10 Şahinefendi, church of the Forty Martyrs. Painting of the Forty Martyrs after the restoration: detail of the lake water

Our investigations had two main objectives. The first was to obtain sufficient knowledge to proceed with the restoration process in the most informed way possible, especially as regards the consolidation and cleaning of the paintings. The second was to integrate knowledge of the materials and techniques of the wall paintings with their art historical study. This combined information is presented in a database, "Cappadocia – Art and Rupestrian Habitat" (Fig. 4.12), which can be consulted online in both Italian and English. In this, data on original materials and execution techniques are presented as constituent parts of an overall structure, termed 'pictorial apparatus', which also includes more usual information, such as iconographic subject matter and classification of decorative motifs. At present the database has 109 entries on religious and lay rock-cut complexes, and includes 3788 images and over 8000

Fig. 4.11 Şahinefendi, church of the Forty Martyrs. The paintings of the vault of the south aisle and of the south apse after the restoration

Fig. 4.12 The database "Cappadocia—Art and Rupestrian Habitat" home page

data items, which can be found under four main search-fields, covering excavated architecture, furnishings and wall paintings (http://bancadati.museovirtualecappadocia.it/ [31, 32]).

Work at the churches at Şahinefendi and New Tokalı allowed particular insights of the development and sequencing of their wall painting technologies. At the Church of the Forty Martyrs in Şahinefendi, previously unknown layers in the complex sequencing of pictorial surfaces of the church were identified for the first time,

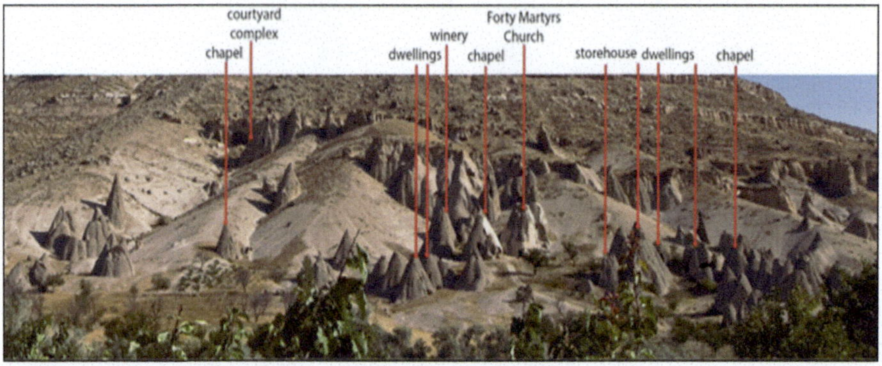

Fig. 4.13 The rock village of Şahinefendi. The arrows indicate the church, the courtyard complex and the annexed rooms

Fig. 4.14 Şahinefendi, church of the forty Martyrs. The south apse niche showing painting of a date-palm (Phase I)

while technical examination and analysis elucidated their constituent materials and application techniques. Four layers correspond to four distinct phases that can be dated to between the proto-Byzantine period and the thirteenth century [4, 5]. At the New Tokalı church, in-depth investigation of the paint materials use, which are of the

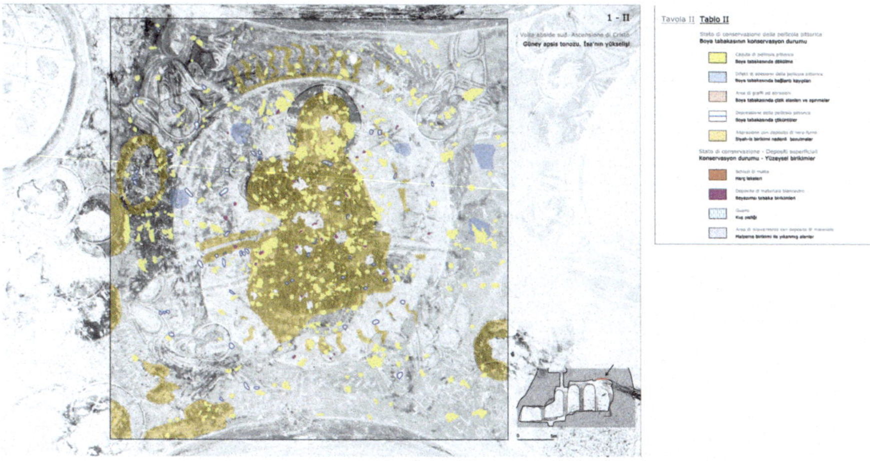

Fig. 4.15 Şahinefendi, church of the Forty Martyrs. Graphic documentation of the southern apse. Areas marked in yellow indicate painting belonging to Phase II (V. Valentini)

Fig. 4.16 Şahinefendi, church of the Forty Martyrs. Detail of painting in the south apse. The dark areas of paint showing through the upper painting belong to Phase II

Fig. 4.17 Şahinefendi, church of the Forty Martyrs. North apse, Deesis (Phase III)

highest quality, and of their application on dry plaster with various binding media, confirming its status as a unique example in Cappadocia.

4.4 The Church of the Forty Martyrs in Şahinefendi and Its Four Phases (Sixth–Seventh—Thirteenth Centuries)

The church of the Forty Martyrs (Kırkşehitler kilise or Altı Parmak kilise) [23, 25, 30] is located in the middle of the oldest rock-cut settlement (Figs. 4.4, 4.5 and 4.13). At least three excavation phases have been identified, based on analysis of the

Fig. 4.18 Şahinefendi, church of the Forty Martyrs. Entrance to the south aisle, the Dormitio Virginis and the inscription belonging to Phase IV

Fig. 4.19 Şahinefendi, church of the forty Martyrs. North nave, niche with the Crucifixion (Phase IV)

morphology of the rock-cut spaces and from evidence left by excavation processes [13, 14, 31].

At the start of the project in 2006, the entire wall painting programme was concealed by accumulated smoke blackening, produced from fires lit during agricultural re-use of the interior, which impeded legibility of the schemes and concealed their chromatic quality (Figs. 4.6 and 4.7). The cleaning operations restored their intense colour and legibility (Figs. 4.9 and 4.10). During this process, close observation of the paintings from the scaffolding made it possible, as mentioned above,

Fig. 4.20 Gomeda valley, church of San Basil

Fig. 4.21 Gomeda valley, church of San Basil. Detail of the ceiling painting

to identify and specify the complex sequence of layers that constitute the internal decoration, which can be dated to between the sixth–seventh and thirteenth centuries [31].

A palimpsest of superimposed painted layers was identified on the basis of distinguishing features and morphological characteristics [3]. In Cappadocia, wall painting stratigraphies are not always related to the sequential processes associated with a single painting, but are also complicated by execution phases of different periods laid dry on earlier painting, which may omit certain expected steps in the painting process. These phenomena have been identified in numerous painted sites: Saint

Fig. 4.22 Gomeda valley, church of San Basil. Detail of apse painting

Fig. 4.23 Gomeda valley, church of San Basil. Detail of the apse painting showing the overlap of the preparatory layers

Stephen in Cemil, St. Basil in the Gomeda valley, St. John the Baptist in Çavusin, Saints Joachim and Anna in Kızıl Çukur. In Şahinefendi, by integrating the study of the paintings with the excavation phases of the church, we gained an understanding of how the evolution of the internal architecture was accompanied by the renewal of the pictorial decoration [32]. The first core of the church was a chapel with a single aisle, flanked by a burial chamber with a decoration of red lines that outline the architecture and incorporate consecration crosses [33]. In the south apse, a niche in the southern wall is painted with a red date palm on a white background, with a band

Fig. 4.24 Kızılçukur valley, church of Saints Joachim and Anna

Fig. 4.25 Kızılçukur valley, church of Saints Joachim and Anna. Detail of the apse painting showing the overlap of the preparatory layers

of green acanthus leaves around its arch (oral communication, Giuseppe Barbera) (Figs. 4.11 and 4.14). This is contemporary with the painting in the original chapel, which is datable to the proto-Byzantine period (sixth–seventh/early-ninth century), and can be designated phase I. Subsequently, the apse of the chapel was painted with a *Maiestas Domini*, dateable to the late ninth century (phase II) (Fig. 4.11). Today, the only recognizable aspects of this painting are some figures that show through a subsequent pictorial layer, which was executed *a secco* (phase III) (Figs. 4.15 and 4.16).

Fig. 4.26 Göreme, church of Karşı becak

Fig. 4.27 Göreme, church of Karşı becak. Detail of the apse painting showing the use of different red pigments

The painting of phase III is the most extensive, and is associated with a major architectural expansion, when the north aisle was excavated to turn the chapel into a church with two aisles covered by a barrel vault and terminated by apses. In the south aisle, the vault is painted with medallions featuring busts of martyred saints and a short Christological cycle, while in the apse, the Ascension of Christ is above a register of figures of bishop-saints (Fig. 4.13). In the north aisle, the vault has a large composition showing the Forty Martyrs of Sebaste immersed in the frozen lake (Figs. 4.8 and 4.9), while in the apse we find the *Deesis* flanked by the Archangels,

Fig. 4.28 Kavaklı Dere, church of Badem

Fig. 4.29 Kavaklı Dere, church of Badem, detail of the painting of the arch

and below isolated figures of bishop-saints (Fig. 4.17). We are inclined to date this phase to the eleventh century, based on our overall studies of the church and its pictorial decoration. A subsequent phase (phase IV) marks the completion of pictorial campaigns, and includes the Dormition of the Virgin (*Dormitio Virginis*) located at the entrance to the south aisle (Fig. 4.18), and the Crucifixion painted in the large niche on the north wall of the north aisle (Fig. 4.19). This phase is uniquely distinguished by a dedicatory inscription, giving a commission date of 1216–1217 and the patron's name, Hieromonaco Macario, which is found on the entrance arch to the south aisle (Fig. 4.18).

Fig. 4.30 Göreme Open Air Museum, church of Yilanli Kilise. The painting of the vault

Fig. 4.31 Göreme Open Air Museum, church of Karanlik. The paintings of the domes and vaults

The first phase of painting was executed on a white lime plaster finished with a thin gypsum skim-coat, on which pigments were applied with the aid of an added binding medium, an execution technique that has parallels in proto-Byzantine painting in Cappadocia [34]. The subsequent phase (phase II) **is now only** partially present, appearing as dark spots on the pictorial layer of the previous phase, and can only be identified through graphic documentation (Fig. 4.15) and stratigraphic analysis. It consists of a thin layer of preparatory whitewash, which provides the base for painting with pigments that were fixed with the aid of an added binding medium.

Fig. 4.32 Göreme Open Air Museum, church of Karanlik. The Nativity

Fig. 4.33 Şahinefendi, church of the Forty Martyrs, detail of the Crucifixion (Phase IV)

The extensive third phase (phase III) comprises an application of a lime-based mortar containing inorganic aggregates and vegetable fibre. This initially provided a base for *fresco* painting, which, after setting, was augmented by rich painting applied with an added medium. The last phase (phase IV) consists of *fresco* painting over a lime-based mortar rich in Fig. 4.1 inorganic aggregates and finely chopped vegetable fibres.

Fig. 4.34 Cemil, church of St. Arcangelos. Detail of an angel figure (1217–1218)

4.5 The Techniques of the Şahinefendi Paintings in Their Context in Cappadocia

The diversity of wall painting materials and techniques identified in the various phases in the Church of the Forty Martyrs in Şahinefendi [35–37] are mirrored in the technological findings of the more than 100 painted churches in Cappadocia surveyed and presented in our database. This overview of Cappadocian paintings and the characterization of their original materials allowed us not only to reconstruct basic working practices, but also to understand the unique technical variations associated with individual workshops. An initial general classification can be made, based on identification and characterization of plasters, pigments and their application techniques, and whether these constitute *fresco* painting, or painting on dry plaster with the aid of an additional binding medium. This then allows us to define two macro-sets of paintings, the first relating to the sixth–seventh—ninth centuries, the second to the tenth–thirteenth centuries; and to specify, within these broad chronological time-spans, sub-groups referable to the activities of the same workshops.

Fig. 4.35 Göreme Open Air Museum, church of Tokalı. Interior from the presbytery

Phase I of the painting phases (Fig. 4.14) belongs to the period of proto-Byzantine painting, and can be linked to a large group of paintings with shared characteristics, such as at the churches of Agios Basilios in the Gomeda valley near Mustafapaşa (sixth–eighth century) (Fig. 4.20); Agios Stephanos at the Keşlik monastery in Cemil (sixth–eighth century); Saints Joachim and Anna in the Kızıl Çukur valley (sixth–seventh—ninth century); Üzümlü, or of the stylite Niceta, in the Kızıl Çukur valley (late-seventh—early-eighth century); Karşıbecak in Göreme (ca. 700); and the Balkan Dere in Ortahisar (sixth–eighth century) [24, 30]. These paintings have in common the fact that they were mostly produced using a *secco* technique with a palette of homogenous colours, made up of reds, yellows, greens, browns and black (Fig. 4.21). Their secondary supports comprise thin layers of white plaster that cover internal surfaces, sometimes in several overlapping layers, following the course of the rock morphology. The palette consists of natural earth colours and more complex pigment mixtures that contain lead-based pigments [35, 36]. For example, the most intense red is obtained by mixing red lead (or minium) and red earth (hematite)

(Fig. 4.27), while some shades of yellow combine yellow ochre (goethite) with lead oxide (litharge) (Fig. 4.22) [18, 35, 36].

Preparatory layers are made of thin layers of gypsum-based plaster, mixed free of aggregates. This was a widespread practice in the region and characterizes above all this phase of Cappadocian painting [35, 36]. The brilliant white of the gypsum plaster provides the background for subsequent paint layers. Surface marks left by spatulas and brushes in the plaster provide evidence of working procedures, which was very likely applied as a malleable and dense paste. At almost all the sites from this period analysed, painting schemes comprise two plaster layers: the first layer was spread mainly with a trowel or spatula, so that it would tenaciously adhere to the rocky support; the second layer is thinner and smoothed on the surface with wide brushes. A variant of this practice is found in the decoration of the north apse of the church of Saints Joachim and Anna in the Kızıl Çukur valley (phase II, north apse and narthex) (Figs. 4.24 and 4.25) [38]. Here, impressions in the plasterwork show that the first plaster layer was applied by brush, which was left intentionally rough to support the second plaster layer. Similarly, in the church of Agios Basilios in the Gomeda valley [39], the preparatory layer of the first painting phase in the apse (Figs. 4.26 and 4.27) is made of a white plaster that was spread by brush, in order to follow the topography of the rock as closely as possible. The second layer, which accommodates the painting, is also a thin layer of white plaster spread by brush (Figs. 4.22 and 4.23).

The plastering evidence testifies to swift working practices, which witnessed architectural modules plastered in broad sequences, progressing from the apse to the vault and the walls. Evidence of interruptions or overlapping seams relating to scaffolding lifts or day-stages is not readily apparent, but may be lost or disguised. Even though the time required to prepare and apply the plaster layers would have been quite short, other operations were carried out simultaneously. Preparatory drawing could be applied to the still wet gypsum layer, for example.

The application of two gypsum plaster layers is a distinctive feature of this group of paintings. Although its definitive recognition requires scientific analyses, it is probably very widespread, and is certainly present in the paintings of phase I and II of the Church of Saint Basil; in the second phase of the church of Saints Joachim and Anna in the Kızıl Çukur valley; in the church of the stylite Niceta in the Kızıl Çukur valley; in the church of Karşı becak in Göreme; and in the second phase of the church of Agios Stephanos at the Keşlik monastery in Cemil [35, 36, 39, 40]. The extensive use of these gypsum plastering procedures denotes the adoption of a technical approach that is associated with a homogeneous group of paintings produced in the same period, or at least in a narrow chronological timeframe, and which, as a consequence, meant that their paintings were executed in *secco* techniques on dry plaster. These technical convergences should not be taken to indicate the practice of a specific workshop, but they do highlight a distinct tradition of mural practice in the region between the sixth–seventh and ninth centuries.

The second phase of painting at Şahinefendi (Figs. 4.11, 4.15 and 4.16), although still belonging to the proto-Byzantine period, shares few technical and material characteristics with the phase I scheme, except with respect to the application of plaster layers. It can be compared with the paintings in the church of Badem near Kavaklı Dere from the ninth century (Fig. 4.28), even recalling the iconographic subject matter in the apse, which can be glimpsed under the Ascension of the third phase of painting here. The preparatory layer of the Badem paintings is made up of two coats of white plaster that combine lime and gypsum, the first spread with a spatula and the second with a brush (Fig. 4.29). The painting is applied in broad brushstrokes with overlapping layers, and its iconographic composition includes, as in Şahinefendi, a *Maiestas Domini* in the apse. The colour range includes ochres (light red, dark red), black and green. The green, which is used as a background colour in the apse, was applied last, and would have required the addition of a binding medium.

The third phase of painting at Şahinefendi (Figs. 4.8, 4.9, 4.11 and 4.17) is linked to one of the richest periods of church decoration in Cappadocia. Churches that are datable to the eleventh century can be divided into two main stylistic categories, respectively centred around the Yılanli Kilise (Fig. 4.30) and the so-called "Column Churches" (Elmali Kilise, Karanlik Kilise, Çarikli Kilise) (Fig. 4.31) [23, 25, 41]. The first group is characterized by paintings created on very thin layers of plaster using a limited palette, while the second group of the column churches, fewer in number, preserve higher quality painting. The phase III painting at Şahinefendi can be directly compared with the paintings of this last group, with which it shares very close stylistic, iconographic and technical features. There is a very close iconographical relationship, for example, with the Nativity scene of the church of Karanlik Kilise within the Open Air Museum of Göreme (Figs. 4.11 and 4.32) [23]. From a technical perspective, plasters in this group were enriched with organic (vegetable fibres) and inorganic aggregates, and some have been found to combine both gypsum and lime [35, 36]. Use of lime allowed *fresco* painting, which was employed to lay in large base areas. After setting, these areas were then additionally painted with pigments mixed with a binding medium (in Karanlik church, analysis carried out in 1982 identified vegetable gum [26]). The colour range is typically much broader, including white, carbon black, yellow and red ochres, and green. Similar technological features are found in phase III of the painting at Şahinefendi. The plasters are likewise made of lime and gypsum enriched with inorganic aggregates and sparse vegetable fibres. The use and sequencing of *fresco* and *secco* painting is also comparable.

The last painting phase at Şahinefendi (phase IV), which is dated to 1216–1217 (Figs. 4.18 and 4.19), differs significantly from the others. It is one of only a few other examples of this painting approach from this era, and from the same area, which also include the Church of the Archangel in Cemil (Figs. 4.33 and 4.34) and the Church of Taşkinpasa. Compositional and stylistic similarities with the Church of Taşkinpasa are also marked. The main technical difference, with respect to the previous phases, is due to the presence in the plaster of material extracts from local rocks and very thin vegetal fibres, which were added to improve its working properties and performance characteristics [42]. This phase is also distinguished by the use of specific pigments, such as Jarosite yellow, which have rarely been found in Cappadocian wall paintings.

4.6 The Wall Paintings of the New Tokalı Church and Tenth-Century Painting

The wall paintings of the New Tokalı church are unique among the painted churches of Cappadocia in terms of their iconographic complexity, stylistic quality, distinct technical execution processes, and the use of painting materials of rare magnificence [2, 43, 44] (Fig. 4.35). They belong to a church dug out in the area beside the apse of the older church (Old Tokalı), which is now the first space when entering the rock-cut complex (Fig. 4.36). The paintings of the New Tokalı, produced in the

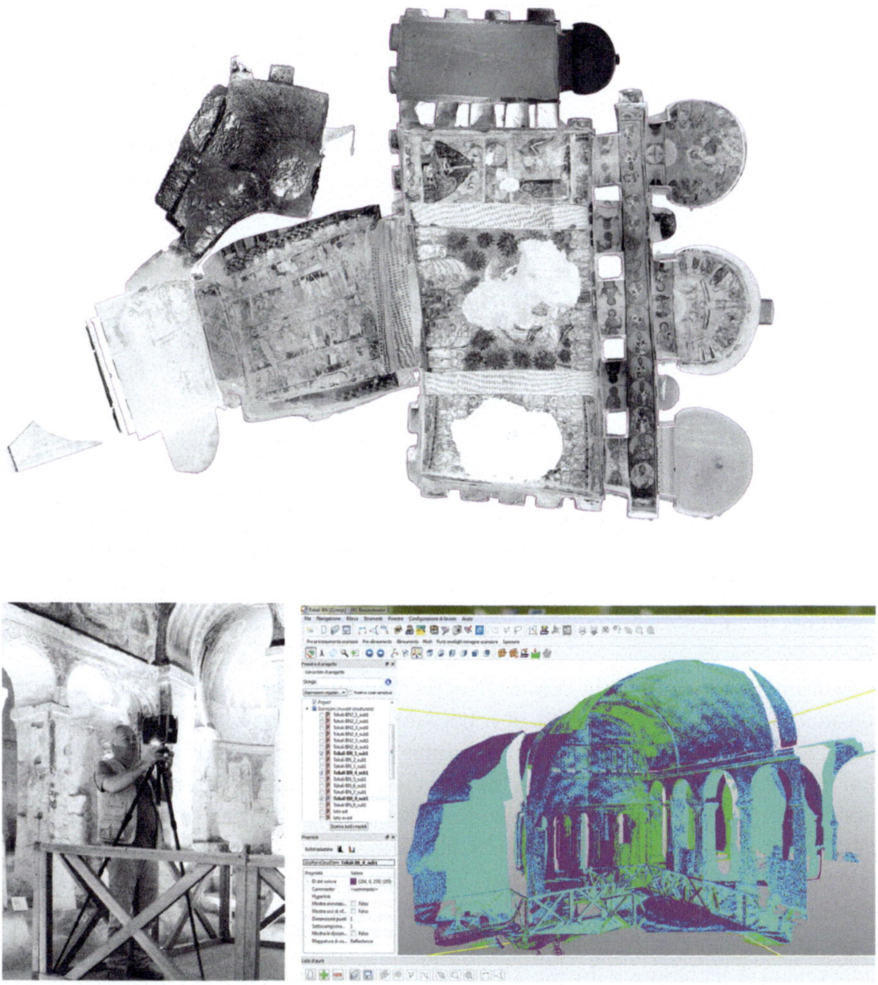

Fig. 4.36 Göreme Open Air Museum, church of Tokalı. Relief and planimetry (M. Carpiceci, Sapienza University)

mid-tenth century, are therefore contiguous with those of the Old Tokalı, which date back to 913–920. While the chronological difference between the two painting complexes is small, immense differences characterize approaches to the creation of their wall paintings. The work produced by the workshop whose task it was to paint the new church is unique among those active in Cappadocia at this time, and finds no comparison elsewhere. It was most likely Constantinopolitan painters who came to this area at the wish of a patron who was closely related, through the Phocas family, to the imperial court of Constantinople in the middle of the tenth century, before Niceforo Phoca became Byzantine emperor (963–969). They had access to high quality precious materials and in considerable quantities. In particular, the use of ultramarine blue obtained by processing from lapis lazuli, and of gold and silver leaf, is both lavish and surprising.

The workshop painted the surfaces of the entire cave-complex using *secco* techniques [27, 35, 36]. The approach is entirely different from the other Cappadocian paintings analysed in the region. Here, a white finishing layer applied over the plaster provides the base for painting using organic binders. Examination using a portable microscope (50x magnification) revealed a phenomenon of micro-cracking (craquelure) which is typical of such paintings (Figs. 4.37 and 4.38).

Analysis indentified casein as a binder at Tokalı, but physical evidence in the form of small bubbles in the paint surface also suggests that glue was used. The palette includes natural earth red (hematite) and yellow ochre (goethite), carbon black and green earth together with more valuable pigments such as ultramarine blue, and the organic colourant, indigo. The extensive use of ultramarine (lapis lazuli), which covers the blue sky of the scenes with a lighter shade and fills the garments of Christ and the Virgin with a darker violet-blue shade, is remarkable (Figs. 4.39 and 4.40). The modulated darker shading of the garments is achieved by the blue being painted over a light or dark grey background. The meticulous cleaning of the paintings restored vigour to the blue backgrounds and made extraordinary details legible once again. Unexpected complexity is evidenced in traces of gilding in the haloes, which were once covered with thin gold-leaf (Fig. 4.41), and in the use of silver for the highlights of Christ's robe in the scene of the Transfiguration painted in the corridor (Fig. 4.42).

Painting practices in tenth-century Cappadocia incorporate a multitude of execution methods that are evidenced in a network of workshops active in the valleys between Göreme and Çavusin. Wall paintings produced by these workshops vary in their technical quality, ranging from those in the New Tokalı church, which we have defined as uniquely rich, to other churches that contain more rudimentary schemes using poorer materials. The Dovecote Church in Çavusin (963–969) is chronologically very close to the New Tokalı church [23, 30] (Figs. 4.43 and 4.44). The painting schemes of both churches share the same subject matter, incorporating the Mission of the Apostles and the Ascension, but they differ in terms of their painting materials and execution techniques. In the Dovecote, for example, use of ultramarine blue (lapis lazuli) is found only in the narthex as the background for the figures of the Archangels and in the scene showing the Vision of St. Eustace [35, 36] (Fig. 4.45). Here, the precious pigment is applied over a red preparation, whereas in the New

Fig. 4.37 Göreme Open Air Museum, New Tokalı Church, south wall. Detail of the painting

Tokalı church, as mentioned, the base colour is light grey and dark grey. Otherwise, the general palette is mainly based on earth pigments (such as iron oxides and green earth), although lead-based pigments, now altered to black, are also present. A very thin layer of white plaster is used, free of aggregates. These technological disparities indicate the work of an entirely different (and inferior) workshop, which is also reflected in the style of the Çavusin paintings. Despite sharing the same iconography as the New Tokalı church, the figures were composed with no sense of proportion and are very elongated (Figs. 4.39 and 4.44).

The paintings in the Old Church of Tokalı (Figs. 4.46 and 4.47)—which are similar to the second pictorial phase of the Church of Saint John in Güllü dere (Ayvali kilise)—are certainly datable to 913–920 (Fig. 4.48) [23, 30]. With the church of the Holy Apostles in Mustafapaşa (Fig. 4.49), they constitute the linchpin of another homogeneous group of paintings, probably produced by the same workshop using broadly similar execution processes and a common palette. Nevertheless, the execution techniques differ. In the church of Tokalı, the painting was executed on a *fresco* base and subsequently with pigments mixed with a binding medium; while

Fig. 4.38 Göreme Open Air Museum, New Tokalı Church, south wall. Detail of micro-cracking (craquelure) in the paint layer (50x magnification)

Fig. 4.39 Göreme Open Air Museum, New Tokalı Church, vault. Mission of the Apostles

Fig. 4.40 Göreme Open Air Museum, New Tokalı Church, vault. Detail of the Adoration of the Magi

Fig. 4.41 Göreme Open Air Museum, New Tokalı Church, vault. Detail of the gold- leaf used to gild the haloes

Fig. 4.42 Göreme Open Air Museum, New Tokalı Church, corridor, west wall. Detail showing the use of silver for the highlights of Christ's robe in the scene of Transfiguration

Fig. 4.43 Çavusin, the church of Dovecote. Interior

Fig. 4.44 Çavusin,Dovecote church. Painting of the vault with the Mission of the Apostles

Fig. 4.45 Çavusin, Dovecote church, narthex. The ultramarine blue here is painted on a red background

in Güllü dere, the preparatory layer—comprising a thin white plaster laid over the oldest pictorial phase—was the base for a painting made entirely using the *secco* technique. The painted decoration of the vault of the nave and of the apse arch of the church of the Holy Apostles is considered to be by the same workshop that was responsible for the Old Church of Tokalı, and shared aspects of painting technology can be identified at both churches [45].

Fig. 4.46 Göreme Open Air Museum, church of Tokalı. Interior with the Old Tokalı church

Fig. 4.47 Göreme Open Air Museum, Old Tokalı church, detail of the painting of the vault

The multitude of execution techniques adopted in the tenth century also includes a group of paintings characterised by the predominant use of local and poorer materials in the production of preparatory layers. Beige coloured plasters, rich in inorganic aggregates (local stones) and finely shredded vegetables, which are particularly crumbly, are a shared feature of this group. An example are the paintings of the Church of the Archangels in Zindanönü (first half of the tenth century) (Fig. 4.50) where the decoration, largely lost today, was produced using an intense chromatic range that played with the contrast between the red and green tones laid out in compact washes. The plasters here comprise two layers, one with very little binder, but rich in straw

Fig. 4.48 Gullu Dere, church of Saint John in (Ayvali kilise). Detail of the painting of the second phase

Fig. 4.49 Mustafapaşa, church of the Holy Apostles. Detail of the painting

Fig. 4.50 Zindanönü, church of the Archangels. Detail of the painting

and tuffaceous aggregates; and the other, thin and white, applied with a brush. The same characteristics can be found in the plaster stratigraphies of the paintings in church 3 in Göreme (start of the tenth century), and in those of church 9 in Göreme (first quarter of the tenth century) (Fig. 4.51), both in the Göreme valley area.

Fig. 4.51 Göreme Open Air Museum, church n. 9. Detail of the painting

4.7 From Techniques to Execution Methods

In the field of wall paintings, the term 'techniques' may be defined as the main procedures utilized by muralists over time, whereas 'execution methods' refers to the specific, often unique, actions and procedures used by craftsmen and painters to bring these techniques to completion. Consequently, a true understanding of wall painting technology requires much more than the knowledge of the constituent materials used. By interrogating their execution methods, we obtain a better understanding of how wall paintings were produced. This provides a bridge between viewing wall paintings as simply material artefacts to appreciating them as unique works of art. A synthetic approach is required in this process, which considers the sequential procedures that contribute to the creation of wall paintings, from their setting-out techniques and preparatory drawing, and the methods and tools used to transpose initial designs to the wall surface, such as the use of stencils, to the ways in which paint layers are gradually built up and superimposed. These components emphasize the fact that often a painted image is not the result of a unitary construction, but is instead an assembly of single elements—faces, hands, arms, drapery folds, etc.—replicated sequentially on the preparatory layers [46–49]. The use of these approaches in medieval painting in the West and East have been the subject of extensive research, but study of their application in the wall paintings of Cappadocia is still ongoing.

The paintings described in this paper extend over a number of centuries, and encompass the output of many workshops, reflecting different geographical areas and cultural influences. It is therefore useful to focus compassion between these diverse paintings on specific replicable features. An analysis of the succession of painting layers used to construct the faces in Cappadocian wall painting schemes provides specific insights of the complexity of the painting process. The examples described below date from the ninth to the eleventh centuries, one of the richest periods of painting in the history of Cappadocia's painted churches.

The faces painted in phase III (eleventh century) in the Church of the Forty Martyrs in Şahinefendi were begun with a light yellow base layer, followed by overlapping colours (Fig. 4.52). The main features were drawn with red strokes, while the principal skin tone, which extends from the face to the neck, was blocked in with a very warm pink (Fig. 4.53). Over this, intense pink brushstrokes define the cheeks and the lips, and underline the nasal area, forehead and chin. Once the basic complexion had been established, the shadows were introduced, in greens (Fig. 4.53) and a dark brown tone, and then the features were defined with decisive brown brushstrokes (Fig. 4.54). Lastly, black strokes highlighting the volume of the face were painted (Fig. 4.55), along with white highlights.

In contrast, the faces of the Marian cycle painted on the vault and walls of the north aisle in the church of Saints Joachim and Anna, in the Kızıl Çukur valley, built approximately two centuries earlier, in the ninth century, have a more schematic structure (Fig. 4.56). A base skin tone of warm pink tending to yellow accommodates subsequent painting. The facial features are then built up in a sequence of brushstrokes applied in red lines followed by white highlights, placed carefully to impart volume,

Fig. 4.52 Şahinefendi, church of the Forty Martyrs, south vault. The Virgin of the Nativity

Fig. 4.53 Şahinefendi, church of the Forty Martyrs. Portable microscopy image (50x magnification) showing the overlay of the green shade on the shades of pink

Fig. 4.54 Şahinefendi, church of the Forty Martyrs. Portable microscopy image (50x magnification) showing the features defined with brown brush strokes

Fig. 4.55 Şahinefendi, church of the Forty Martyrs. Potable microscopy image (50x magnification) showing how the black lines complete the construction of the face

such as in the use of vertical lines under the eyes. The highlights were probably made with two different types of white: lime white and lead white, the latter now blackened due to the alteration of the pigment

Fig. 4.56 Kızıl Çukur valley, church of Saints Joachim and Anna. Detail of an angel in the vault

An interesting departure from these general approaches to facial painting is encountered in the New Tokalı church (mid-tenth century), where a variety of methods can be attributed to different painters. In particular, analysis of the paintings in the corridor between the apses and the *naos*, conducted during the restoration process in 2018, allowed us to explore this aspect further, and to distinguish with precision the work of various painters, who worked in uniquely individual ways. The work of three distinct groups has been identified. The first group includes the faces found in the central part of the walls and ceiling, in turn divided into two sub-groups ('a' and 'b'), which are based on a common template, but are then separated by the manner in which subsequent paint layers are applied. The facial types of the two sub-groups, constructed as usual by adding fields and linear strokes, have a different basic colour scheme. Those positioned between the central apse and the northern apse (I-a) are characterized by a dark red skin tone, the features of which are then defined by additional brushstrokes and highlights (Fig. 4.57). In contrast, in sub-group I-b, the dark red skin tone appears to be absent, and the features are built up in a sequence of more precise chromatic steps (Fig. 4.58). This latter type seems to be very widespread at the New Tokalı church, with the faces of the figures on the north and south walls of the *naos* and, most likely, also those painted on the vault probably belonging to it.

The faces of the figures in the north apse and in the part of the corridor facing it are part of the second group (Fig. 4.59). They maintain distinctive traits found in the facial painting of the first group (I-b), but employ a different colour range. In particular, they are characterized by the presence of an ochre-green base that is used sparingly to define the shadows. This is also used with a pink skin tone that underlies the lines that define the features. The third group includes the faces of the figures arranged on the south side of the corridor. Here, facial complexions are built up by means of slight and nuanced chromatic shifts, from pink to darker shades, on top

Fig. 4.57 Göreme Open Air Museum, New Tokalı Church, corridor between naos and apses. Face type I-a

Fig. 4.58 Göreme Open Air Museum, New Tokalı Church, corridor between naos and apsis. Face type I-b

Fig. 4.59 Göreme Open Air Museum, New Tokalı Church, corridor between naos and apsis. Face type II

of which are dark passages and highlights that provide definition without specifying the anatomical details (Fig. 4.60). The eyes, in fact, are depicted with only a black pupil and a comma-shaped highlight that indicates the sclera of the eye.

The wealth of technical information contained in these few comparative examples indicates the potential for further study to deepen our knowledge of technical methods, which could then help identify the different workshops that painted in Cappadocia and clarify unresolved chronological issues [P. P.].

4.8 Conclusion

The technical data presented in this paper, as already mentioned, derives from the different but complementary outcomes offered by a survey of multiple churches, and the focus provided by the restoration work carried out at two specific painted sites. The survey allowed us to compile information from more than 100 churches and to build a general framework. The restoration work allowed us to conduct an in-depth study of original materials and execution techniques, and to extrapolate from these findings different execution methods used in Cappadocian painting.

In summary, findings indicate that the use of different materials and execution techniques can be related to four macro-phases into which we have divided painting practice in Cappadocia. In the overall development of painting up to the eleventh century, there was a general tendency to expand from simplified forms to more complex approaches, which is partly related to the development of more ambitious

Fig. 4.60 Göreme Open Air Museum, New Tokalı Church, corridor between naos and apsis. Face type III

architectural spaces. In the tenth and eleventh centuries, painting procedures proliferated, which can be associated with the presence of different workshops coexisting in the extended territory between the Open Air Museum and Çavusin. This diversity encompasses a broad range, from the most modest wall painting schemes up to those of an exceptional nature, the latter exemplified by the painting in the New Tokalı church or the painting cycles of the so-called 'Column Churches'.

Acknowledgements This research was made possible thanks to the General Directorate of Monuments and Museums of the Turkish Republic, which authorised the survey, research and restoration activities of Tuscia University in Cappadocia. The work was carried out in collaboration with the Nevşehir Archeological Museum and the Nevşehir Conservation and Restoration Regional Laboratory.

Our thanks also go to Tuscia University and the Italian Foreign Affairs Ministry for their continuous support.

We would also like to thank Taniguchi Yoko for the friendship and scientific exchange that started in Cappadocia where she is involved in the University of Tsukuba's conservation projects; and to Prof. Aoki Shigeo for inviting us to participate at the International Colloquium on the Conservation of Asian Wall Paintings and the Painting Technology Exchange (Tokyo 2019).

References

1. Andaloro, M.: The project on the rock paintings in Cappadocia (2012). Research and preservation: 1. the Göreme Open Air Museum; 2. the church of the Forty Martyrs in Şahinefendi. In: XXXVI International Symposium of Excavations, Survey and Archaeometry, Muğla, 27–31 May 2013 (Conference Proceedings), Ankara, pp. 120–135 (2014)
2. Andaloro, M.: Yeni Tokalı Kilisesi: Bir Diğer Tabir ile Kapadokya'nın Sistin Şapeli/The New Tokalı kilise, or the Sistine Chapel of Cappadocia. In: Öger, A. (ed.) 1. Uluslararasi Nevşehir Tarih Ve Kültür Sempozyumu, Nevşehir Üniversitesi, 16–19 Kasım 2011 (Conference Proceedings), Nevşehir, vol. 5, pp. 5–26 (2012)
3. Andaloro, M.: Painting on rock-hewn settlements in Cappadocia: knowledge, preservation and enhancement (2016 campaign). In: 35. Araştırma Sonuçlan Toplantısı, Bursa, 22–26 Mayıs 2017 (Conference Proceedings), Bursa, vol. 2, pp. 99–112 (2017)
4. Andaloro, M., Borghini, S.: The restoration of the Church of the Forty Martyrs in Şahinefendi in Cappadocia. 2009 campaign, in 26. Arkeometri Sonuçları Toplantısı, Istanbul, 24–28 Mayıs 2010 (Conference Proceedings), Ankara, pp. 147–160 (2011)
5. Açıkgöz, F., Alberti, L., Andaloro, M., Caldi, C., Dikilitaş, G., Pogliani, P., Valentini, V., Yalçınkaya, U.: Kayalık Kapadokya'daki Resimli Süslemelerin Restorasyon Çalışmaları/The restoration of the rock hewn paintings in Cappadocia. In: Arkeoloji ve Sanat, vol. 154, pp. 190–191 (2017)
6. Andaloro, M. (ed.).: La Teoria del Restauro nel Novecento da Riegl a Brandi, Atti del Convegno Internazionale di Studi, Viterbo, 12–15 Novembre 2003 (Conference Proceedings), Firenze (2006)
7. Bon Valsassina, C.: Restauro made in Italy, Milano (2006)
8. Brandi, C.: Teoria del Restauro, Torino (1963)
9. Shufuko no riron (trad. Hidehiro Ikegami e Idemi Otake; rev. a cura di Shigetoshi Osano), Sangensha Publishing Inc., Tokyo 2005
10. Pogliani, P.: Restorasyon Çalışmaları Aracılığı ile Kapadokya Kayalı kiliserenine Diğer Kazandırma Çalışmaları/The improvement of the rock churches of Cappadocia through the restoration. In: Öger, A. (ed.) 1. Uluslararasi Nevşehir Tarih Ve Kültür Sempozyumu, Nevşehir Üniversitesi, 16–19 Kasım 2011 (Conference Proceedings), Nevşehir, vol. 6, pp. 289–302 (2012)
11. Ozil, R.: The objectives of the International mural painting conservation project (Göreme, 1973–1990). In The safeguard of the rock-hewn churches of the Göreme valley, International Seminar, Ürgüp, Cappadocia, Turkey, 5–10 September 1993 (Conference Proceedings), Rome, pp. 163–170 (1995)
12. Andaloro, M.: Natura e figura in Cappadocia. In: Quintavalle, C.A. (ed.) Medioevo, natura e figura, Associazione Italiana Storici dell'Arte Medievale, Parma 20–25 Settembre 2011 (Conference Proceedings), Milano, pp. 189–213 (2015)
13. Benucci, M., Crescenzi, C., Giustiniani, C., Romagnoli, G.: Survey on a monastic complex in the area of Göreme (Nevsehir, Turkey). In: Parise, M., Galeazzi, C., Bixio, C., Yamac, A. (eds.) Hypogea 2017. International Congress of Speleology in Artificial Cavities, Goreme, 6–8 March 2017 (Conference Proceedings), Istanbul, pp. 404–412 (2017)
14. Benucci, M., Carpiceci, M., Colonnese F., Inglese, C., Romagnoli, G.: The rock-cut settlement and the courtyard complex of Şahinefendi (Nevşehir, Turkey). In: Parise, M., Galeazzi, C., Bixio, C., Yamac, A. (eds.) Hypogea 2017. International Congress of Speleology in Artificial Cavities, Goreme, 6–8 March 2017 (Conference Proceedings), Istanbul, pp. 251–260 (2017)
15. Andaloro, M.: La chiesa di Haghios Basilios e la pittura delle origini in Cappadocia. Linee di un progetto. In: Quintavalle, C.A. (ed.) Medioevo, natura e figura, Associazione Italiana Storici dell'Arte Medievale, Parma, 20–25 Settembre 2011 (Conference Proceedings), Milano, pp. 215–227 (2015)
16. Bordi, G.: La chiesa di Haghios Basilios e la pittura delle origini in Cappadocia. Palinsesti pittorici. In: Quintavalle, C.A. (ed.) Medioevo, natura e figura, Associazione Italiana

Storici dell'Arte Medievale, Parma, 20–25 Settembre 2011 (Conference Proceedings), Milano, pp. 229–238 (2015)

17. Bordino, C.: La chiesa di Haghios Basilios e la pittura delle origini in Cappadocia. Fonti e iscrizioni. In: Quintavalle, C.A. (ed.) Medioevo, natura e figura, Associazione Italiana Storici dell'Arte Medievale, Parma, 20–25 Settembre 2011 (Conference Proceedings), Milano, pp. 239–256 (2015)
18. Pogliani, P.: La chiesa di Haghios Basilios e la pittura delle origini in Cappadocia. Materiali e stesure pittoriche. In: Quintavalle, C.A. (ed.) Medioevo, natura e figura, Associazione Italiana Storici dell'Arte Medievale, Parma, 20–25 Settembre 2011 (Conference Proceedings), Milano, pp. 257–264 (2015)
19. Bordi, G.: El Nazar a Göreme. Sulle tracce delle chiese delle origini in Cappadocia. In: Paribeni, A., Pedone, S. (eds.) Di Bisanzio dirai ciò che è passato, ciò che passa e che sarà. Scritti in onore di Alessandra Guiglia, Roma, pp. 113–128 (2018)
20. Bordino, C.: La chiesa di sant'Eustachio a Göreme e la sua decorazione pittorica attraverso il database "Cappadocia. Arte e habitat rupestre". In: Jurkovic, M. (ed.) CARE (Corpus architecturae religiosae Europae, IV-X saec.)—Meaning and Use of Corpora, 24th Annual International Scientific Symposium of IRCLAMA, Hortus Artium Medievalium, vol. 24, pp. 80–89 (2018)
21. Bordino, C.: El Nazar a Göreme. Una rara rappresentazione pittorica di Longino Martire di Cappadocia? In: Paribeni, A., Pedone, S. (eds.) Di Bisanzio dirai ciò che è passato, ciò che passa e che sarà. Scritti in onore di Alessandra Guiglia, Roma, pp. 129–147 (2018)
22. Coşkuner, B.: Wall-painting methods of Byzantine art: examples from Cappadocia. In: Ödekan, A., Akyürek, E., Necipoğlu, N., Türkoğlu, I. (eds.) On ikinci ve on üçüncü yüzyıllarda Bizans dünyasinda değişim/Change in the Byzantine world in the twelfth and thirteenth centuries, 25–27 June Istanbul 2007 (Conference Proceedings), pp. 495–504 (2010)
23. Jolivet-Lévy, C.: La Cappadoce un siècle après Guillaume De Jerphanion, avec la collaboration de Nicole Lemaigre Demesnil, 2 vol., Paris (2015)
24. Thierry, N.: Une école de peinture greco-orientale du haut Moyen Age. In: Byzantine East, Latin West. Art Historical Studies in Honr of Kurt Weitzmann, Princeton, pp. 299–305 (1995)
25. Restle, M.: Byzantine Wall-Painting in Asia Minor, I-III, Greenwich, CT (1967)
26. Dangas, I.: Conservation of mural paintings of the Karanlik Church, Göreme (Turkey). In: The safeguard of the rock-hewn churches of the Göreme valley, International Seminar, Ürgüp, Cappadocia, Turkey, 5–10 September 1993 (Conference Proceedings), Rome, pp. 171–180 (1995)
27. Schwartzbaum, P.M.: The conservation of the mural paintings in the rock-cut churches of Göreme. In: Epstein, A.W. (ed.) Tokalı Kilise. Tenth-century metropolitan art in Byzantine Cappadocia, Washington, DC, pp. 52–57 (1986)
28. Schwartzbaum, P.M.: The ICCROM project for conservation of mural paintings in the rock churches of the Göreme valley (1971–1983). In: The safeguard of the rock-hewn churches of the Göreme valley, Proceedings of an international seminar, Ürgüp, Cappadocia, Turkey, 5–10 September 1993 (Conference Proceedings), Rome, pp. 187–202 (1995)
29. Jolivet-Lévy, C.: La Cappadoce médiévale. Images et spiritualité, St Léger Vauban (ed. It. Milano 2001) (2001)
30. Thierry, N.: La Cappadoce de l'antiquité au moyen âge, Turnhout (2002)
31. Andaloro, M., Valentini, V.: La Banca Dati Cappadocia - Arte e Habitat rupestre. Dai corpora cartacei alle applicazioni web. In: Jurkovic, M. (ed.) CARE (Corpus architecturae religiosae Europae, IV-X saec.)—Meaning and Use of Corpora, 24th Annual International Scientific Symposium of IRCLAMA, Hortus Artium Medievalium, 24, pp. 74–79 (2018)
32. Andaloro, M., Pogliani, P.: La banca dati Cappadocia-Arte e habitat rupestre. Per un progetto di conoscenza, conservazione e valorizzazione. In: Sogliani, F., Gargiulo, B., Annunziata, E., Vitale, V. (eds.) VIII Congresso Nazionale di Archeologia Medievale SAMI, Matera 12–15 Settembre 2018 (Conference Proceedings), Sesto Fiorentino, pp. 239–242 (2018)
33. Andaloro, M., Bordi G., Bordino C., Pogliani P., Valentini, V.: Signs on the rock. Red paintings in the valleys of Göreme and Kılıçlar, In: Parise, M., Galeazzi, C., Bixio, C., Yamac, A. (eds.) Hypogea 2017. International Congress of Speleology in Artificial Cavities, Cappadocia, March 6/10, 2017 (Conference Proceedings), Istanbul, pp. 11–19 (2017)

34. Pelosi, C., Santamaria, U., Agresti, G., Castro, F., Lotti, D., Pogliani P.: Project on the rock paintings in Cappadocia. Analytical investigation of the church of Forty Martyrs in Şahinefendi and other sites (Report 2008). In: 25. Arkeometri Sonuçları Toplantısı, Denizli, 25–29 Mayıs 2009 (Conference Proceedings), Ankara, pp. 535–552 (2010)
35. Pelosi, C., Agresti, G., Andaloro, M., Baraldi, P., Pogliani, P., Santamaria, U.: The rock hewn wall paintings in Cappadocia (Turkey). Characterization of the constitutive materials and chronological overview. e-Preservation Sci. **10**(2013), 99–108 (2013)
36. Pelosi, C., Agresti, G., Andaloro, M., Baraldi, P., Pogliani, P., Santamaria, U.: Micro-Raman and internal micro-stratigraphic analysis of the painting materials in the rock-hewn church of the Forty Martyrs in Sahinefendi, Cappadocia (Turkey). In: 7th International Congress on the Application of Raman in Art and Archaeology, 2–6 September 2013 (Conference Proceedings), Ljubljana, pp. 164–165 (2013)
37. Santamaria, U., Agresti, G., Pelosi, C., Pernella, A.: Rock's painting materials in Cappadocia: Şahinefendi Forty Martyrs Church. In: XXX International Symposium of Excavations, Survey and Archaeometry, Ankara, 26–30 May 2008 (Conference Proceedings), Ankara, pp. 307–316 (2009)
38. Thierry, N.: Haut Moyen Ăge en Cappadoce. Les églises de la region de Çavusin, II, Paris (1994)
39. Agresti, G., Andaloro, M., Baraldi, P., Pelosi, C., Pogliani, P., Santamaria, U.: Mural paintings in Cappadocia (Turkey). Characterization of the constitutive materials and chronological overview. e-Preservation Sci. **10**, 99–108 (2013)
40. Taniguchi, Y.: Scientific Studies on Conservation for Üzümlü Church and its wall paintings in Cappadocia, Turkey, Annual report on the activities in 2014 (2014) http://rcwasia.hass.tsukuba.ac.jp/contents/content%20images/publication/uzumlu2014.pdf
41. Ousterhout, R.G.: Visualizing community: art, material culture and settlement in Byzantine Cappadocia, Washington (Dumbarton Oaks Studies, 46) (2017)
42. Pelosi, C., Agresti, G., Andaloro, M., Baraldi, P., Pogliani, P., Santamaria, U., La Russa, M.F., Ruffolo, S.A., Rovella, N.: Micro-Raman and internal micro-stratigraphic analysis of the painting materials in the rock-hewn church of the Forty Martyrs in Şahinefendi, Cappadocia (Turkey). Archaeometry **58**(4), 659–672 (2016)
43. Jolivet-Lévy, C.: Tokali Kilise revisited: new considerations on Christ's Ministry in the New Church. In: Öger, A. (ed.) 1. Uluslararasi Nevşehir Tarih Ve Kültür Sempozyumu, Nevşehir Üniversitesi, 16–19 Kasım 2011 (Conference Proceedings), Nevşehir, vol. 6, pp. 289–302 (2012)
44. Epstein, A.W.: Tokalı Kilise. Tenth-century metropolitan art in Byzantine Cappadocia. Washington, DC (1986)
45. Thierry, N.: Un atelier de peinture du debout du Xe s. en Cappadoce, l'atelier de l'ancienne église de Tokalı, in Bulletin des Antiquaires de France, pp. 170–178 (1971)
46. Andaloro, M.: Archetipo, modelli, sagome a Bisanzio. In: Quintavalle, A.C. (ed.) Medioevo: i modelli, Associazione Italiana Storici dell'Arte Medievale, Parma 27 Settembre – 1 Ottobre 1999 (Conference Proceedings), Milano, pp. 567–574 (2002)
47. Andaloro, M.: La parete palinsesto; 1900, 2000. In: Osborne, J., Brandt, J.R., Morganti, G. (eds.) Santa Maria Antiqua al Foro Romano cento anni dopo, Colloquio Internazionale, Roma, 5–6 Maggio 2000 (Conference Proceedings), Roma, pp. 97–111 (2005)
48. Pogliani, P.: Pittori e mosaicisti nei cantieri di Giovanni VII (705–707). In: Bordi, G., Carlettini, I., Fobelli, M., Menna, M.R., Pogliani, P. (eds.) L'officina dello sguardo. Scritti in onore di Maria Andaloro, Roma, vol. II, Immagine, memoria, materia, pp. 443–450 (2014)
49. Pogliani, P.: Le pitture del complesso di San Magno. Fra le stesure pittoriche: analisi delle tecniche d'esecuzione. In: Gianandrea, M., D'Onofrio, M. (eds.) Fondi nel Medioevo, Roma, pp. 113–120 (2016)
50. De Jerphanion, G.: Une nouvelle province de l'art byzantin. Les églises rupestres de Cappadoce, 5 vol., Paris (1925–1942)

Chapter 5
At the Western Edge of the Silk Road: Challenges of Conserving a Unique Nabataean Wall Painting in Petra, Jordan

Stephen Rickerby

Abstract The ancient city of Petra in Jordan, declared a UNESCO World Heritage Site in 1985, is famous for its monuments and buildings carved out of sandstone rock. A rock-cut interior in Siq al-Barid, a suburb of Petra, preserves the only *in situ* wall painting with figurative subject matter, dating to the first century AD. Given our limited knowledge of Petra's Nabataean art and culture, the importance of this unique survival cannot be underestimated. From 2006-10, the Petra National Trust, a NGO, collaborated with the Courtauld Institute to conserve and clean the painting. Emerging from beneath obscuring dirt and graffiti, its extremely high quality and stylistic indebtedness to Hellenistic art were recognised. Scientific examination revealed a sophisticated original technology. This paper considers the techniques of the painting and its conservation, including issues of balancing site protection requirements against the demands of tourism, and the need to integrate local communities in these efforts.

Keywords Petra · Nabataean · Wall painting · Conservation · Original technology · Site management and tourism

5.1 Introduction

The ancient city of Petra, in Jordan, is famous for its monuments and buildings cut out of sandstone rock. These were once richly painted, but centuries of weathering have revealed the underlying "rose-red" stone for which the site is best known today. One cave complex uniquely preserves a 2000-year old figurative painting. Located in a suburb of the city that was originally also a centre of viticulture, its chambers were a discrete meeting place for a male fraternity, whose activities revolved around ritualised wine drinking and feasting dedicated to the cult of Dionysos. Fittingly, the interior was sumptuously painted in imitation of a bower heavily laden with

S. Rickerby (✉)
Consultant, Conservation of Wall Painting Department, Courtauld Institute of Art, University of London, Vernon Square, Penton Rise, King's Cross, London WC1X9EW, UK
e-mail: stephen.rickerby1@btinternet.com

S. Aoki et al. (eds.), *Conservation and Painting Techniques of Wall Paintings on the Ancient Silk Road*, Cultural Heritage Science,
https://doi.org/10.1007/978-981-33-4161-6_5

ripened grapes. By modern times the site and its painting had fallen into a desperate state. Used as a shelter by shepherds, the painting was entirely covered by soot from their fires. Gunshot damage, attempted thefts, and ancient and modern graffiti and iconoclasm, had all taken an appalling toll.

From 2006–2010, the Petra National Trust (PNT), a Jordanian NGO, collaborated with the Courtauld Institute (University of London), under the aegis of the Department of Antiquities of Jordan (DoA) and the Petra Archaeological Park (PAP), to study, conserve and clean this unique painting [1]. Efforts were also made to address difficult site management and protection requirements.

5.2 Archaeological Context

Declared a UNESCO World Heritage Site in 1985, Petra is one of the world's most spectacular archaeological sites. Located in mountainous terrain east of the Rift Valley in southern Jordan, its remarkable monuments and facades are carved into multicoloured Cambrian and Ordovician sandstone ridges and gorges over a vast area. The urban centre is about 10 km^2, while the Petra Archaeological Park, which also incorporates the surrounding suburbs, covers 264 km^2. First settled in about the sixth century BC by Arab traders known as Nabataeans, Petra subsequently developed into a powerful city-state, maintaining its independence until 106 AD, when it was annexed by the Roman Empire. The thriving capital that the Nabataeans created was supplied with water by an ingenious network of cisterns, channels and dams, and from here they controlled the spice trade from the south, the Silk Road trade from the east, and the main trade routes to the Mediterranean in the north [2].

Approximately 10 km to the north of Petra is Beidha, which in the Nabataean period was a fertile vine-growing and wine-making region, as evidenced by the large number of wine-presses that have been found in the area [3]. In this region too is a narrow gorge, now known as Siq al-Barid—and sometimes referred to as a suburb of Petra— which preserves numerous Nabataean rock-cut monuments and chambers, ancillary rooms, cisterns and channels for transporting water. One of these complexes (Brünnow number 849), erroneously known as 'the painted house', preserves the most outstanding example of Nabataean in situ painting [4, p. 43].

5.3 Description of the Site and Its Painting

The site originally comprised a number of features and elements, some of which survive intact, while others can be inferred from archaeological evidence (Fig. 5.1). A flight of stone-cut steps on the south side of the Siq leads up to a platform in front of a large square chamber. This was originally served by a gated entrance, and by a small side door. The chamber is fashioned as a *biclinium*, having two rock-cut benches, on its east and south sides, while a smaller vaulted recess is excavated in

Fig. 5.1 The painted cave complex at Siq al-Barid originally comprised a number of features and elements, some of which survive intact, while others can be inferred from archaeological evidence. A flight of stone-cut steps on the south side of the Siq leads up to a platform in front of a large square chamber, the *biclinium*, which is served by ancillary rooms, cisterns and channels for transporting water (© *Petra National Trust and Courtauld Institute*)

the back, south wall (Fig. 5.2). The principal chamber is decorated on its south wall with painted stuccowork, which consists of an orthostat dado surmounted by fictive ashlars. Although other examples of this type of architectural decoration survive at Petra, this is the most extensive.

The most stunning painting, however, is found in the recess. Its vault and walls are covered with an exquisite decoration of intertwining vinescrolls and flowers, populated by small winged boys (*erotes*), other figures, and a variety of birds and insects (Figs. 5.3 and 5.4). The *erotes* gather grapes into baskets, wielding pruning hooks and balancing on ladders to do so; others fend off birds with bow and arrow, and spear. The painting was once complemented by richly decorated stuccowork. Remains of an entablature with dentil decoration survive at the springing of the vault, and a stuccowork medallion originally adorned the vault's centre.

Dating of Nabataean monuments is made difficult by a lack of documentary evidence. Precise dating of the painted cave complex at Siq al-Barid is further complicated by the fact that it was adapted over time. But it can be confidently ascribed originally to the first half of the first century AD, based on the similarity of its tooling styles and architectural features to other datable monuments at Petra [4, p. 62]. Most importantly, it is certain that its figurative painting is Nabataean, not Roman, and therefore has immense art-historical significance.

Fig. 5.2 The main chamber (the *biclinium*) has two rock-cut benches, on its east and south sides, while a smaller vaulted recess is excavated in the back, south wall, as shown here. The *biclinium's* south wall is decorated with painted stuccowork, which consists of an orthostat dado surmounted by fictive ashlars. The vaulted recess is painted in imitation of a bower heavily laden with ripened grapes. This image shows the cave complex before conservation (© *Petra National Trust and Courtauld Institute*)

Fig. 5.3 The painted vault in the recess, before conservation. The exquisite decoration of intertwining vinescrolls and flowers, figures, and a variety of birds and insects was obscured and disfigured by smoke blackening, deliberate damage and graffiti (© *Petra National Trust and Courtauld Institute*)

Fig. 5.4 Detail of two birds on the vault, before conservation. The naturalistic depiction of the birds and plants represented in the painting allowed many of them to be identified (© *Petra National Trust and Courtauld Institute*

5.4 Significance and Meaning

Technical findings demonstrate that the majority of Petra's rock-cut monuments were originally painted [5]. Although notable survivals of in situ painting are relatively few, trace evidence is copious and convincing. But with the exception of the vaulted recess at Siq al-Barid, these examples are all architectural in character, imitating fine materials such as marble revetments and *opus sectile*, or representing illusionistic architectonic features such as doorways and pilasters [6, 7]. Very few examples of figurative painting have been discovered, and in comparison with these, that at Siq al-Barid is the finest in quality. Moreover, the other documented paintings are known either from fallen fragments discovered during archaeological excavations, or were found in situ and were subsequently detached.[1] This makes the wall painting at Siq al-Barid unique, as the only figurative Nabataean painting to survive in situ.

Most of what is known of Nabataean culture is extrapolated from the standing architectural evidence. Relatively few contemporary documents or artefacts are known, and references to Petra's original inhabitants are scant, and are also usually second-hand and somewhat unreliable. The painting at Siq al-Barid is of inestimable value for the information it conveys of Nabataean life, and of the influences that shaped its customs and religious practices. The vintaging theme places the painting's influences firmly in a Greco-Roman tradition, while the stunning naturalism recalls Hellenistic painting practice, other examples of which are now extremely rare [8, 9, pp. 1–5]. There is convincing evidence that these influences were disseminated via Ptolemaic Alexandria [4, p. 152].

While partial survival makes it impossible to know how usual figurative painting was at Petra, indications are that is was reserved for the most sumptuous and important settings. The exquisite painting at Siq al-Barid suggests particular patronage and use of the cave complex. The painting has been described many times in the literature, but owing to its fragmentary condition, small-scale detailing, and dirty appearance prior to cleaning, its iconography and meaning were misinterpreted [9–11]. Features clarified as a result of cleaning led to a major reinterpretation. The vintaging motif suggested a Dionysian context, and a fresh interrogation of the painting confirmed this association. The vinescroll consists of three intertwining vines—the grape-vine, ivy, and bindweed—all of which are associated with Dionysos. Music and dance, essential elements of Dionysian cult activities, are represented by a flute-player and a fragmentary dancing female (*maenad*?) figure, whose head is wreathed with vines. A thriving cult of Dionysos existed in Nabataean Petra, and in wine-producing Beidha in particular [3, p. 495]. There seems little doubt that the painted cave complex at Siq al-Barid was dedicated to Dionysos, and was used for exclusive fraternity meetings, or *symposia*, whose cultic activities focused on ritualised wine drinking [12, pp. 165–169]. Indeed, there is documentary evidence elsewhere in Beidha that *biclinia* were used for such purposes [4, p. 108].

5.5 Conservation Programme

The conservation programme was developed collaboratively over a 4-year period. Reflecting concern for the painting's unique significance—and its extremely vulnerable condition—a prioritised, cautious and gradual approach was adopted for all treatment components, based on investigations and analysis, and the results of trials. Wider aims of the project included efforts to improve site protection and address access issues. The programme was also an opportunity to highlight the plight of other survivals of wall painting in the Petra region. A comprehensive survey of painted plasters was carried out, with the aim of developing a wall painting conservation and management strategy for the area.

5.6 Diagnostic Investigations

5.6.1 Condition Assessment

Human activity has had the greatest adverse impact on the paintings. Habitation and use of the cave complex by local shepherds resulted in fire damage, including near comprehensive soot blackening and heat-related phenomena such as preferential blistering and loss of specific paint passages. Although measures were put in place to prevent these activities in 1985, more damage continued to occur, including targeted attacks on figurative elements of the painting. A metal grille installed across the painted recess for its protection had the unintended consequence of providing a means of climbing up to its vault, resulting in two damaging theft attempts. Graffiti, both incised and applied, was extensive, and there was evidence that this was continuing to accrue.

Various types of environmental deterioration were evident. Weathering, insect burrowing, and localised salts activity had caused pitting, decohesion, and severe undercutting of plaster layers, threatening the survival of the painting. It was also clear from examination of historic photographs that loss of blistering painting was gradual but ongoing (Fig. 5.5). Complex inherent paint changes and degradation effects were prominent features of the original painting technology. As is often the case, defining the precise nature and rates of change of these phenomena was problematic.

Fig. 5.5 Environmental deterioration included loss of blistering painting, which was gradual but ongoing. Complex inherent paint changes and degradation effects were prominent features of the original painting technology, too (© *Petra National Trust and Courtauld Institute*)

5.6.2 Environmental Assessment

As an exposed painted site, it was recognised from the outset that possibilities of environmental control were extremely limited. However, macro- and micro-environmental conditions were monitored over a 1-year period to assess the contribution of these factors to deterioration. Micro-cores were also taken to investigate salt types and their origins, and to assess the potential for salt reduction procedures.

As expected, monitoring established that significant seasonal and diurnal fluctuations of relative humidity and temperature in the macro-environment were only slightly buffered internally, providing conditions for regular activation of some soluble salts and possible destabilisation of sensitive paint materials. Large quantities of sodium chloride salts originating in salt-rich strata of the sandstone support were found in localised areas of deteriorated plasterwork.

5.6.3 Original Technology

Non-invasive technical investigations included visual assessment, multi-spectral imaging, and portable microscopy. Paint and plaster samples were also obtained for microscopic examination and immunological analysis, respectively.

Previous research had established that the secondary support consisted of three layers of lime plaster [5, pp. 145–148]. The conservation programme additionally identified a thin calcium sulphate ground-layer with a proteinaceous component (Fig. 5.6). Identified inorganic paint materials include a range of iron-oxide pigments, carbon black, Egyptian blue, atacamite green (possibly altered from malachite) and lead white, some bound in a proteinaceous medium. The presence of red, blue, and yellow organic colorants was established in paint layers affected by blistering and flaking. A stunning technological feature was the lavish use of gold, which in Petra had previously only been found on architectural mouldings. While only minute traces of gold leaf now remain, close examination demonstrated that the autumnal vine leaves and stems were extensively gilded (Fig. 5.7). Extremely sophisticated techniques of colour mixing and layering were used to subtle effect, for example in conveying a translucent, iridescent quality to the depiction of birds' wings [1, pp. 216–217].

The painting at Siq al-Barid is certainly indebted to Hellenistic painting practice [13, 14], but it is also technically distinct, notably in the use of a gypsum ground. This, and the extensive use of organic paint materials and media, indicated particular circumstances of technological transfer and development within the Nabataean context. Worryingly, the painting had been assumed to be a *fresco* painting, and cleaning proposals were made on this basis as recently as 1995 [15]. Had this occurred, the painting would have been irreversibly damaged.

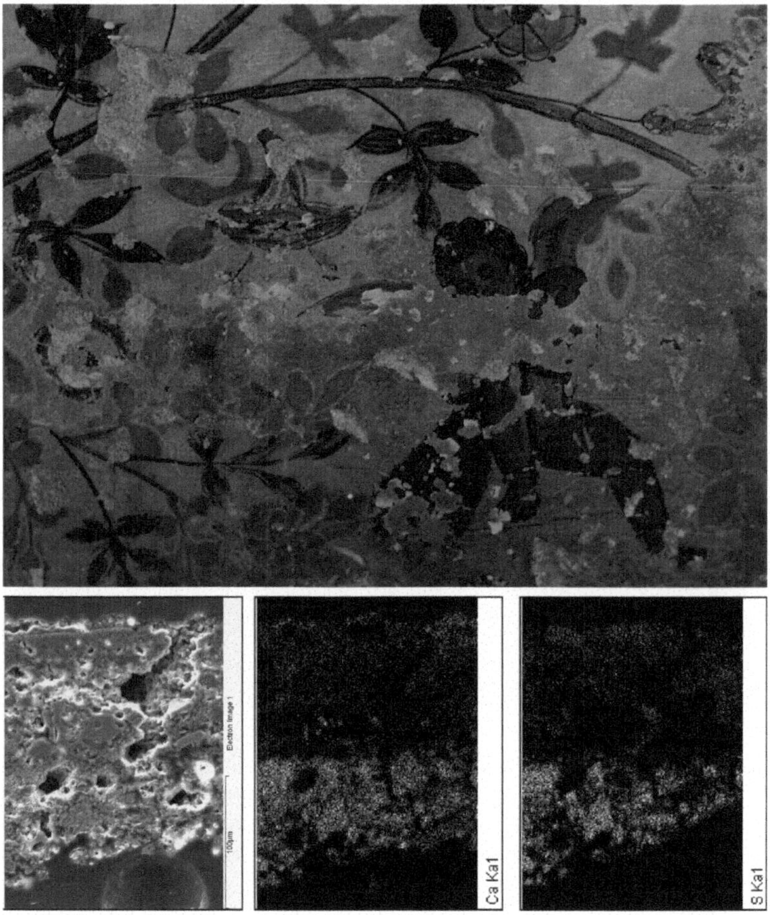

Fig. 5.6 The use of a gypsum ground is a particularly unusual aspect of the painting technology. SEM-EDX elemental mapping was used to detect inorganic components in the ground (bottom), while uv-induced fluorescence revealed extensive organic components in the paint materials (top) (© *Petra National Trust and Courtauld Institute*)

5.7 Remedial Treatment

The technology findings had far-reaching implications for remedial treatment in terms of its organisation and approach. Stabilisation of the unpainted and painted plasterwork was prioritised in the first two years of the programme, and cleaning was deferred as a treatment option, only to proceed if a safe method could be established. This was an important precedent considering that previous cleaning proposals had been advocated on a mistaken understanding of the painting technology.

During the stabilisation phase, extensive cleaning trials were also carried out and carefully evaluated. Some of the pigments used were incompatible with alkaline

Fig. 5.7 A stunning technological feature is the lavish use of gold. While only minute traces of gold leaf now remain, close examination demonstrated that the autumnal vine leaves and stems were extensively gilded. This detail shows one of the gilded vine leaves (© *Petra National Trust and Courtauld Institute*)

cleaning agents, while the water-sensitive nature of the gypsum ground, paint media, and gold mordant precluded prolonged aqueous treatments. The heterogeneous and often extremely fragile condition of the painting prevented a general approach. A policy of selective and modulated cleaning was eventually implemented, avoiding highly water-sensitive areas of the painting (Figs. 5.8 and 5.9). The cleaning system utilised hydrated magnesium silicate/aluminium silicate sorbents to break down dirt deposits, possibly through adsorption by these clays of neutral and cationic organic compounds [16].

The safe and effective cleaning of the painting required intensive resources, which the project partners both recognised and supplied. Emerging from beneath obscuring dirt and graffiti, the extremely high quality of the painting was recognised for the first time, generating international attention. Consideration next turned to a range of site management and protection issues.

5.8 Site Management Issues

During the course of remedial treatment, site management issues were the subject of regular on-site discussions between project partners. Considerations relating to the long-term preservation of the uniquely important painting in its fragile archaeological environment, while improving access and information for increasing numbers of

Fig. 5.8 The cleaning of the painting was transformative, allowing it to be fully appreciated as an outstanding work of art. It is now recognised as one of the most precious survivals of wall painting from the ancient world (© *Petra National Trust and Courtauld Institute*)

Fig. 5.9 The cleaning of the painting was transformative, allowing it to be fully appreciated as an outstanding work of art. It is now recognised as one of the most precious survivals of wall painting from the ancient world (© *Petra National Trust and Courtauld Institute*)

visitors, and bringing the benefits of tourism to the local population, presented a range of challenges.

5.8.1 Risks of Intentional Damage

Local populations have experienced many changes since they were removed from Petra's main archaeological areas to nearby purpose-built settlements in 1985. Jordan's peace treaty with Israel in 1994 was followed by increases in tourism and archaeological activity, which have had significant effects, both positive and negative [17, p. 99]. While some communities have benefited economically from these developments in Petra's urban centre, those around Beidha have been largely excluded from them. Myths of hidden treasure still abound, and archaeological activity, which is often interpreted locally as treasure hunting, has encouraged clandestine and often destructive activities to compensate for economic exclusion. Cases of important archaeological artefacts being destroyed in an attempt to find treasure believed to be concealed within are not uncommon [17, pp. 133–136].

The recent history of the painted cave complex at Siq al-Barid demonstrates these concerns and local preoccupations. Although much past and ongoing damage has been casual and opportunistic—such as the problem of graffiti—other recurrent damage is differently motivated. The targeted attacks on the painting—including with gunshots—and the two theft attempts have already been mentioned. In spring 2009, the plaster rim of the lost stuccowork medallion on the vault of the recess was also attacked and badly damaged. The impetus for this was probably treasure hunting, ironically spurred by the attention on the site that the conservation programme had itself created.

5.8.2 Tourism and Site Protection

Most tourists to Petra omit a visit to Siq al-Barid from their itinerary. But increased awareness of the site—partly as a consequence of its conservation—appears to be reflected in growing visitor numbers. While this is potentially beneficial for the local community, visitor-related graffiti is still a problem, and the extreme vulnerability of the site is a concern [Fig. 5.10].

Its main rock-cut features—the access stairs, the threshold and floor of *biclinium* chamber, and the tops of its benches—are all susceptible to visitor-related attrition. As the steps are steep, eroded and dusty, and completely open on their north side, there are also visitor safety issues. Since the metal grille closes the painted recess, the *biclinium* is the main area for visitor viewing, which cannot safely accommodate more than 10-15 people. Data collected during the conservation programme recorded many groups of more than 40 people. Controlling access is therefore key not only to protecting the painting, but also for safeguarding the site and its visitors.

Fig. 5.10 Visitors climbing the rock-cut steps to the painted cave complex. Increases in visitor numbers, partly in response to publicity surrounding the conservation project, raised concerns about visitor safety, and the protection of the painting in its fragile archaeological environment (© *Petra National Trust and Courtauld Institute*)

5.9 Outcomes

Almost 10 years after the completion of the conservation project, its outcomes can be critically assessed. Its immediate achievements—the stabilisation and cleaning of the painting—were successful, as were the organisational structures that enabled this. A close and productive collaboration between project partners allowed the programme to adapt as new information and issues emerged. This flexible and iterative approach was essential considering the uniquely important status of the painted site.

The vast increase in understanding of the painting in its context was another positive outcome, enhancing the site's exceptional artistic and archaeological significance. Created with the utmost skill and sophistication using expensive paint materials, including gold leaf, the cave complex is now recognised as preserving one of the most precious survivals of wall painting from the ancient world. In the absence of written records, it is also a remarkable testament to Nabataean religious beliefs and practices in this desert region at the beginning of the first millennium.

The environmental exposure of the site presents a challenging legacy. The open nature of the cave complex imposes severe constraints on what if anything can be achieved with respect to its environmental protection. Indeed, it was necessary to acknowledge these circumstances and modify expectations accordingly. Hence, it was decided not to implement major environmental interventions—such as blocking the openings to reduce the effects of temperature and humidity fluctuations—considering both the uncertain outcomes of such measures and their potentially adverse impact on the archaeological fabric. These decisions were, however, also based on information acquired from condition studies, which indicated that rates and effects of environmental deterioration, while not negligible, were relatively slow acting.

The site management issues proved more challenging still. It became clear during the conservation programme that protecting the site from intentional damage required a dual approach, combining improved physical protection with efforts to change local attitudes towards the site and its painting through mediation and inclusion. These aims were difficult or impossible to achieve in practice for a variety of reasons. Options for physical protection are extremely limited by the nature and configuration of the site. No one proposed solution solved all problems, with each option having certain drawbacks, including issues of restricted viewing, inadequate safety provision and intrusion on the archaeological fabric. The optimal physical solution—introducing a gate at the top of the stairs—required the presence and supervision of a designated key-holder. Administrative and resources issues at both national and local levels prevented this from being implemented, however. A more radical option—to keep the site permanently closed to visitors—would avoid all these problems, but preventing tourism would inevitably create local resentment, placing the painting under threat.

For now, the only improved site protection measure is an extension on top of the metal grille that separates the inner recess from the outer *biclinium* chamber. While this does protect the unique vintaging painting from theft and other damage, the rest of the site remains vulnerable to visitor-related attrition, and safety issues for visitors also continue to be neglected. Despite recognition that integrating local communities into protection and preservation efforts is the most effective use of scarce resources, and is also the best means of maintaining a positive connection between the local population and the site, these efforts were stalled by wider events and circumstances.

Within a year of the completion of the conservation programme, the 'Arab Spring' engulfed the region. Although Jordan avoided serious upheaval, direct and indirect effects were numerous. Jordan absorbed over 650,000 Syrian refugees, at a cost of $2.5 billion per year (equal to 6% of Jordan's GDP). Combined with Jordan's Palestinian and Iraqi inhabitants, refugees reportedly made up 28% of the country's total population of 10 million.[2] These problems overshadowed concerns for the care of cultural heritage. Meanwhile, climate studies carried out between 1961 and 2005 have tracked increasing temperatures and decreasing precipitation, which are having a negative impact on Jordan's ecology, agriculture and nomadic populations.[3] Increasing tourism at sites such as Petra is inevitably seen as a way of diversifying diminished economies and benefiting adversely affected communities. But since Petra's outlying areas are not included under UNESCO's listing, sites such as the painted cave complex at Siq al-Barid remain at the margins of administrative oversight and protection. Promoting more tourism brings as many potential threats as benefits.

Measures to ensure the protection of the site and its wall paintings are still desperately needed, but they must be decided and implemented with sensitivity and be supported by appropriate resources, if they are to succeed. In the context of conserving wall paintings in situ, these issues and dilemmas are not unusual. But for the remarkable paintings at Siq al-Barid, there is perhaps added poignancy. Their particular insight of a lost culture and aesthetic is of incalculable value. It is to be hoped that despite all the challenges and obstacles that stand in the way of achieving their lasting preservation, the conservation programme's longer-term objectives will be eventually realised.

Notes

1. Fragments of figurative painting were found and collected during excavations at al-Zantur, none of which are on public display [18]. The most significant examples discovered in situ were detached in the 1970s from the Winged Lions Temple, but their current location and condition are not known [19].
2. http://www.jordantimes.com/news/local/syrian-refugees-cost-kingdom-25-billion-year-report.
3. https://www.ncbi.nlm.nih.gov/pmc/articles/PMC5576883/.

Acknowledgements The conservation project was commissioned by the Petra National Trust (PNT). Particular thanks are due to its then Executive Director, Aysar Akrawi, for her dedication to the project, and to her staff in facilitating it. Thanks are also due to the late Dr Fawwaz al-Khraysheh (Director-General, Department of Antiquities of Jordan), and Mr. Suleiman Farajat (former Director, Petra Archaeological Park), for their support. Tahani al-Salhi (PAP) provided on-site logistical help and organisation. The project's development, including research presented in this paper, is indebted to the professionalism and dedication of Lisa Shekede, the author's co-conservator. At the Courtauld Institute, Professor David Park and the late Professor Sharon Cather provided unstinting direction and support. Thanks also go to Dr. Giovanni Verri for multispectral imaging work and to Jennifer Porter for ELISA organic analysis. Finally, thanks are due to Maram Na'es for conservation assistance.

References

1. Akrawi, A., Shekede, L.: A unique Nabataean wall painting in Petra: Conservation in situ. In: Rozeil, C., Roy, A., Saunders, D. (eds.) Conservation and the Eastern Mediterranean, IIC 2010 Istanbul Congress, 20–24 September (Conference Proceedings), London, pp. 214–219 (2010)
2. Graf, D. F., Sidebotham, S. E.: Nabataean trade. In: Markoe, G. (ed.) Petra Rediscovered: Lost City of the Nabataeans, London, pp. 65–73 (2003)
3. Bikai, P.M., Kanellopoulos, C., Saunders, S.L.: Beidha in Jordan: a Dionysian hall in a Nabataean landscape. Am. J. Archaeol. **112**, 465–450 (2008)
4. McKenzie, J.: The Architecture of Petra, British Academy Monographs in Archaeology No. 1, Oxford (1990)
5. Shaer, M.: The Nabataean mortars in Petra area: investigation of types and applications, MA thesis, Institute of Archaeology and Anthropology, University of Yarmouk (unpublished) (1977)
6. Zayadine, F.: Recent excavations and restorations of the Department of Antiquities (1979–80). Annu Dep Antiq Jordan **25**, 341–355 (1981)

7. Kolb, B.: Petra—from tent to mansion: living on the terraces of Ez-Zantur. In Markoe, G. (ed.) Petra Rediscovered: Lost City of the Nabataeans, London, pp. 230–237 (2003)
8. Kakoulli, I.: Greek Painting Techniques and Materials From the Fourth to the First Century BC, London (2009)
9. Dalman, G.: Petra und Seine Felsheiligtümer, Hinrichs, Leipzig (1908)
10. Horsfield, G., Horsfield, A.: Sela-Petra, the Rock, of Edom and Nabatene. Q. Dep. Antiq. Palest. **7**, 21–24 (1938)
11. Glueck, N.: A Nabataean painting. Bull. Am. Sch. Orient. Res. **141**, 13–23 (1956)
12. Healey, J.F.: The Religion of the Nabataeans: A Conspectus, Cologne (2001)
13. Brekoulaki, H., Perdikatsis, V.: Ancient Painting on Macedonian Funerary Monuments, 1v – 111 c. B.C.: a comparative study on the use of colour, in Colour in Ancient Greece: The Role of Color in Ancient Greek Art and Architecture (700–31 B.C). In: Proceedings of the conference held in Thessaloniki, 12th–16th April, 2000 organised by the J. Paul Getty Museum and Aristotle University of Thessaloniki, Thessaloniki, pp. 147–154 (2002)
14. Rhomiopoulou, K., Brekoulaki, H.: Style and painting techniques on the wall paintings of the "tomb of the Palmettes" at Lefkadia, in Colour in Ancient Greece: The Role of Color in Ancient Greek Art and Architecture (700–31 B.C): Proceedings of the conference held in Thessaloniki, 12th–16th April, 2000 organised by the J. Paul Getty Museum and Aristotle University of Thessaloniki, Thessaloniki, pp. 107–116 (2002)
15. Franchi, R., Pallecchi, P.: The wall paintings at Petra and Beidha in Jordan: mineralogical-petrographic study of the causes of deterioration and proposals for their restoration. Stud. Hist. Archaeol. Jordan **5**, pp. 99–104 (1995)
16. Shariatmadari, H., Mermut, A.R.Benke, M.B.: Sorption of selected cationic and neutral organic molecules on palygorskite and sepiolite. Clays Clay Miner. **47**(1), 44–53 (1999)
17. Hejazeen, E. G.: Tourism and local communities in Jordan: perception, attitudes and impacts, in Tourism Studies from the University of Eichstätt-Ingolstadt, 8, Profil, Vienna (2007)
18. Kolb, B., Gorgerat, L., Grawehr, M.: Swiss-Liechtenstein excavations on Az-Zantur in Petra, 1998. Annu. Dep. Antiq. Jordan **43**, 261–372 (1999)
19. Hammond, P.C.: Excavations at Petra 1975–1977. Annu. Dep. Antiq. Jordan **22**, 81–101 (1977–1978)
20. Burn, L.: Hellenistic Art from Alexander the Great to Augustus, London (2004)

Chapter 6
pXRF and FTIR Spectrometry Applied to the Study of Azurite and Smalt in Romanian Medieval Wall Painting

Olimpia Hinamatsuri Barbu

Abstract The theme of this paper is the analysis of medieval Romanian mural paintings, specifically the characterization of the pictorial layers containing blue pigments and the identification of their degradation products. The equipment used for this purpose included a portable X-ray fluorescence system (pXRF) and a Fourier transform infrared spectrometer (FTIR). The pXRF analyses were performed *in situ* in order to determine the pigments' elemental composition and to identify the sampling areas for FTIR analyses. This study presents the identification of two blue pigments. The first is azurite, which was found at Arbore church; the church of St. George, Voroneț monastery; the church of Humor monastery; St. George church of St. John the New Monastery, Suceava; the Princely Church of St. Nicholas, Curtea de Argeș; and the Doamnei church, Bucharest. The second is smalt, both well-preserved and discoloured examples of which were identified at the church of Sucevița Monastery from Suceava. Calcium oxalate is the most frequently found alteration product identified in the pictorial layer with azurite, as well as in smalt pigment layers. In many cases the smalt pigment was not well-preserved, showing pronounced discolorations especially on exterior paintings. The detailed study of smalt at Sucevița, based on samples taken from twelve areas, demonstrates that the alteration of this pigment can be correlated with the formation of oxalate and carboxylate. FTIR spectrometry highlighted the occurrence of potassium carboxylate and of organic material in the smalt-containing pictorial layer, suggesting either *a secco* technique or a later intervention.

Keywords Wall painting · Smalt · Degradation · FTIR · pXRF · Azurite · Fresco technique

---

O. H. Barbu (✉)
National Museum of Romanian History, Centre of Research and Scientific Investigation, Calea Victoriei 12, 030026 Bucharest, Romania

S. Aoki et al. (eds.), *Conservation and Painting Techniques of Wall Paintings on the Ancient Silk Road*, Cultural Heritage Science,
https://doi.org/10.1007/978-981-33-4161-6_6

6.1 Introduction

In wall painting conservation, copper-based pigments demand special care due to their sensitivity to alkaline solutions. In the Romanian medieval paintings from Northern Moldavia, the most frequently used copper-based blue pigment was azurite, utilized in the representation of garments and for backgrounds. The blue paint layer is likely to have been applied *a secco* (with an organic binder) over colour applications (usually black, red or brown) painted *a fresco*, to enhance its resistance to environmental deterioration [1]. Alteration of azurite has been observed in association with factors such as excessive humidity and environmental fluctuations, and also previous restoration treatments involving the use of alkaline solutions, for example, ammonium carbonate, ammonium oxalate or barium hydroxide [2, 3]. These alterations can make it difficult to identify the pigment by visual observation alone.

Smalt is another blue pigment whose identification can be impaired by degradation. This pigment, applied both *a fresco* and *a secco*, was frequently used in the medieval painting of Muntenia. In many cases the pigment is not well preserved, showing pronounced discoloration especially on exterior paintings. Unlike natural azurite, which was used mostly for the representation of garments, smalt was usually employed for covering large background surfaces depicting the sky. It was usually also applied over a layer of carbon black, as smalt has poor hiding power.

There have been few studies to date in the identification through instrumental analysis of azurite or smalt in Romanian mural painting [4–10]. The intention of the current paper is therefore to address this lack in the literature. In order to understand the nature and causes of the alterations observed *in situ*, painted models of azurite and smalt with different binders were prepared according to traditional methods, for analysis alongside samples taken from the paintings themselves [11–16]. The present research provides analytical identification of azurite and smalt pigments from Romanian wall paintings together with their alteration products using FTIR. pXRF was also used for *in situ* identification of chemical elements and for the selection of representative areas for FTIR analysis.

6.2 Experimental

6.2.1 Paint Samples

In order to confirm azurite and smalt occurrences, paint samples were taken from a range of thirteenth to eighteenth century paintings at several monuments in Romania (Table 6.1).

Table 6.1 Pigments and degradation products identified in samples from Romanian mural paintings containing blue pigment. Notations: c = century (date of painting), CaOx = calcium oxalate, az = natural azurite

Historical monuments	County	c	Results		
			az	CaOx	Other compounds
First Princely church of Saint Nicholas in Curtea de Argeș – ruins of old Argeș (archaeological painting fragments)	Argeș	13	+	–	Carbon-based black
Princely Church of Saint Nicholas in Curtea de Argeș	Argeș	14	–	+	Natural ultramarine (lapis lazuli), carbon-based black
			+	+	Carbon-based black
Sântămărie – Orlea church	Hunedoara	14	+	+	Nitrate, gypsum, calcium carbonate
Old Humor, ruins of old church (archaeological painting fragments)	Suceava	15	+	–	Carbon-based black, silicates
Church of Humor monastery	Suceava	16	+	+	Gypsum, carbon-based black
			+	+	Gypsum, malachite, nitrate, calcium carbonate
St. George church of St. John the New monastery	Suceava	16	+	+	Carbon-based black, nitrate
			+	+	Malachite, carbon-based black
St. George church of Voroneț monastery	Suceava	16	+	+	Carbon-based black, silicate
Arbore church (exterior painting)	Suceava	16	+	+	Silicate
St. Nicholas - Popăuți church	Botosani	16	+	+	Nitrate, calcium carbonate
Church of Tismana monastery (painting by Dobromir)	Gorj	16	–	+	Smalt, gypsum, carbon-based black
Church of Sucevița monastery	Suceava	16	–	+	Smalt, carbon-based black
Doamnei church (painting by Constantinos and Ioan)	Bucharest	17	+	+	Gypsum, nitrate, calcium carbonate
			–	+	Smalt, carbon-based black
Colțea church (painting by Pârvu Mutu, porch)	Bucharest	17	–	+	Smalt, carbon-based black, artificial ultramarine (repaint)

(continued)

Table 6.1 (continued)

Historical monuments	County	c	Results		
			az	CaOx	Other compounds
Fragments of mural painting found in 2014 at St. George the New church (painting by Pârvu Mutu)	Bucharest	18	+	+	Malachite, carbon-based black
Church of Văcărești monastery (painting fragments extracted by *stacco* technique; the historical monument was demolished during the communist regime)	Bucharest	18	+	+	Smalt, carbon-based black, proteins (glue used for *stacco* extraction), gypsum
			–	+	Artificial azurite (repaint), gypsum, proteins (glue used for *stacco* extraction), carbon-based black
Stavropoleos church (exterior painting, porch)	Bucharest	18	–	+	Smalt, gypsum, carbon-based black

6.2.2 *Reference Materials*

Data interpretation was carried out using commercially available reference materials, samples from our laboratory collection, as well as painted models consisting of new laboratory-prepared azurite and smalt applied both *a secco* and *a fresco*. In order to simulate conservation-related alterations of the pigments, the naturally aged samples were subjected to various conservation treatments. These were then subjected to instrumental analysis using pXRF and FTIR. The organic binders used were egg, casein, gum arabic, flour paste, hide glue and linseed oil. Suppliers of pigments, binding media and materials for conservation treatments were as follows: Kremer Pigmente, Germany (smalt, lead white, minium, cinnabar, red ochre, green earth, gum arabic, hide glue, linseed oil); La Verdure, France (azurite and malachite); Merck, Germany (casein, ammonium oxalate, oxalic acid and barium hydroxide); Chimopar, Romania (ammonium carbonate); Rocks and minerals, O. Edelstein Collection (kaolinite, calcium carbonate and gypsum).

6.2.3 *Equipment*

A portable XRF system from InnovX Systems-Alpha Series with W anti-cathode, operating at 35 kV and 40 μA, acquisition time of 30 s and equipped with a Si-PIN detector, was utilized for the determination of chemical elements in the pictorial layer. FTIR analyses were performed by means of a Bruker Tensor 27 spectrometer, operating in the spectral range 4000–400 cm^{-1}, with 4 cm^{-1} resolution on samples ground with KBr (pellets of 3 mm diameter). An optical microscope, Nikon Eclipse LV100 equipped with VIS/UV light source and Nikon D90 digital camera, a portable microscope Dino-Lite AD4113T-I2V, and an application written in C#, PigmentX, were used as additional methods to identify some pigments [17]. For statistical data analysis the MATLAB program was employed, which helped in developing calculation methods for Principal Component Analysis (PCA).

6.3 Results and Discussion

This paper is the result of several years of painting research, the investigations being performed mainly in the preliminary stages of the restoration process, where the identification of the painting constituents assisted intervention decisions. Table 6.1 presents a selection of samples of blue pigments identified in Romanian medieval wall paintings.

6.3.1 Painting Support

The Romanian mural paintings dated between the thirteenth and eighteenth centuries which were studied for this paper appear to have been executed *a fresco*, occasionally completed with *a secco* additions to the lime plaster support. The presence of plaster joins corresponding to scaffolding lifts (*pontate*) and day-work (*giornate*), with visible traces of compaction of the plaster with a trowel, incised preparatory drawing, and impressions of brushstrokes, indicate painting onto fresh lime plaster, *a fresco*. The painting was applied on to a maximum of three layers of lime plaster. In some cases the wall was first prepared with a coarse layer of lime and sand (*arriccio*). More usually the first plaster layer (*intonaco*) was between 3 and 30 mm thick and composed of lime putty reinforced with chopped straw, according to the tradition of Byzantine *fresco* technique. Occasionally, this layer was covered by a thinner one (*intonachino*), in which the lime plaster was reinforced with hemp tow. The painting was then quickly applied in successive layers before complete carbonation of the lime plaster.

6.3.2 Natural Azurite

Natural azurite, Cu3(CO3)2(OH)2, a copper-based pigment, is one of the most frequently found pigments in the mural painting of churches from Northern Moldavia between the fifteenth and sixteenth centuries. Wall paintings containing azurite from several Romanian medieval monuments were characterized using pXRF and FTIR techniques. The results obtained are shown in Table 6.1 and some of the FTIR spectra of these samples are compared in Fig. 6.1. From the eighteenth century the naturally-occurring mineral azurite is progressively replaced by other blue pigments, including artificial azurite (blue verditer). In the painting at Văcărești monastery artificial azurite is used for large zones of blue painting, specifically the background areas depicting the sky, while the natural azurite is confined to the representation of garments. Most frequently, the azurite used in Romanian mural paintings has infrared absorption frequencies at 3428, 1464, 1417, 1092, 953, 837, 815, 498, and 455 cm^{-1}. Only at Arbore and the Princely Church at Curtea de Arges, where the blue pigment preserves its deep blue colour, was an absorption band of azurite at 1489 cm^{-1} detected. This band was absent in the spectra of many azurite samples, including all the laboratory-prepared samples, where the azurite was applied a fresco directly onto lime or on an intermediate layer of carbon black; this finding implies that the band may have been hidden by the calcium carbonate formed by lime carbonation. The absence of this band was also observed for azurite samples which had been previously treated with ammonium carbonate and barium hydroxide solutions, both of which have been used in the past for cleaning and consolidation interventions.

Calcium oxalate is the most frequently identified alteration product associated with azurite. Calcium oxalates (whewellite, IR: 1619–29 ν_a(COO), 1318–24 ν_s(COO), 775–781 δ(OCO) cm^{-1}, or weddellite, IR: 1640–47 ν_a(COO), 1325–30 ν_s(COO), 781–785 δ(OCO) cm^{-1}) were identified in paint layers containing pigments such as azurite, malachite, natural earths, smalt, lapis lazuli, lead white and minium. The highest ratio of oxalate to azurite was found on the exterior mural paintings at

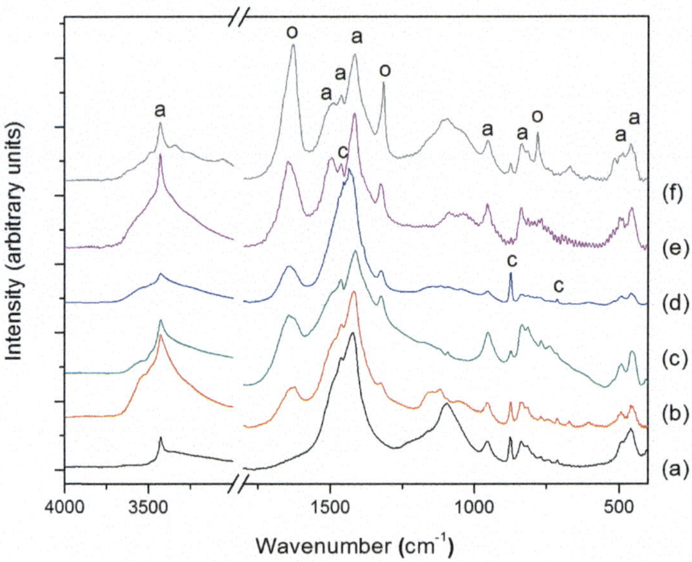

Fig. 6.1 FTIR spectra of samples containing azurite from Old Humor (a), Voroneț (b), St. John the New (c), Doamnei (d), Princely church Curtea de Argeș (e) and Arbore (f); Notations: a—azurite, o—calcium oxalate and c—calcium carbonate

Arbore church. Several hypotheses have been advanced in the literature concerning the factors responsible for the formation of oxalates in paint films: these include the use of organic binders (proteinaceous materials, oils or carbohydrates), the presence of atmospheric pollutants, and the use of organic materials in conservation treatments [18–20]. A biological origin has also been postulated, in which oxalates are produced as a result of fungi metabolism [21]. Although the source of this degradation product is not clear, it is worth mentioning that, in the analyzed samples, the oxalates are mostly present on exterior wall paintings. Despite expectations, no copper oxalates were found in any of the medieval samples containing copper-based pigments. Copper oxalates (IR: 1660 νa(COO), 1365–70 νs(COO), 1320–25 νa(CO2 wag) and 820–825 δ(OCO) cm^{-1}) were identified on laboratory-prepared layers containing azurite and malachite, applied on a carbon black layer, in *fresco* technique, treated with cellulose poultices impregnated with ammonium oxalate solution 10%. Gypsum, with characteristic bands at 3548, 3408, 1685, 1622, 1145, 1117, 671 and 602 cm^{-1}, was also identified in superficial layers with azurite and malachite (Fig. 6.2).

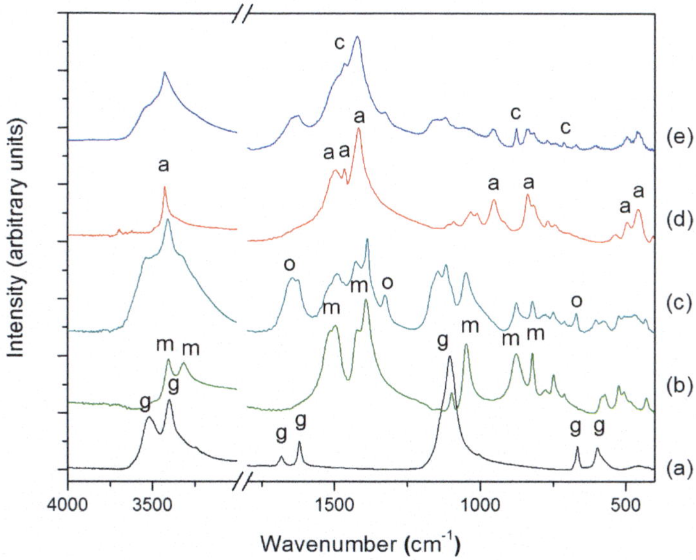

Fig. 6.2 FTIR spectra for samples from Humor church containing azurite (e) and malachite (c) in comparison with FTIR reference spectra for azurite (d), malachite (b) and gypsum (a) Notations: a— azurite, m—malachite, o—calcium oxalate, c—calcium carbonate, g—gypsum

6.3.3 Smalt

Smalt was largely used in Romanian mural paintings between the fifteenth and eighteenth centuries, after which it was gradually replaced by other blue pigments. Smalt is a potassium salt (in some cases also with sodium) colored with cobalt oxide derived from various cobalt ores [22]. Smalt degradation was frequently observed on the murals, its color changing to yellowish or gray in the painting. Loss of the blue color has been mentioned in the literature as an alteration process due to a pigment-binder reaction, resulting in the formation of potassium carboxylates [23, 24] The presence of a higher amount of potassium in the center of deeply colored smalt granules compared to the discoloured edges, and the detection of potassium in the organic binder (oil) surrounding the pigment, suggests the migration of potassium from the pigment to the binder, resulting in a decrease of alkalinity [25]. The chemical bonding of potassium to acidic groups of fatty acids found in oils increases water solubility of the paint layer, further facilitating degradation. Chemical analyses revealed variations in the chemical composition of the pigment, specifically in ratios of Si, Co and K (Na) oxides. The discoloration was explained by a change in the coordination system of cobalt from tetrahedral toward octahedral coordination, and the diminution of potassium content of the smalt granules [26, 27] Deterioration in the colour of smalt pigment was also observed at the mural painting of Sucevița monastery (sixteenth century) (Fig. 6.3). Here this alteration can be correlated with the formation of potassium carboxylate and calcium oxalate.

To shed more light on the processes relating to the presence of oxalate and carboxylate in painting layers containing smalt, pXRF and FTIR spectra were processed using principal component analysis (PCA). The aim was to classify the samples into groups characterized either by the presence of particular chemical elements (pXRF data) or by the presence of compounds, based on the identification of chemical bonds (FTIR data). PCA is a data analysis tool involving linear reduction of the dimension of a system with many variables by identifying the orthogonal directions

Fig. 6.3 Sampling areas of smalt pigments from Sucevița monastery labelled S1, S2, S4, S7 and S12 (see Table 6.2 for identification)

of maximum variance and projecting the data into a reduced dimensional space, formed by the components' subspaces. PCA is a mathematical procedure that transforms the number of possible correlated variables into a smaller number of uncorrelated variables, which are called principal components (PC). The first component concentrates the maximum variability of the data, and the next component retains as much as possible of the remaining variability. PCA methods applied to XRF data (PCA-XRF) were useful for the selection of samples of interest, characterized by the presence of smalt (with As, Co, Fe, Ca, Pb, Sr and K content) (Fig. 6.4). Other samples, containing Cu and Zn, were excluded from the statistical analysis applied to FTIR spectra. The FTIR data confirmed the presence of smalt by the identification of absorption bands at 1022 and 440 cm^{-1}. All the smalt samples selected for PCA-FTIR, showed these bands, the differences being variation in calcium oxalate, calcium carbonate, gypsum, carboxylate or other compounds.

The first six PC components computed using the FTIR data (corresponding only to smalt samples) accumulated 99.5% of variability, from which a percent of 64.35 refers to the first component, PC1. It was found that the samples exhibiting extreme positive values of PC1 (Fig. 6.5), namely the points 8, 9 and 10, correspond to smalt samples taken from the visibly discolored smalt, from the church Exonarthex (samples labeled S7 and S12, Fig. 6.3. Spectra denomination and their associated number for principal components representation are shown in Table 6.2. The first component differentiated the smalt samples containing calcium oxalate from those with calcium carbonate (Fig. 6.6a). The second principal component PC2 (26.15%)

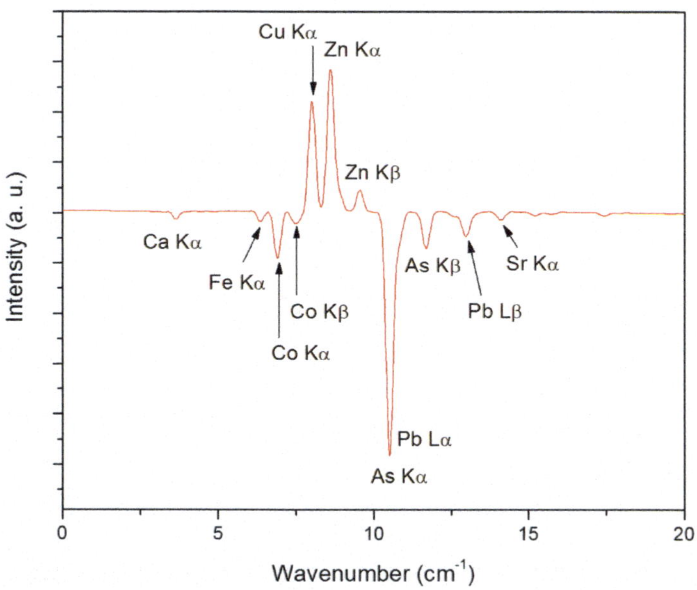

Fig. 6.4 Loading plot of the principal component PC1 for the pXRF spectra of blue samples from Sucevița monastery

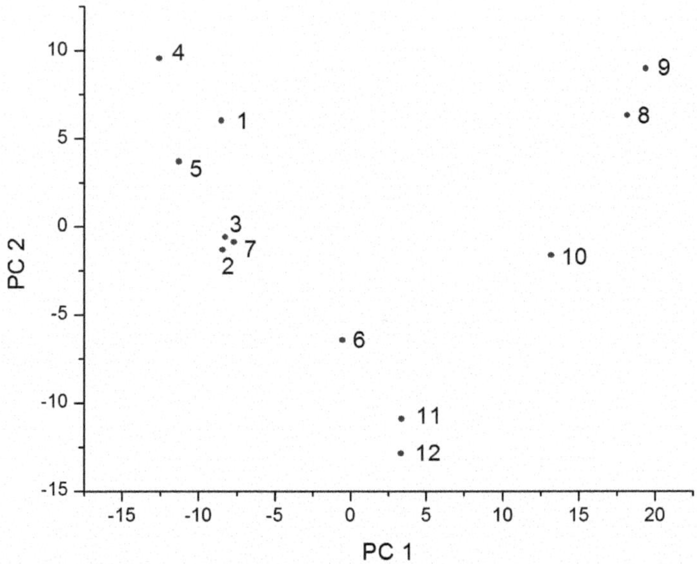

Fig. 6.5 Score plot of the two principal components, PC1 and PC2, for FTIR data of smalt samples

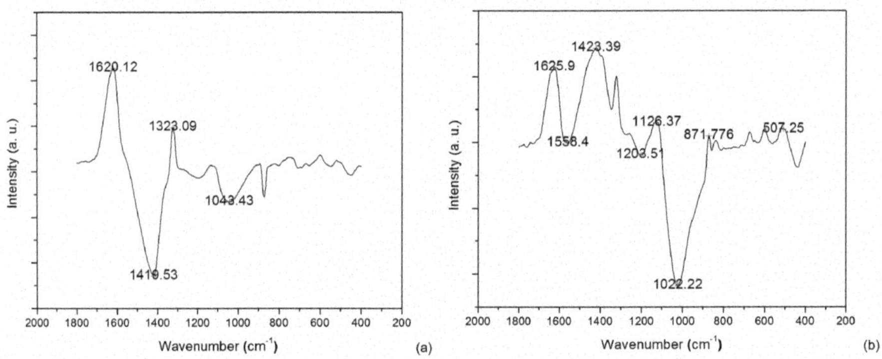

Fig. 6.6 Loading plots of the first component, PC1 (a) and of the second component, PC2 (b) for the FTIR data of smalt samples

separated the spectral bands of calcium oxalate, calcium carbonate, gypsum and nitrate (traces) from the characteristic bands of smalt pigment (Fig. 6.6b). The highest negative values for PC2 (points 11 and 12) correspond to the well-preserved smalt samples, where the colour is still blue (samples labeled S8 and S13, Table 6.2). The asymmetric form of the smalt band, exhibiting a minimum at nearly 1022 cm^{-1} and a small shoulder at 1200 cm^{-1}, was decomposed in the two component bands. At the same time, a small intensity band at 1558 cm^{-1} was separated from the oxalate peak, this band showing the presence of potassium carboxylate, which can form in smalt on contact with a carboxyl containing compound. The other noticeably

Table 6.2 Sampling areas of smalt samples from Sucevița Monastery and the associated number for principal components (PC) representation of FTIR data

Point in PC space	Sample code	Sampling area	Sample appearance
1	S1	Burial chamber, North-West, background	Dark gray
2			
3			
4			
5	S2	Pronaos, South, blue background	Light blue
6	S3	Burial chamber, North, background	Light gray
7	S4	Burial chamber, North, St. Luca, background	Light gray
8	S7	Exonarthex, West, black background	Light gray
9			
10	S12	Exonarthex, East, background	Light gray
11	S8	Pronaos, South, blue background	Deep blue
12	S13	Pronaos, South, blue background	Deep blue

discoloured smalt samples contain significant amounts of calcium oxalate, calcium carbonate, gypsum and nitrate. Using pXRF spectrometry, it was found that the strongly discoloured samples contain less potassium than those which preserve a deep blue colour. Further examination of the spectrum obtained from an area with discoloured smalt, sample S4 (point 7 in PC space, Fig. 6.3), showed some additional bands (2921, 2851 and 1733 cm^{-1}) which may be attributable to an organic material (oil). This may indicate that in altered smalt, potassium soaps form by potassium reacting with the fatty acids in the oil, suggesting *a secco* application of the pigment. However, there is also the possibility that an organic material has been used in past conservation treatments.

6.4 Conclusion

The presence of azurite and smalt in Romanian mural paintings dating between the thirteenth and eighteenth centuries was confirmed through analysis of blue pigments from sixteen historical monuments. Among the degradation products affecting them, the most prominent is calcium oxalate. FTIR spectroscopy attested that oxalates are present in almost all the investigated samples of both azurite and smalt pigments. Considerable variations in the appearance of the smalt layers were observed, from well-preserved areas retaining the original blue coloration to highly discoloured areas with a grey or yellow appearance. Principal component analysis (PCA) was used to correlate differences in FTIR spectra of smalt samples with these differences in physical appearance. PCA-FTIR analysis attested the presence of potassium carboxylate in areas where the blue colour survives. *In situ* examination in combination with data obtained by physical and chemical analyses led to the conclusion that the paintings were executed for the most part using the *fresco* technique. It is highly probable that some pigments, like smalt, were applied with an organic binder addition, for example, the painting from the church of Sucevița monastery, in which smalt samples were found to contain potassium carboxylates and oil. The identification of chemical elements in the pictorial layer by pXRF was useful in differentiating areas of smalt blue, as well as revealing the compositional differences between the discoloured and well-preserved smalt areas.

Acknowledgements I would like to express my gratitude to Dan Mohanu, Oliviu Boldura, Maria Dumbravician, Silviu Petrescu and Gheorghe Niculescu for their support, and for kindly providing samples from the historical monuments. I would also like to thank the following colleagues and friends who helped me in data collection, analysis and interpretation, offering their experience and knowledge to ensure clear understanding of conservation issues and degradation processes in paintings: Migdonia Georgescu, Ileana Mohanu, Zizi Baltă, Irina Petroviciu, Elena Martin, Paula-Gina Baranovschi, John Sarcoe and Ana-Maria Vlad.

References

1. Istudor, I. și Balș, I.: Contribuții la cunoașterea materialelor folosite în pictura murală exterioară a bisericilor din secolul al XVI-lea din Bucovina și la unele probleme de tehnică, Revista Muzeelor, nr. 6, pp. 491–496 (1968)
2. Boldura, O.: Mural painting in the north of Moldavia. Aesthetic modifications and restoration. Editura ACS, 2nd edn, Bucharest (2013)
3. Dei, L., Ahle, A., Baglioni, P., Dini, D., Ferroni, E.: Green degradation products of azurite in wall paintings: identification and conservation treatment. Stud. Conserv. **43**(2), 80–88 (1998)
4. Barbu, O.H., Zahariade, A.: Noninvasive in situ study of pigments in artworks by means of VIS, IRFC image analysis, and X-ray fluorescence spectrometry. Color Res. Appl. **41**, 321–324 (2016)
5. Hidaka, M., Hidaka, K., Junichi, M.: Technique of the mural paintings of the Hospital Church in Horezu Monastery. Eur. J. Sci. Theol. **11**(3), 231–239 (2015)
6. Mohanu, D., Poll I., Baltă, Z., Barbu, O.H., Georgescu, M.: Noi fragmente de pictură murală descoperite în anul 2014 la Biserica Sf. Gheorghe Nou. Observații tehnice. In: Istoria Oraşului București (5): Mărturii Brâncoveneşti, Editura Agir, pp. 403–424 (2015)
7. Buzgar, N., Buzatu, A., Apopei, A.I., Cotiugă, V.: In situ Raman spectroscopy at the Voroneţ Monastery (16th century, Romania): new results for green and blue pigments. Vib. Spectrosc. **72**, 142–148 (2014)
8. Merticaru, C., Istudor, I., Cimpeanu, G.: Investigation concern the outdoor wall painting on the 16th century monasteries from Bucovina, Romania, Art '05—Nondestructive Investigations and Microanalysis for the Diagnostics and Conservation of the Cultural and Environmental Heritage, Lecce (Italy), May 15–19, Book of Abstracts, p. 184 (2005)
9. Ionescu, O.H., Mohanu, D., Stoica, A.I., Baiulescu, G.E.: Analytical contributions to the evaluation of painting authenticity from Princely church of Curtea de Arges. Talanta **63**(4), 815–823 (2004)
10. Istudor, I.: Considerații tehnice asupra unor picturi murale din Romania (Epocile Antică, Medievală, Modernă), Revista Monumentelor Istorice, Anul LXXII, nr. 1, pp. 76–100 (2001–2003)

11. Dionisie din Furna: Erminia picturii bizantine. Editura Sophia, București (2000)
12. Cennini, C.: Il libro dell'arte (traducere, note şi indice de Toscani, N. Al.), Editura Meridiane, Bucureşti (1977)
13. Kiplik, D. I.: Pictura monumentală, Editura de stat pentru literatură şi artă (1953)
14. Vinner, A. V.: Materialele şi tehnica picturii monumentale decorative, vol. 1–2, Editura Iskusstvo (1953)
15. Vinner, A. V.: Cum să folosim tempera, Editura Arta (1951)
16. Verona, A. G.: Pictura – studiu tehnic – fresco, secco, tempera, pictura în ulei, encaustica; întreţinerea şi restaurarea picturilor, Editura şi tiparul Sfintei Monastiri Neamţu (1943)
17. Barbu, O. H., Dima, C., Stefănică, A. and Nicolau, H.: "PigmentX" application for pigment identification, IRUG12 - The twelfth Biennial Infrared & Raman Users Group Conference, Ormylia (Greece), May 23–25, Book of Abstracts, pp. 87–88 (2016)
18. Higgitt, C., White, R.: Analyses of paint media: new studies of Italian paintings of the fifteenth and sixteenth centuries. Natl. Gall. Tech. Bull. **26**, 88–104 (2005)
19. Salvadó, N., Butí, S., Tobin, M., Pantos, E., Pradell, T.: The nature of medieval Synthetic Pigments: The Capabilities of SR-Infrared Spectroscopy. IRUG 6 Conference, pp. 296–301 (2004)
20. Cariati, F., Rampazzi, L., Toniolo, L., Pozzi, A.: Calcium oxalate films on stone surfaces: experimental assessment of the chemical formation. Stud. Conserv. **45**, 180–188 (2000)
21. Monte, M., Bianchini, L.: Formation of calcium oxalate deposits on marble specimens in fungal culture. In: 10th International Congress on Deterioration and Conservation of Stone, vol. II, pp. 241–246 (2004)
22. Mühlethaler, B., Thissen, J.: Smalt. In: Roy, A. (ed.) Artists' pigments: a handbook of their history and characteristics, Vol. 2, National Gallery of Art, Washington, DC, pp. 113–131 (1993)
23. Keune, K.: Binding medium, pigments and metal soaps characterised and localised in paint cross-sections. Ph.D. Thesis, University of Amsterdam (2005)
24. Spring, M., Higgitt, C., Saunders, D.: Investigation of pigment-medium interaction processes in oil paint containing degraded smalt. Natl. Gall. Tech. Bull. **26**, pp. 56–70 (2005)
25. Boon, J. J., Keune, K., van der Weerd, J., Geldof, M. and van Asperen de Boer, J. R. J.: Imaging microspectroscopic, secondary ion mass spectrometric and electron microscopic studies on discoloured and partially discoloured smalt in crosssections of 16th century paintings, Chimia **55**, 952-960 (2001)
26. Santopadre, P., Verita, M.: A study of smalt and its conservation problems in two sixteenth-century wall paintings in Rome. Stud. Conserv. **51**, pp. 29–40 (2006)
27. Robinet, L., Spring, M., Pagès-Camagna, S., Vantelon, D., Trcera, N.: Investigation of the Discoloration of Smalt Pigment in Historic Paintings by Micro-X-ray Absorption Spectroscopy at the Co K-Edge. Anal. Chem. **83** (13), pp. 5145–5152 (2011)

Chapter 7
Technical Characteristics of Church Wall Paintings in the Balkans

Midori Hidaka

Abstract This paper deals with a technical study of medieval wall paintings in churches located in the Balkans, focusing on Sopoćani monastery, an eminent example in present-day Serbia. By analyzing their painting materials and techniques, the paper aims to clarify the technical and cultural characteristics that defined Balkan wall painting practices in the medieval period.

The naos of the Sopoćani, the largest space in the church, belongs to the oldest construction and is one of the best-preserved parts of the interior. The practical devices used by painters in the medieval period in the creation of wall paintings are discussed through comparison of findings shown in this paper with other related examples in Serbia, Kosovo and Romania. The research was achieved by two different methods: optical documentation with architectural recording using 3D scanning system; and analytical research of the materials and technical approaches used in extensive gold leaf decoration. The survey by 3D scanning revealed (1) the quality of the original construction which is different from that of the later extended structure, and (2) the unique devices used by medieval painters to compensate for viewing the murals at distance, which suggests some comparison with oriental wall paintings, too.

The investigation of the gold leaf decoration disclosed the existence of a red mordant layer under the gold leaf. Findings of the survey indicate that painters modulated the gold leaf decorations by varying the colour of the mordant layer, which included compositional adjustments and use of different pigment combinations. Material analysis also revealed the possibility of silver chloride being used as a painting material in medieval Serbia.

Keywords Wall painting · Balkan · Church · Medieval · Painting technique

M. Hidaka (✉)
School of Conservation, Tokyo National University of the Arts,
12-8, Ueno-Park, Taito, Tokyo 110-8714, Japan
e-mail: makadih3531@gmail.com

S. Aoki et al. (eds.), *Conservation and Painting Techniques of Wall Paintings on the Ancient Silk Road*, Cultural Heritage Science,
https://doi.org/10.1007/978-981-33-4161-6_7

7.1 Introduction

Due to its intricate historical, religious and political backgrounds, the meaning of what is implied by the region of the Balkans has shifted from time to time. In this paper, the Balkans implies the geographical region that is usually recognized today, which is also historically defined as the area that came both under the strong influence of the Roman Empire and the transitional influence from the rule of the Ottoman Empire afterwards.

The Balkans are located at the south-eastern edge of the European continent, forming a bridge to the Middle East, Africa and Asia. For this reason, the area was geologically extremely important throughout the medieval period. Its medieval art was formed through the migration of peoples: Persian and Roman art, Christianity and a variety of other cultural influences were all introduced to Europe through the Balkans. Waves of influence and occupation have been facilitated by its porous geography, which is without steep mountain chains, and is bordered by the Great Hungarian Plain in the north, and the Mediterranean Sea and the Black Sea in the south. Consequently, the Balkans have also experienced many conflicts since ancient times, and have been disputed by the Great Powers in its modern history, too [1]. In the medieval period, combined cultural influences are probably reflected in the wall paintings, informing the identity of the painters who tried to maintain faithfully their Byzantine mural traditions for centuries [2], until the influence of Renaissance art also spread over the area [3].

The Christianization of the Balkans is thought to have been completed by the tenth century, which is complemented by the contiguous establishments of monasteries on Mount Athos during second half of tenth century [4]. Indeed, monasticism spread rapidly in the area, embedding the influence of the Eastern church.

The Grand Prince Stefan Nemanja (reign: 1168–1196) successfully extended his domain to establish the Grand Principality of Serbia by taking advantage of the political confusion in the Eastern Roman Empire triggered by the Crusades. He and his successors, who ruled the region of present-day Serbia until the fifteenth century, fostered a golden age of religious art, abundantly funded from the profits of international trade [2]. They richly endowed many religious foundations, whose interiors were decorated with brilliant wall paintings. Monastic life had previously been established in Egypt, Syria and Asia Minor as an ascetic practice, located in places far from human dwellings, and structured according to vows of chastity, submission and honourable poverty [5]. By the time of its diffusion into the Balkans, it had become an integral and essential part of human affairs. Monasteries often functioned as centres of politics and society, providing important places of meeting and intellectual exchange [4]. In this context, wall paintings were developed not only for decorative purposes, but also as a visual device to illustrate and interpret Christian theology clearly and easily to all visitors.

Medieval wall painting practice in the Balkans is thought to have been exclusively influenced by the traditional materials and techniques of painting production in Ancient Rome. A previous study carried out by the author indicates that painters, while faithfully applying some aspects of traditional Roman practices, such as the use of a limited number of painting materials, inevitably retained some regional characteristics and techniques, clearly different from those of Roman tradition [6]. Following this result, this paper discusses characteristic features of Balkan wall paintings in respect of their iconographical planning and the use of gold leaf decoration.

7.2 Objectives

The main object of this study is the church of the Holy Trinity, Sopoćani Monastery, located in southern Serbia, 15 km from the border with Kosovo. It was one of the most important religious centres in medieval Serbia.

The monastery was established by the Serbian king, Stefan Uroš I, whose reign was recorded to be between 1243 and 1274/1275 [7]. An inscription recording the date of the wall paintings has long been indecipherable. From complementary historical, archaeological and iconographical point of views, it can be reasonably suggested that the work was carried out between 1263 and 1274/1275. The construction of the original church was finished by 1263, which gives a terminus post quem for the production of the wall painting. The terminus ante quem, 1274/1275, is based on the year of Sava II's death (1273/1274), uncle of the founder Stefan Uroš I. A canonized portrait of Sava II was painted in the apse of the church by the year of the end of Stefan Uroš I's reign (1276), and historical records tell that the original wall paintings had been completed before King Uroš I was dethroned by his eldest son [7, 8].

The history of the Holy Trinity Church after its completion in the thirteenth century was marked by a series of destructive episodes, which started with structural damages during the Battle of Kosovo (1389). The most serious attacks were the continuous destruction done by the Ottoman army in the Late Middle Ages, including the violent removal of the entire lead roof in 1689. The church, since then, had been left abandoned and roofless for more than two hundred years, which inevitably accelerated the decay of the wall painting as well as the structure itself. This period of neglect lasted until the 1920s, when a series of interventions were planned and started with insufficient resources to address all the damages.

The Holy Trinity church consists of eleven architectural spaces: the apse; the Diaconicon; the Prothesis; the Naos; the Pronaos; the four small chapels; and the porch with its massive bell-tower. Though the architectural style of the church belongs to the so-called Raška school in general [7], its neat proportions and simple visual impact show some influence of the Romanesque period as well. After the completion of the original construction, the church underwent several periods of renovation, extension and repair-working, especially in the thirteenth–fourteenth centuries (when two of the four small chapels, the porch and the bell tower were added), the sixteenth–seventeenth centuries and finally in the twentieth century [7].

This paper focuses on the west wall of the naos, the greater part of which is preserved in good condition, unaffected by later additions or interventions. Wall paintings cover the entire interior surface following Byzantine tradition. An outstanding decorative feature is the extensive use of gold leaf in their backgrounds. Although the gold backgrounds now appear as either black or dark brown in colour due to degradation, traces of remaining gold leaf can still be detected by careful observation with the naked eye.

Findings from this study of the monastery of Sopoćani were compared with other medieval complexes in the region, which have also been fully or partially measured and analysed applying the same methodological approaches. These are: Studenica monastery (Serbia), Žiča monastery (Serbia), Đurđevi Stupovi monastery(Serbia), Ravanica monastery (Serbia), Gračanica monastery (Kosovo), Dečani monastery (Kosovo), Peć monastery (Kosovo), Bogorodica Ljeviška church (Kosovo), Horezu monastery (Romania), Humor monastery (Romania) and Comana monastery (Romania).

7.3 Methodology

The greater part of the west wall of the naos at Sopoćani is painted with the Dormition of the Virgin, while full-length portraits of saints, founders and donors occupy the lower part.

Two types of surveys were carried out on this wall:

A. optical documentation to study the surface topography and structure of the wall painting; and
B. analysis to study the original materials and technique of the wall painting.

7.3.1 Optical Documentation

Notable features of medieval Balkan churches are their sophisticated architectural proficiency to create an expanse of large space, and the presence of high-quality wall paintings to cover the vast interior surfaces. It is not a straightforward task to record simultaneously the objects of two different scales; the architectural volume reaching to a height of 20 m, and the minute technical detail charactering the wall paintings. Two different 3D scanners were used to achieve the structural research and documentation: a laser scanner for the larger scale; and a LED scanner for the smaller. Photogrammetry using a high-resolution digital camera was also utilized for the accurate record of the colours of the wall painting.

Equipment specifications:

- Laser scanner

  Light source: near-infrared ray
  Measuring distance: 0.6–600 m
  View field: 360° × 100°

Accuracy: 5 mm

- LED scanner

  Light source: blue LED
  Measuring distance: 0.17-0.3 m
  View field: 30° × 21°
  Accuracy: 0.05 mm

From the view of practical use and margin of accuracy, the laser scanner is to be applied for much longer distances than the stated measuring capacity given above (0.6-600 m). The LED scanner, on the other hand is fully effective at recording minute surface details by keeping it at a working distance between 0.17-0.3 m with the view-field limited to 180 × 140 mm. This research ranges over a wider scale through the combination of these two scanners to be applied respectively for measuring the historical structure and its depicted surface detail. The survey itself conditioned the type of the instrument. Given the demands of recording wall paintings in situ, equipment should be easy to carry up and down the narrow path of the scaffolding, be operationally sustainable, and facilitate rapid work under difficult conditions. 23 laser scans with scanning angles ranging from 20° to 360° were carried out on the interior and exterior of the structure, and 477 full- angle LED scans were made of the painting on the west wall of the Naos. The sizes of the data obtained by the 3D scanning are: 315 GB (minimum 363 M, maximum 10 GB) for the Laser scanner; 310 GB (minimum 60 KB, maximum 450 KB) for the LED.

7.3.2 Analysis of Original Materials

Small samples were taken from two locations, one from the background of the wall painting, and the other from the halo of a figure, to analyze the chemical properties of the painting layer which once composed the gold leaf decoration and is now decayed. Cross-sections, produced from the samples, were examined under a polarizing microscope in order to observe their detailed construction and stratigraphy. The unmounted samples were also examined under an optical microscope. Both the cross-sections and the samples were then analysed by SEM-EDX (Scanning Electron Microscopy-Energy Dispersive X-ray spectography), to investigate the composition of the material layers in further detail and determine their constituent elements. Through SEM-EDX analyses the constituent elements are recognized and located precisely in the stratigraphy of the samples. This analytical process is of considerable advantages for gaining a fuller understanding of the original painting techniques.

The samples were examined with an Optical Microscope (OM) and analysed by SEM-EDX, followed by detail measurement with an X-ray Micro Diffracted Goniometer (MDG) to determine their compounds. Instrumental specifications were:

- SEM-EDX analysis: two instruments were used, a JSM-5400 (JEOL Ltd.) observing SEM image, BS (back-scattered electron) image; and a JSM-6360

(JEOL Ltd.) attached energy dispersive X ray spectrometer INCA x-sight (O x f o r d Instruments Ltd.); accelerating voltage: 15 kV.

- MDG analysis: RINT2100 with PSPC-MDG2000 and RINT rapid(RIGAKU (Corporation); target & X-ray: CuKα, tube voltage: 40 kV, tube current: 30 mA, colimeter: 100 μmφ, measurement time: about 2000s and 3000 s. In PSPC-MDG2000 diffraction patterns were detected with a Position Sensitive Proportional Counter combined with multi-channel analyzer and in RINT rapid they were detected as a figure using an imaging plate. For both using a computer measurement dates were converted to 2theata dates for search compounds.

7.4 Results

7.4.1 Optical Documentation

Results of the 3D scanning show significant differences between the original structure and the later extended addition. The original walls vary relatively little in thickness (between 0.957–1.001 m), whereas the thickness variation in the later extension is wider (0.457–1.132 m). The ground-plan and elevation images produced from the laser scan data also revealed the accurate planning of the original structure, where architectural units were based on geometrical proportions (Fig. 7.1). The same tendency was recognized in comparison sites as well.

In the later-added porch, the difference in width along the north-south axis was measured 0.784 m, and the largest outward inclination of the columns is 7 degrees. The leaning columns can be clearly seen in old photographs, but it is not possible to get the inclination measurement from these slightly distorted and partially non-dated images.

Further survey and long-term monitoring are urgently required to determine whether these structural displacements are increasing or can be considered stable within the limit of certain safety margins.

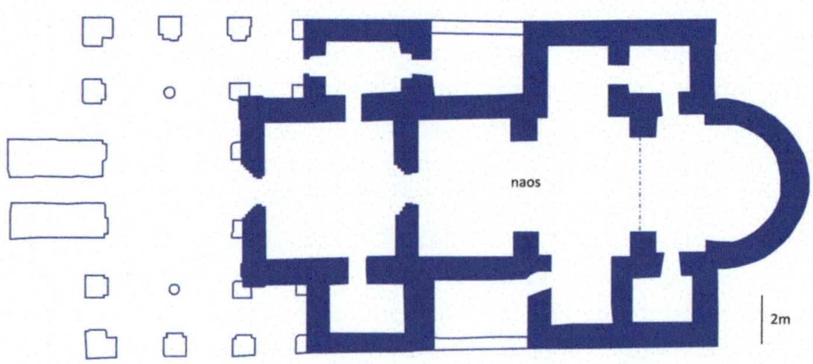

Fig. 7.1 Grond-plan of Sopoćani church

Analysing the data given by the 3D LED scanning, the author discovered interesting compositional devices that had been adopted in the work of painting on the west wall of the naos (Fig. 7.2). The painted surface can be divided into 3 zones (lowest, middle and uppermost) corresponding to pontata, or scaffolding lifts. In the production of medieval wall paintings in the Balkans, lime plaster was applied and painted in sequential horizontal sections. By observing and mapping the plaster joins between these sections, the height of the original scaffolding lifts is determined. As

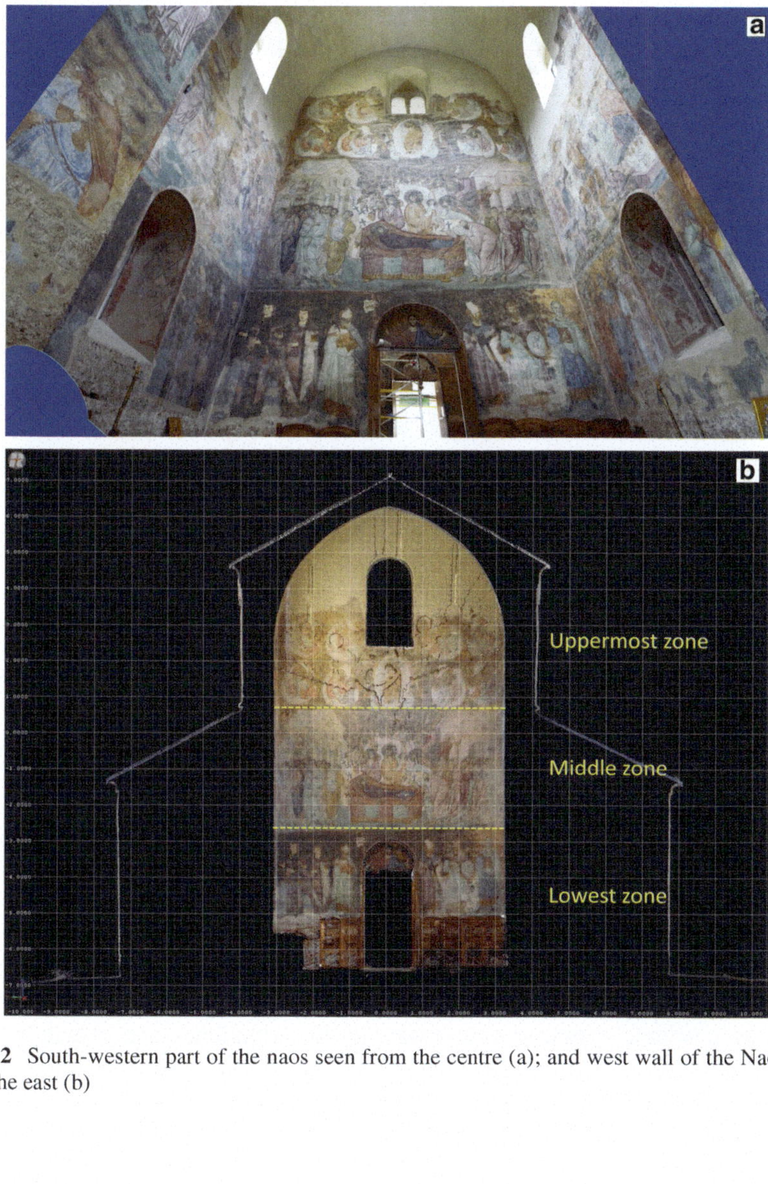

Fig. 7.2 South-western part of the naos seen from the centre (a); and west wall of the Naos seen from the east (b)

Table 7.1 Relative figure-sizes according to elevation height in Balkan Churches

Element	Sopoćani (13c)	Đurđevi Stupovi (12c)	Gračanica (14c)	Dečani (14c)
Upper part	1.229 m	0.818 m	0.998 m	0.515 m
Middle part	1.761 m	1.160 m	1.229 m	0.676 m
Lower part	2.168 m	1.469 m	1.681 m	0.835 m

is generally supposed, 92 figures painted on the wall of this area differ in size, orientation and posture, but the difference in height is notably consistent. Their average height, from head to foot, in each of the horizontal levels is: 2.168 m in the lowest zone ($\sigma = 0.06$); 1.761 m in the middle zone ($\sigma = 0.27$); and 1.229 m in the uppermost zone ($\sigma = 0.11$). Namely, the higher the figures are positioned, the shorter they are depicted making them appear very far from floor-level. This finding leads us to assume that the original painters intended to convey a certain visual illusion to produce a sense of higher volume for those standing on the ground level of the church. This typical illusionistic device is also found in other comparative examples, at Đurđevi Stupovi monastery, Gračanica monastery and Dečani monastery [Table 7.1]. Although figure-sizes vary according to the overall dimensions of these churches and the corresponding scale of their wall paintings, the decreasing rate of figure-size according to elevation is 79–85% from the lowest to middle zones, and 70–76% from the middle to uppermost zones. It is to be noted that these examples show relatively little divergence.

In other regional centres of medieval painting production, such as in Constantinople and Thessaloniki, figure-sizes in Orthodox churches differed much according to their importance in the overall painting scheme. For example, dedicatory saints and local patron saints, reflecting their particular significance or contribution to their church history, were depicted much larger than other figures. In the central and northern Balkans, on the other hand, depictions of figures developed along more realistic lines [9]. If figures are almost equal in height as a human body, it is possible to introduce the illusionistic technique to make people feel a wider inner space. Thus, 'naturalism' in painting practice of the Balkans might have led to the birth of the perspectival device that we observe to have been used at Sopoćani and other churches in the region.

Regarding the connection between setting-out and the painting technique at Sopoćani, LED scanning data give us another fact of interest: all 58 saints (figures are 92 in total) are given haloes which were drawn by compass, leaving a detectable centre-point and accurate circle to define the halo circumference. Significantly enough, the positioning of the circles shows slight differences between the three scaffolding zones. In the lowest zone, the circles were centred at a point on the face of each saint between his eyes and temple, whereas in the uppermost zone, the centre points are positioned higher on the forehead. In the middle zone, the centres are mostly placed around the forehead, and some at the height of the temple (Fig. 7.3). In addition, the faces of the figures in the middle and uppermost zones are painted slightly longer, compared with those in the lowest zone. All this integrated evidence

Fig. 7.3 Differences in the positioning of the haloes (left: lowest zone, middle: middle zone, right: uppermost zone)

demonstrates how the painters consciously as well as deliberately applied the technical device of heightening to compensate for the inherent foreshortening problems of viewing a large wall painting from the ground floor. The skillful painters were ingenious enough to shift the inscribed haloes slightly up without distorting their perfect circular form, revealing their highly sophisticated awareness of spatial representation. Comparative evidences from other churches in the region suggest that this practice was widely shared among painters.

It is interesting to consider the significance of the compositional techniques used at Sopoćani in the context of the development of linear perspective in art. Perspective is a technical device employed on a flat surface to represent three-dimensional effects, to give the impression that the depicted image appears as it does to the human eye: objects close to the observer are depicted relatively large, while those further away in depth are shown proportionately smaller. Ancient Greek mathematicians and theatrical producers are thought to have been aware of the technique [10], though perspective drawing is generally regarded as having been established in western art in the early Renaissance in Italy [10, 11]. Thus, in wall paintings executed during and after this period, figures are usually shown smaller in the lower registers and increasingly larger higher up the wall. This device allows every figure, regardless of its location on the wall, to appear the same size to the viewer at floor-level.

The perspective device employed in medieval wall paintings in Balkan churches, on the other hand, were intended to produce quite different effects, and for different reasons: figures close to the viewer were deliberately painted larger than those appearing farther away, in order to intensify the illusion of their spatial context. In viewing the wall painting from the ground floor, the imagined space would be felt higher and wider than the reality: for example, angels painted on a vault would be seen by the observer as if they were flying high in the heavenly sphere. This spatial device was practiced and adopted successfully by Late Medieval Balkan painters. Ordered to work in an Orthodox context, they were already fully conscious of the manipulation of space and illusion for certain iconographic and theological reasons, even though they were not yet aware of the perspective method that was to come with the Renaissance.

It is of further interest to note that the technique of using diminishing figure-sizes in Balkan wall paintings has earlier parallels in wall painting along the Eastern Silk Road, such as those at the grotto sites of Kizil and Dunhuang. While medieval Serbian rulers faithfully followed Byzantine orthodox traditions of painting, they were also tolerant of influences from outside, including artistic styles from western Europe. Influences from Asia were probably also transmitted via trade links, although only a few scholars have explored the possibilities of iconographical-technical similarities between the medieval wall paintings of the Balkans and of Central and Southern Asia [12]. Comparative research in the wider context of the Eurasian continent to be prepared by the author may open new perspective in the study of medieval wall painting technique.

7.4.2 Material Analysis

Microscopic examination and analysis of paint samples revealed the presence of a red mordant layer under the gold leaf decoration at Sopoćani. The mordant, applied over a layer of yellow ochre which was probably applied in the buon-fresco technique [13], functions not only as an adhesive base for the gold, but also to modulate and emphasize its optical effects.

As previously mentioned, although much of the gold leaf decoration is lost or degraded, it can still be identified through careful observation, especially under magnification by instrument, and its presence can be also detected by elemental mapping (Fig. 7.4). The elemental ratios of the gold leaf by atomic percentage based on the averaged results of two analyzed samples are shown in Table 7.2: Au comprises more than 90%, Cu and Ag are the remaining constituents. The high ratio of gold suggests typical composition of gold leaf in the pre-modern era [14].

Interestingly, the colour of the mordant was different in different parts of the wall painting: in the background, a light yellowish-red colour was used, while a darker bluish-red colour was employed in the haloes. This difference in colour is clearly observed in the microscopy images (Fig. 7.5). To determine the rationale behind these differences, a careful scientific examination was carried out.

In the SEM images of one sample of gilding taken from the background of the painting (Fig. 7.4), the detected elements show a high concentration of Fe on the left side of the figure. The brightest part in the back-scattered image is mainly composed of Ag and Cl, which are present over a layer of drying oil. Cross-section examination of the sample clearly shows these oil-containing layers [Fig. 7.6]. The underlying yellow layer contains Fe, Si and Al. Ag and Cl are also seen in the mapping images, which are probably accidentally transferred from other layers.

Examination and analysis of the haloes were carried out on two different sample types: one still preserving the gold leaf surface; the other, now reduced to an altered or degraded black appearance [Fig. 7.4]. The SEM image of the preserved gold shows tiny cracks on the surface, and the drying oil below; in the back-scattered image,

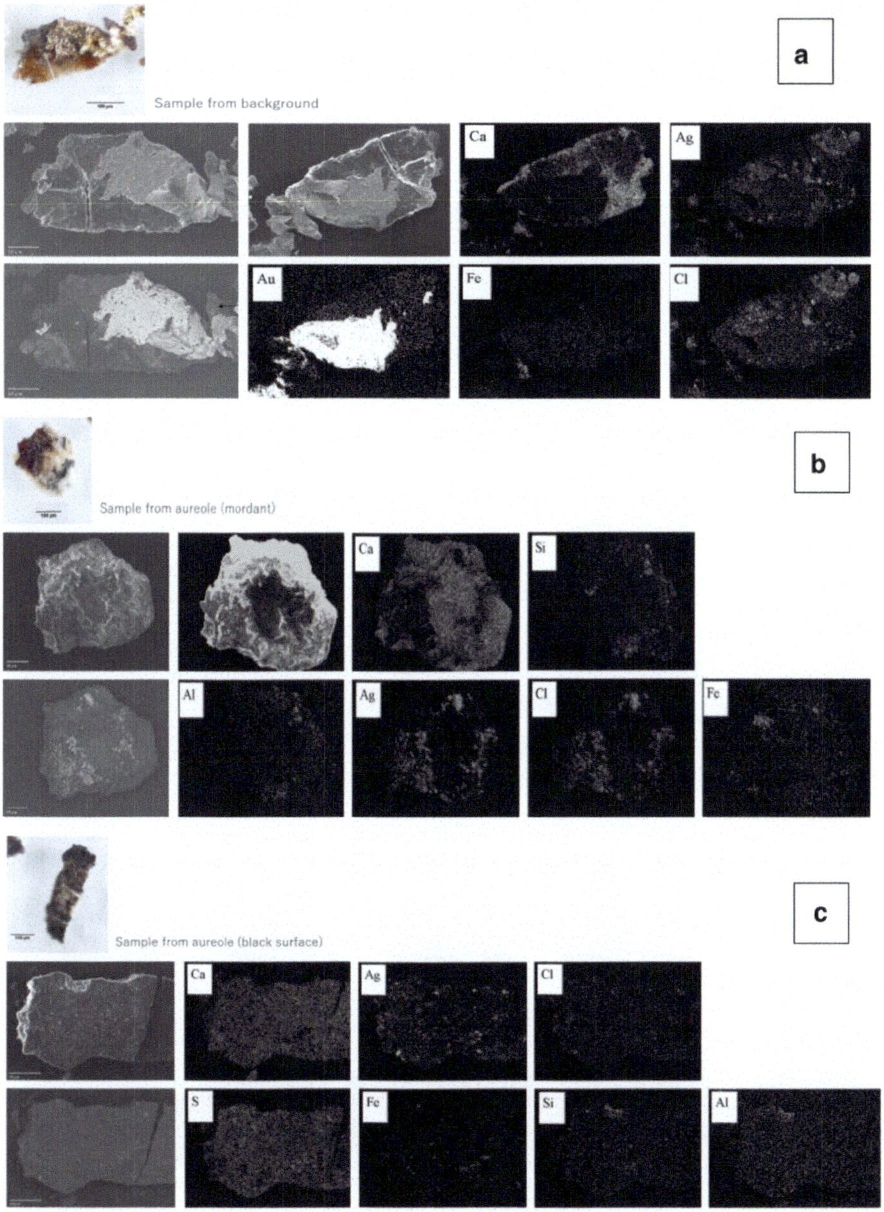

Fig. 7.4 A, B, C: SEM-EDX elemental mapping of gold samples from the background and haloes

Table 7.2 Elemental components of the gold leaf used at Sopoćani church

Element	Atomic ratio (%)
Au	93.0–91.1
Cu	4.3–4.8
Ag	3.2–4.2

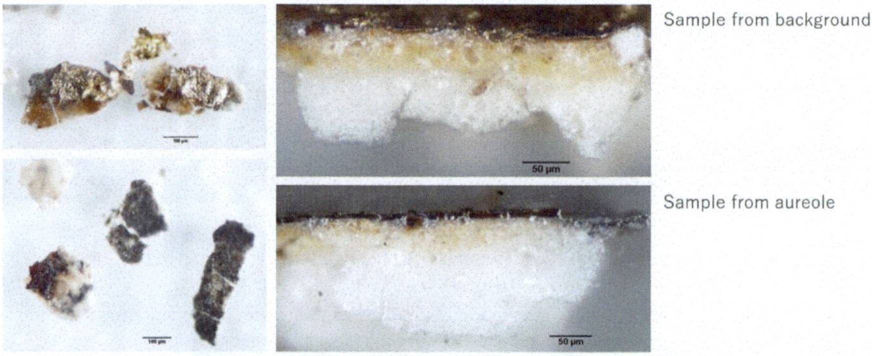

Fig. 7.5 Microscope images of gold samples from the background and haloes

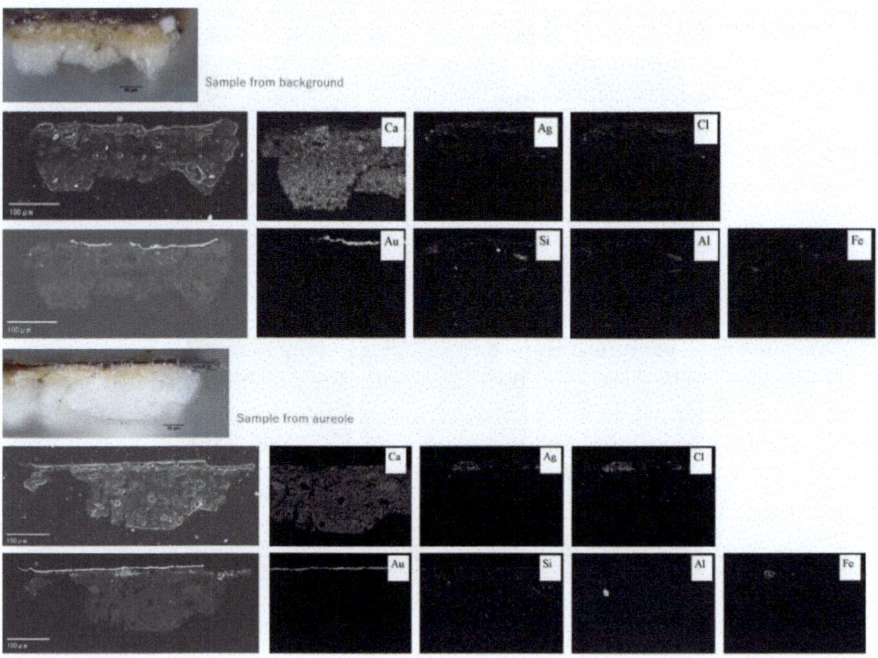

Fig. 7.6 SEM-EDX elemental mapping of cross section samples taken from the background and haloes

random brighter points of Ag and Cl are observed. Interpreting the mapping results as a whole, Fe, Si and Al are present as much as Ag and Cl.

For the other sample, the SEM image of the black surface is quite homogeneous, and, compared with the first sample, there is relatively low contrast in the back-scattered image. The elemental mapping of this sample shows that its dark surface contains a large amount of S, while other elements are unevenly distributed. In the cross-section of this sample [Fig. 7.6], Ag and Cl were detected, especially on the right side of the sample, and Al is also present in small amounts. The underlying preparatory layers and their constituent materials are the same as those found in samples from the background of the painting.

From these analytical results, it is clear that the main component of the mordant is iron oxide red, but that silver chloride (AgCl) is also present in almost similar amounts (ICDD-JCPDS card number, 31-1238 by MDG) (Figs. 7.7 and 7.8). Phosphine (PH^3), Sulphur (S) and iodine were also detected as minor components, although related compounds were not identified.

Iron oxide red is one of the oldest pigments used in human history. In contrast, silver chloride has not so far been detected as a distinctive painting material. However, at Sopoćani, the analysis indicated the presence of silver chloride as a distinct compound in the mordant layers of both the background painting and the haloes. The most likely explanation for this is that it is present as an alteration product. In the technique known as 'Zwischgold' (also called part-gold), gold leaf is layered over silver. In instances of the use of 'Zwischgold' in thirteenth century icon paintings,

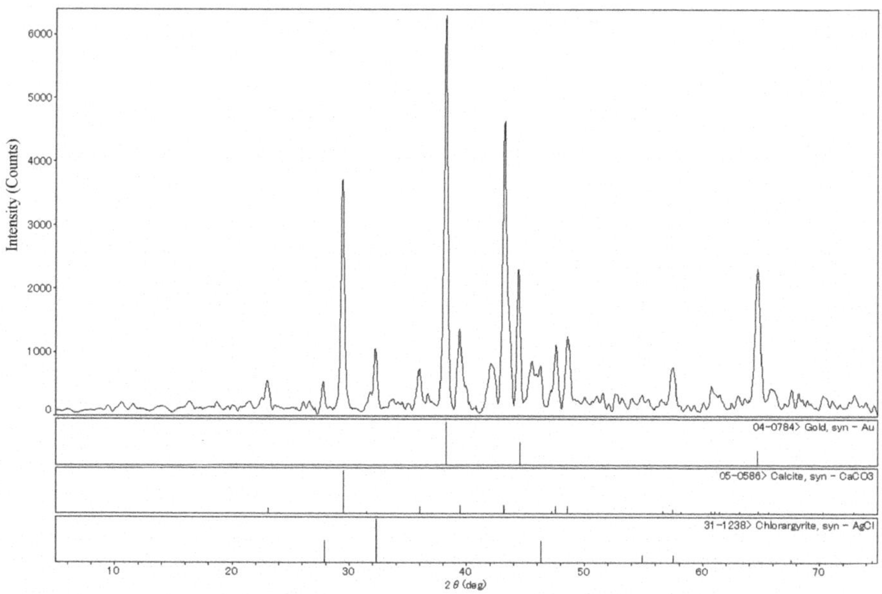

Fig. 7.7 Element mapping of cross section of gold leaf surface

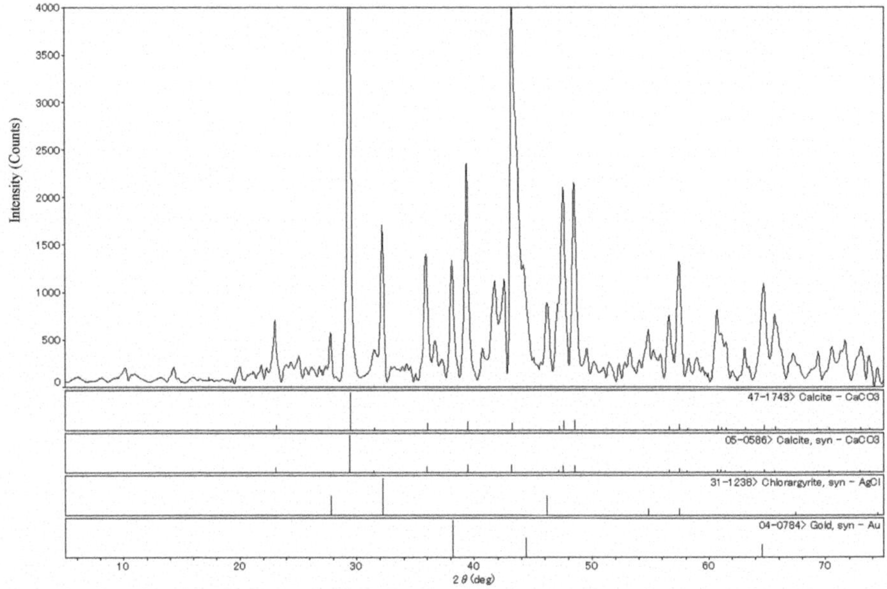

Fig. 7.8 Element mapping of cross section of black surface

the silver foil can be found in an oxidized state and Ag/Cl is a common corrosion by-product. At Sopoćani, the detected AgCl could therefore originate from the corroded silver base of a part-gold application process. However, its distribution patterns in the analyzed samples do not allow for this interpretation.

It is known that the Medieval Kingdom of Serbia (1217–1346) expanded its power through the international trade of indigenous silver and lead [15]. In fact, silver mining occurred close to the Sopoćani monastery [16]. Substantial deposits of silver chloride must have been accumulated as by-products of silver mining and purification activities. As a material, silver chloride is known as silver horn, having a grey appearance, and soft and malleable physical properties. The findings of this study suggest its use as a painting material in the medieval period in the Balkans. The detection of silver chloride in the medieval wall paintings at Horezu monastery, in Romania, also supports this hypothesis [13]. On the tentative examinations carried out by the author, however, silver chloride turned out not to be a very suitable paint material. Though its natural grey colour is affected by red and yellow tones, it becomes almost colourless when crushed for use as a pigment, losing its metallic luster and covering power. The colour grey can be produced by alternative technical means that are both simpler and cheaper, for example by combining black and white pigments. In contrast, silver chloride is not easy to crush nor to process into pigment form, because of its ductility.

Despite these drawbacks, two reasons are taken up to support the idea that it was used as a paint material at Sopoćani. The first is the huge scale of the wall paintings. Each inner wall at Sopoćani covers at least 80 m^2, and in order to paint these immense surfaces, large amounts of paint materials had to be prepared. Silver chloride, which was readily available at the time, could have been mixed with iron oxide red as an extender material.

Secondly, we can reasonably suppose that silver chloride might have been chosen by painters to play a role of modifying the mordant layer under the gold leaf. As already mentioned, in addition to the background of the painting, gold leaf was used on the haloes. The presence of gold is also observed in other places, such as crowns and details of clothes. The original painters must have technically differentiated their use of gold so that adjacent gilded areas could be distinguished. In this regard, it is significant that a substantial amount of silver chloride was detected in samples taken from the background of the painting compared with those taken from the haloes. The observed colour differences in the mordant may have been achieved by combining varying ratios of iron oxide red and silver chloride. These hypotheses need further consideration and analysis.

7.5 Conclusions

Regarding the construction and painting processes at Sopoćani monastery, the author concludes her research in the following three points.

1. Results of the laser scanning demonstrated the superior quality of the original building.
2. LED scanning data revealed a sophisticated awareness on the part of the original painters of spatial planning and manipulation, which was achieved by a variety of compositional devices.
3. Material analyses provide significant insights of the painters' use and application of gold leaf, which may further indicate the use of silver chloride as an original painting material.

Additional remarks should be made as more general conclusions.

Modern procedures of 3D- measurement provide valuable tools for examining and monitoring heritage structures, but there is scope for improving scanning implementation in the field. Even taking into account rapid developments in scanning techniques and performance of equipment, interpretation of scan-data requires both skill and expertise. Here, interdisciplinary collaboration is essential and indispensable.

Regarding the devices and techniques used by painters to determine large-scale mural compositions, further research is needed to elucidate the possible cultural exchanges between West and East that might have mutually informed or exchanged these processes. There is abundant historical evidence that Hellenic culture spread across the Middle East, ultimately extending into Asia, but cultural flows in the northern part of central Eurasia have not been studied sufficiently so far. There is considerable potential for highlighting new cultural interactions between East and West by examining technical aspects of their wall paintings as artistic symptoms of such interactive flow.

Regarding the use and application of gold leaf, certain synthesis of analytical and art-historical studies are highly required. Future work by the author will include additional analysis of original samples and a series of experiments, based on producing replicas according to the instructions contained in medieval painting manuals and treatises. It is hoped that these further studies will provide new insights of thirteenth century painting practice in the Balkan area.

Lastly, it is especially important that scientific information on the medieval painting materials and techniques at Sopoćani not only improves technical and art-historical understanding, but also constitutes an effective base for future conservation efforts.

Acknowledgements I am heartily thankful to all of those who supported and encouraged me in this study, which would not have been possible without the generous permission and assistance of the Institute for the Protection of Historical Monument of Serbia. For the materials analysis, I am grateful for the support of Mr. Jun-ichi Miyata. This study is supported by the Grants-in-Aid for Scientific Research, KAKENHI (JP16H05901, JP16J10608), of the Japan Society for the Promotion of Science.

References

1. Obolensky, D.: Il Commonwealth Bizantino – L'Europa Orientale dal 500 al 1453. Translated by M, Sampaolo. Editori Laterza. Bari. September 1974, pp. 3–36 (1974)
2. Krautheimer, R.: Architettura Paleocristiana e Bizantina. Translated by R. Federici Einaudi, pp. 456–457 (1986)
3. Hidaka, M.: Present state and obstacles of cultural properties in post-conflict Kosovo. Journal for area studies on Eastern Christianity, vol. 1, May 2017, pp. 113–123 (2017)
4. Speake, G.: The Orthodox Christian World. Routledge, pp. 325–330 (2012)
5. Knox, John S.: The monastic movement: origins & purposes. Ancient history encyclopedia, August 2016. Web (2016)
6. Hidaka M., Miyata, J.: Technique of the wall paintings of the hospital church in Horezu monastery. Eur. J. Sci. Theol. **11**(2) (April), 241–252 (2015)
7. Kandić, O.: Istorija I., Manastira, A.: Sopoćani, History and Architecture of the Monastery. Društvo Konzervatora Srbije. Belgrade, p. 20 (2016)
8. Millet, G.: L´ancien art Serbe. les églises. Paris, pp. 24, 64 (1919)
9. Rice, D.: Art of the Byzantine Era. Thames and Hudson Ltd. London, pp. 187–213 (1963)
10. Camerota, F.: La prospettiva del Rinascimento-Arte, architettura, scienza. Architettie architetture. Mondadori Printing S.p.A. Verona, pp. 258–270 (2006)
11. Roccasecca, P.: Dalla prospettiva dei pittori alla prospettiva dei matematici. Il Contributo Italiano alla Storia del Pensiero Scienze. Istitvto della Enciclopedia Italiana. Rome, pp. 137–144 (2013)
12. Upadhya, Om D.: The Art of Ajanta and Sopocani: A Comprehensive Study. Motilal Banarsidass (1994)
13. Hidaka, M., Hidaka, K., Miyata, J.: Technique of the mural paintings of the hospital church in Horezu Monastery. Romania. In: 6th European Symposium on Religious Art. Restoration and Conservation. Florence, pp. 122–125 (2014)
14. Miyata, J.: Bulletin of Art Restoration Studio 21, vol. 15. Tokyo (2001)
15. Medic, M., Todic, B.: Stari Ras sa Sopocanima. Platoneum. Novisad, pp. 82–92 (2014)

16. Melcher, F., Reichl, C.: Economic Geology of the Eastern and South-eastern European (ESEE) Region. Berg-und Huttenmannische Monatshefte. **162**(July), 238–244 (2017)
17. Camerota, F.: Ibid. pp. 7–8 (2006)
18. Hidaka, M., Hidaka, K., Miyata, J.: Technique of the mural paintings of the hospital church in Horezu monastery, Romania. In: 6th European Symposium on Religious Art. Restoration and Conservation, Florence, p. 1 (2014)
19. Kandić, O.: Ibid. p. 247 (2016)
20. Kandić, O.: Ibid. pp. 49–52 (2016)
21. Kandić, O.: Ibid p. 1 (2016)
22. Krautheimer, R.: op. cit. p. 456 (1986)

Chapter 8
Painting Materials and Techniques of the Ajanta Wall Paintings

Yoshiko Shimadzu

Abstract The Ajanta Caves were excavated in two periods: the early caves between the second century B.C. and first century A.D.; and the late caves between the late fifth and sixth centuries A.D. The caves were decorated with polychromed sculpture and wall paintings, both on the interior and exterior. Although general technical investigations of the wall paintings were made as early as the 1930s, more in-depth studies have been produced in recent years. This paper reviews these studies of the painting materials and techniques of the Ajanta paintings, and supplements them with the results of recent investigations carried out in Cave 2, which was excavated in the late fifth century A.D.

The rock-cut caves at Ajanta were leveled with earthen plasters in preparation for painting. The earthen plasters are made of local soil derived from weathered basalt mixed with fine plant materials. Over the upper, finer plaster, a white ground was applied, either of lime or kaolin. The main identified pigments include red and yellow ochre, orpiment, red lead, green earth, and ultramarine blue. The earth pigments were locally available, including green earth, which is celadonite. Natural ultramarine blue (lapis lazuli) was imported from Afghanistan. The yellow pigment, orpiment, survives in an altered state. The use of organic colourants is suspected in the Ajanta paintings but has not been previously confirmed. The recent scientific analysis in Cave 2 identified the red resinous colourant, lac, derived from lac scale insects.

In Cave 2, paint application techniques show evidence of gradation of colours from dark to light, and vice versa; only scant evidence was found for mixing colourants or superimposing paint layers to obtain secondary colours.

Keywords Ajanta wall paintings · Organic red · Lac · Celadonite

---

Y. Shimadzu (✉)
Research Department, National Museum of Japanese History,
117 Jōnai-Chō, Sakura-city, Chiba 285-8502, Japan
e-mail: yshimadzu@rekihaku.ac.jp

S. Aoki et al. (eds.), *Conservation and Painting Techniques of Wall Paintings on the Ancient Silk Road*, Cultural Heritage Science,
https://doi.org/10.1007/978-981-33-4161-6_8

8.1 Ajanta Overview and Previous Technical Studies

The Ajanta Caves are located in central western India, about 100 km from the city of Aurangabad, in the state of Maharashtra (Fig. 8.1). Geologically, the region is a part of the Deccan Traps and the bedrock is basalt. The number of officially recognized caves is 30, including some that are unfinished. The date when the earliest caves were excavated is uncertain, but it is assumed to be sometime between the second century B.C. and the first century A.D. These are referred to as the early caves. A second period of excavation occurred around the late-fifth century to the sixth century A.D., when the greatest number of caves were created; these are referred to as the later caves.

Polychrome reliefs, which were developed in the later period, and wall paintings decorate Caves 1, 2, 6, 7, 9, 10, 11, 16, 17, 19, 20, 21, 22 and 26 [1] (Fig. 8.2). The

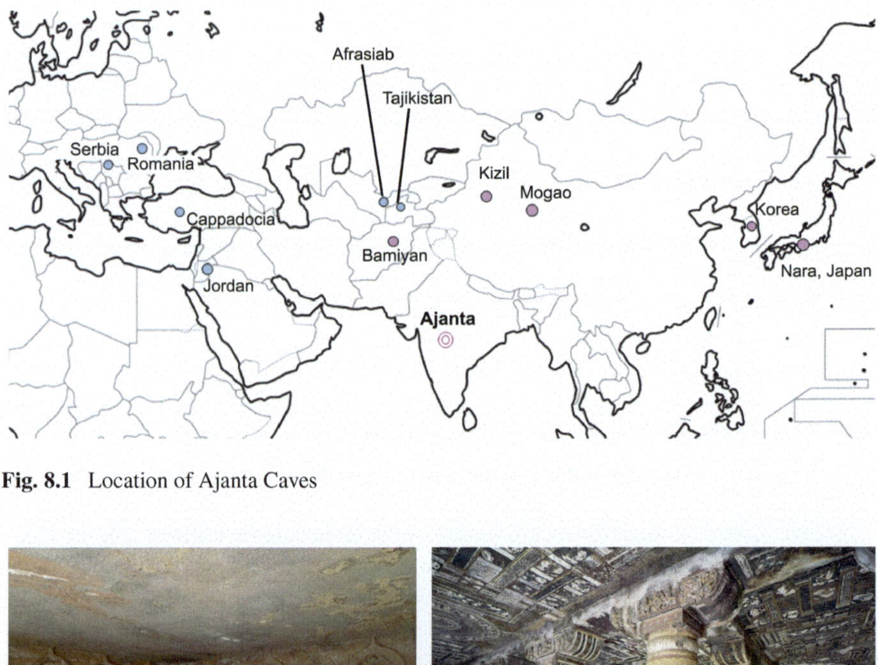

Fig. 8.1 Location of Ajanta Caves

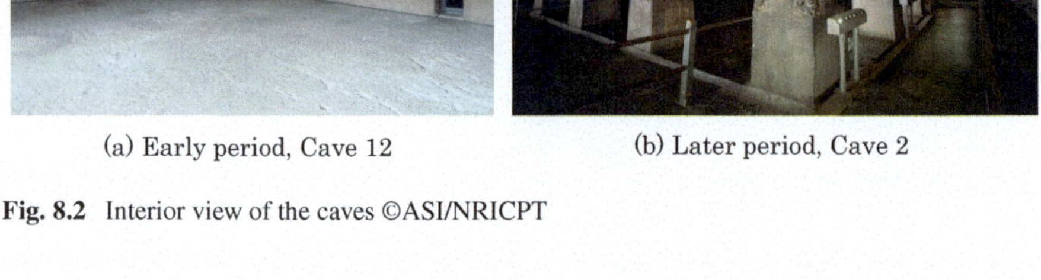

(a) Early period, Cave 12 (b) Later period, Cave 2

Fig. 8.2 Interior view of the caves ©ASI/NRICPT

Fig. 8.3 The early painting of Cave 10 covered with glass panels ©ASI/NRICPT

wall paintings in Caves 9 and 10, which are early caves, were painted over in the later period, and are also covered with glass panels for their protection (Fig. 8.3). These conditions make it difficult to carry out detailed investigations of the original materials and techniques of the early wall paintings. The later wall paintings and polychrome reliefs have been investigated and reported on by the Archaeological Survey of India (hereafter the ASI) on several occasions.

The earliest scientific study of painting materials was published in 1939 by Paramasivan. Since the original publication is not readily available outside of India, its content is referenced from subsequent publications in which it is quoted [2]. Following Paramasivan, Lal [3] and Bharadwaj published the results of their technical studies [4]. Dabhade published a book in 1973 [5] on the techniques of wall paintings in India, which summarized previous studies, including Paramasivan's. In 2007 Lal also published a monograph describing each cave, but its content is mostly the same as his publication of 1967. In Japan, in 1971, Takada and Yamazaki made an abridged translation of Lal's 1967 publication, adding their own interpretation [6, 7]. Sadakane reviewed previous publications and described the paintings based on direct observations [8] (Table 8.1).

The ASI and the Tokyo National Research Institute for Cultural Properties (hereafter NRICPT) conducted a joint project for the conservation of the Ajanta paintings. As a component of this, scientific investigation was carried out and the results were summarized in a submitted report [10]. At almost the same time, the ISCR (Istituto Superiore per la Conservazione ed il Restauro: High Institute for Conservation and Restoration, Italy) published their studies on the Ajanta paintings, mainly from Cave 17 [11].

Table 8.1 List of technical studies (mainly India and Japan)

Year of published	Source	Secondary references and translations
1939	Paramasivan	Set [2]
1967 and 2007	Lal (also in 2007)	Takada and Yamazaki [6, 7] and Sharma [9]
1983	Bharadwaj	
		Sadakane [8]
2014	ISCR (and ASI)	[11]
2015	ASI/NRICPT	[10]

This paper reviews the original materials and techniques of the wall paintings at Ajanta by summarizing reports made by the ASI and presenting results of investigations carried out in Cave 2 by the ASI and NRICPT. Findings of the ISCR are also referenced for comparison.

8.2 Materials and Techniques

8.2.1 Workshop Practice

In Ajanta, although caves were basically prepared immediately after their excavation so that their interiors could be used, walls and ceilings were not necessarily painted right away. Within the same cave, more than two teams of painters appear to have worked simultaneously or in close sequence on the same wall. This practice is evident in Cave 2, where features of figures, or uses of colour, indicate the work of different painters. For instance, on the left wall, the work of more than two painter-teams is detectable in the difference in figure painting between the 'Buddha's biography', 'Buddhas with a worshipping monk', and the 'Thousand Buddhas' (Fig. 8.4). Judging from the manner in which a part of the 'Buddha's biography' was left unfinished, it seems that the other paintings were added some time later.

8.2.2 Rock Support and Plaster Layers

The uneven rock-cut cave interiors were plastered with earthen plaster ('mud plaster' in Indian publications) to provide a smooth surface on which to paint. In general at least two layers of plaster were applied. A rough plaster mix containing plant materials such as seeds or chaff forms the first layer. On top of this, a second smoother layer mixed with finer particles was applied, and on top of this is a white ground (Fig. 8.5).

Fig. 8.4 Paintings of different themes on the same wall, the left wall in Cave 2 ©ASI/NRICPT

Fig. 8.5 Schematic section showing the stratigraphy of the Ajanta paintings

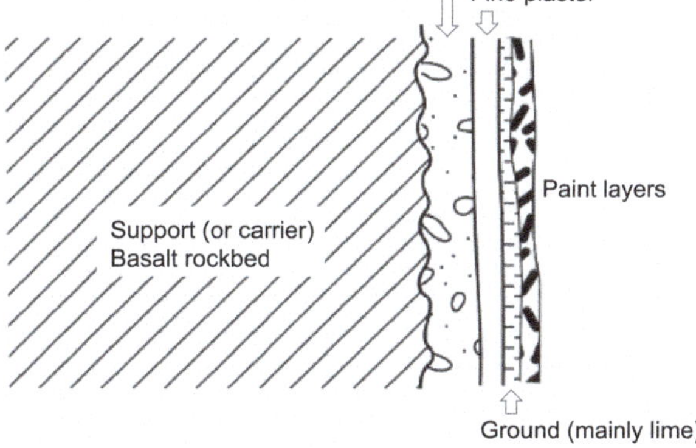

Similar plastering techniques can be found in the wall paintings at Bamiyan in Afghanistan, or at the Kizil grottoes in China [12].

However, at both these sites, the lower earthen plasters contain much more straw, whereas in the Ajanta paintings straw was little used and the size of the plant materials is rather small (Fig. 8.6).

The thicknesses of the earthen plasters differ depending on the roughness of the rock-cut surface to which they were applied. Paramasivan reported use of only a rough plaster layer, which could be more than 50 mm thick [13], superimposed by a finer 'plaster' of about 0.1 mm, which he described as being composed of lime, kaolin, or gypsum. In fact, this fine upper layer must be the white ground rather than a plaster: examination by the ASI and NRICPT, and by the ISCR, confirmed

(a) Ajanta Cave 2 ©ASI/NRICPT

(b) Bamiyan example ©NRICPT

Fig. 8.6 Plant materials in the earthen plaster

that the typical thickness of the white ground layer is about 0.1–0.2 mm. Possibly, Paramasivan mistakenly described the two earthen plaster layers as a single layer.

Lal reports the thickness of the fine earthen plaster as about 2 to 3 mm [3]. In his technical study, the composition of the earthen plaster is reported as 60% silica, 27% iron and alumina, and 2-3% lime and magnesium oxide (Table 8.2).

In contrast, analysis of the rough plaster carried out by the ASI and the Geological Survey of India indicates relatively low silica content, and higher amounts of CaO and MgO, which were interpreted as originating from added granite powder [13]. However, granite does not contain a particularly high amount of calcium or magnesium; and grinding granite into powder form is not a very practical undertaking. Moreover, having adjusted the data to accommodate for a large amount of carbon dioxide derived from organic components and water, these results are consistent with the average data of Deccan basalt [14] (Table 8.2). The major components of basalt rock include pyroxene (containing both calcium and magnesium), forsterite (Mg_2SiO_4), and plagioclase ($NaAlSi_3O_8$ and/or $CaAl_2Si_2O_8$). Depending on occurrence conditions, relatively high ratios of calcium and magnesium are usually present. These components were likewise detected in the earthen plasters, indicating their source material in the local basalt.

8.2.3 White Ground and White Paint

It is likely that the ground-layers employed in the paintings changed over time at the hands of different painters. It would be useful to compare these differences to gain a better understanding of the technological development of the Ajanta paintings. However, at present, there is not enough detailed data to make effective comparison, so in this paper discussion of the ground-layers is based on information contained in published reports and on the findings of our investigations carried out in Cave 2.

Table 8.2 Composition of the rough earthen plaster (weight ratios) compared with Deccan basalt

	Average in Lal [3]	Cave 2*	Cave 17*	Cave 20*	Cave 20*2 (recalculated)	Average of Deccan basalt*3
SiO_2	ca. 60	44.4	45.1	38.5	43.2	48.97
Al_2O_3		12.4	12.0	10.6	11.9	13.66
FeO	ca.27	Trace	2.7	1.8	2.0	9.02
Fe_2O_3		13.2	11.2	9.6	10.7	4.15
TiO_2		2.8	2.8	1.7	1.9	2.49
CaO	2 ~ 3	9.0	8.3	11.0	12.4	9.88
MgO		3.4	3.8	3.1	3.5	5.9
K_2O		1.6	1.0	0.9	1.0	0.72
Na_2O		3.3	3.0	2.5	2.8	2.5
MnO		0.2	0.1	0.1	0.1	0.2
P_2O_5		1.2	0.3	0.3	0.4	0.29
$-H_2O$		1.8	3.1	4.0	4.5	0.98
$+H_2O$		4.5	2.8	3.7	4.1	1.47
CO_2(inorganic)		0.4	0.4	1.3	1.5	–
CO_2(organic)		2.1	4.0	11.2	(0)	–
S		0.4	trace	Trace	–	–

*ASI/NRICPT (2010)
*2 Data for Cave 20 re-calculated to remove the influence of organic components and water
*3 Source: Ghose (1976)

According to Paramasivan, lime, kaolin or gypsum were used for ground-layers, while lime was used as a white paint [2]. Lal [1, 3] reported lime as the ground-layer, and calcium sulfate and a compound of white silicates as white pigments. The identification of the white ground-layers—and of white paint layers generally—may be confused by modern repainting. In Cave 17, for example, zinc was detected by XRF analysis indicating that zinc oxide, a white pigment first used in the nineteenth century, was used at some time after this date. After the rediscovery of the caves in 1819, the paintings were restored on several occasions. On the other hand, Lal reported that baryte (barium sulfate; $BaSO_4$), zinc oxide, lime, or lead white were not used as white paints at Ajanta.

The ISCR reported in 2014 that kaolin was used as a white pigment, a finding also previously made by Sadakane; the ISCR also noted that kaolin was not used as a ground-layer. In contrast to findings made by the ASI, which reported the presence of both lime and gypsum, Sadakane asserts that kaolin is the only white pigment used as white paint.

In Cave 2, the situation is quite varied. A lime ground-layer was found on the paintings of the rear wall, which serves as a base for the background colours. But for the figures, a separate kaolin ground-layer was applied as a base. The kaolin was also allowed to remain selectively visible in white areas of painting (Fig. 8.7). On the other hand, for the painting of the "Thousand Buddhas" in the left aisle of the cave, the lime layer functions as both a ground and a white paint.

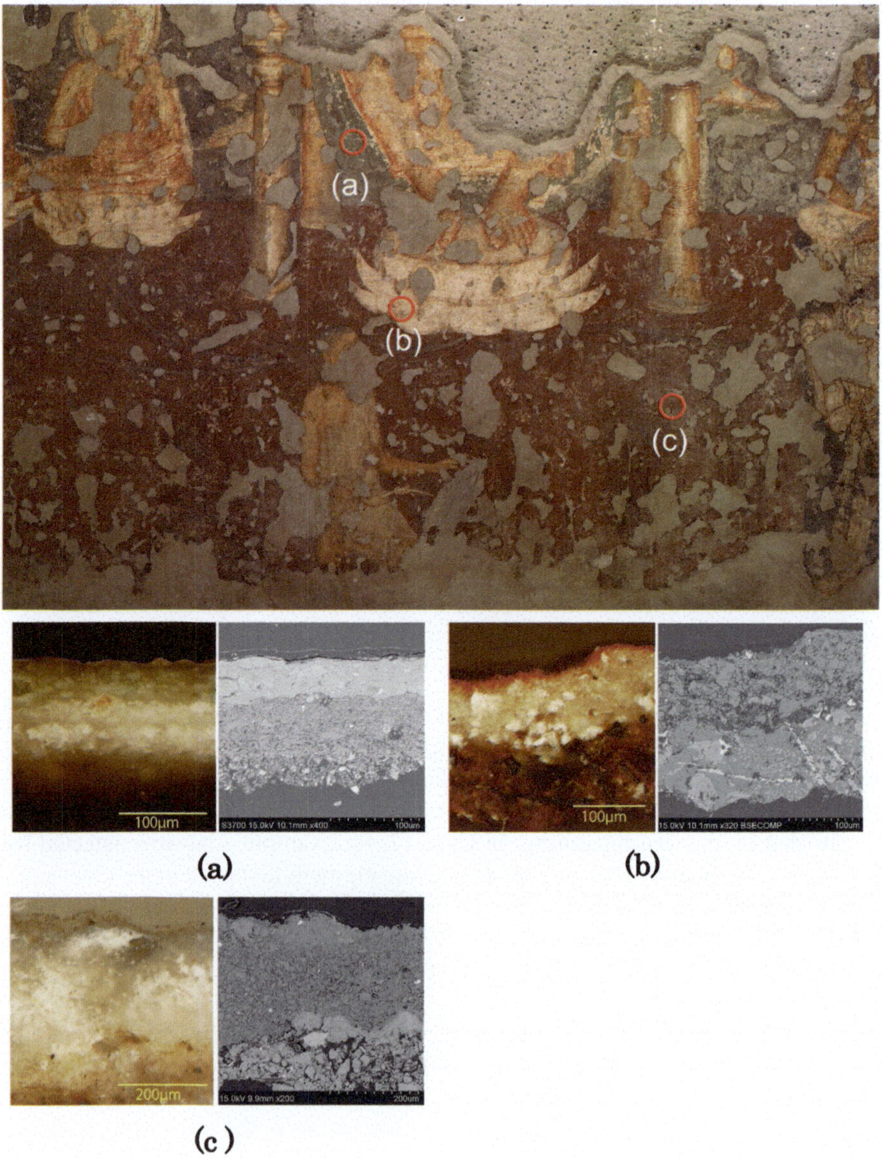

Fig. 8.7 Stratigraphic comparison of the white ground layers in three areas of Cave 2 ©ASI/NRICPT (a) Green area painted on a kaolin white ground over a lime ground (b) Red background painted on a lime ground (c) Kaolin white area painted on a lime ground

There is no clear evidence that gypsum (calcium sulfate) was used as a ground-layer in Cave 2. Detected gypsum may instead be related to past restorations, or be present as a degradation compound.

In Cave 17, the ISCR reported that both the ground-layer and areas of white paint contain calcium, and are likely to be lime, based on examination of cross-sections. A kaolin white was also occasionally found applied as a pigment.

8.2.4 Pigment and Colourants

Paramasivan and Lal report that the pigments used in the Ajanta paintings are red and yellow ochre, green earth, ultramarine blue, and lamp black [1–3]. The thickness of the paint layer is reported by Lal as 0.1 mm. Recent examination of paint cross sections [10, 11] indicates that for fine particulate pigments such as red ochre, the layer thickness is about 10–20 μm, and for green earth, about 30–40 μm.

8.2.4.1 Red

For red colours, it is reported that different shades of red ochre were used. Lal reports that the inorganic reds, red lead or vermilion, were not used at all [1]. In Japan, Takada [6] and Yamazaki [7] mostly adopted Lal's account, although Takada also suggested the use of red lead and vermilion, based on observation of vivid red on the lips of the Buddha figures; while the white appearance of many other lips were thought to be related to the degradation of an organic red. Sadakane [8] assumed the use of red lead based on observation of characteristic colour alteration in areas of the red painting. In addition, he pointed out the probable use of the organic red colour, lac.

In the ISCR's report, the use of an organic colourant was also suggested. This was because certain key elements associated with inorganic reds—iron for red ochre, lead for red lead, and mercury for vermilion—remained undetected by the limited XRF analysis that was carried out. However, the ISCR also went on to suggest that an organic colourant was used in combination with red ochre to obtain the desired shade of red; it was also assumed that red lead and vermilion were used.

In Cave 2, XRF analysis identified the presence of lead, strongly suggesting the use of red lead, which appears also to have undergone alteration (Fig. 8.8). On the other hand, vermilion was not found.

Even though use of organic colourants in the Ajanta wall paintings has often been suggested, firm analytical data has not been published. In the joint project of the ASI and NRICPT in Cave 2, a strong indication of the presence of organic colourants was found on the columns of the left shrine [13]. The colours on the columns facing the hall are almost entirely absent, exposing the white ground or rock surface beneath, whereas on the shrine side, some colours—including red—have survived relatively unchanged. Using ultraviolet (UV) light examination, a characteristic red-pink fluorescence was observed, indicating the presence of organic

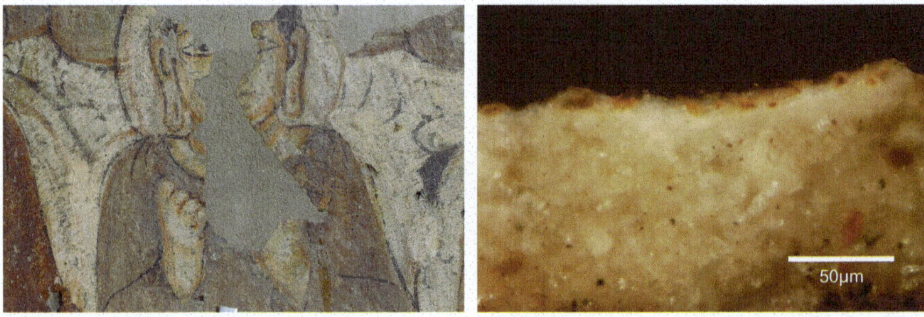

(a) characteristic alteration of red in Buddha's outer garment

(b) photomicrograph of cross-section of red layer showing presence of altered red lead particles

Fig. 8.8 Use of red lead in Cave 2 ©ASI/NRICPT

paint materials (Fig. 8.9). Scientific analysis by HPLC (High Performance Liquid Chromatography) of the red paint detected laccaic acids, proving the use of lac insects (a type of scale insect) as a painting colourant. In cross-section, the lac-containing paint was present as a translucent glaze-like layer, with no observable particles, which precludes the use of a lake form (Fig. 8.10). If purified lac dye powder had been used, the thin paint layer would be either distinctly evident, or dispersed particles of lac dye would be visible in it, neither of which were the case.

Lac insects are known as a source of red dye, and have been widely used for dying fabric—particularly in India and eastern Asia. Lac dye is a general term for the red colourants derived from stick lac, which is the resinous secretion of lac insects (Fig. 8.11). Another product of stick lac is shellac resin, which has been used as a coating material for objects including furniture and musical instruments.

After bibliographical research, a Jain treatise on painting materials and techniques was found, the *Jain Chitra Kaladruma* (tenth century), which describes a recipe for red ink derived from lac insects [16]. Since the treatise may have been originally used in and around the Ajanta region, and the recipe does not requires that lac be applied onto a substrate, which is also the case in Cave 2, it can be suggested that this recipe—or a similar one—was used at Ajanta.

According to the recipe, stick lac is first crushed into small pieces, then dissolved into a basic solution by being warmed. Taniguchi et al. [17] carried out experiments to make the red paint following the recipe's instructions. The resulting paint (hereafter called 'red lac resin') had moderate viscosity, and did not require an added binder or other vehicles. After drying, it became a translucent red film.

The FTIR spectra of both the red lac resin and the red paint from Cave 2 were almost identical (Fig. 8.12), and demonstrated the presence of laccaic acids by HPLC. The spectra were similar to that of shellac, suggesting that a large quantity of resinous components remained. Furthermore, the IR spectrum of the organic red paint analysed by the ISCR [11] is also similar to the red lac resin.

(a) Front of shrine showing near absence of colour

(b) Detail on the rear of the right column showing red and green paint

(c) Same detail showing red-pink UV-fluorescence in the area of red painting

Fig. 8.9 Cave 2, left shrine ©ASI/NRICPT

At Ajanta, it is well known that shellac resin was used in the nineteenth and early twentieth centuries as a protective coating on the wall paintings, and it is therefore possible that IR spectra might indicate this rather than the use of lac as a colourant. In order to rule out this possibility, FTIR analysis was performed on various stick lac samples. This confirmed that the spectra of red lac resin showed a characteristic absorption band around 1565 cm^{-1}, which is not found in the shellac resins.

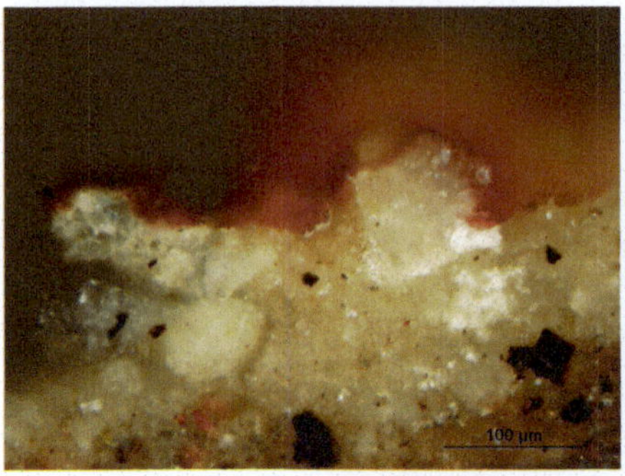

(a) Photomicrograph of cross-section of red paint in normal light

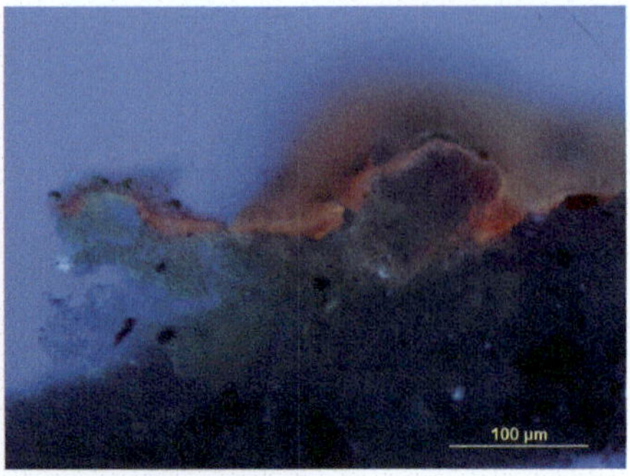

(b) Photomicrograph of cross-section of red paint in UV light

(c) SEM back-scattered image

Fig. 8.10 Evidence for use of an organic colourant in the red paint on the column ©ASI/NRICPT

Fig. 8.11 Stick lac

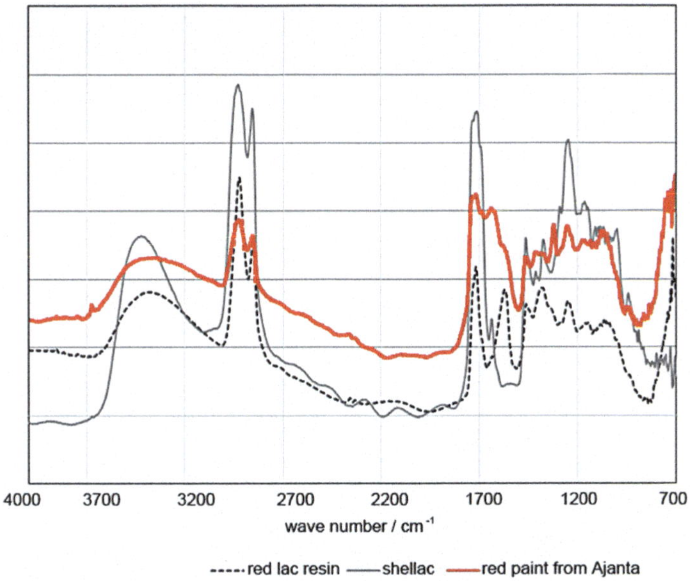

Fig. 8.12 Comparison of FTIR spectra of the organic red paint layer in Cave 2 with shellac and red lac resin

8.2.4.2 Yellow

Beside yellow ochre, the use of orpiment is indicated by the investigations of the joint project of the ASI and NRICPT in Cave 2, and also by the independent findings of the ISCR. Orpiment is a bright yellow mineral with the chemical composition, arsenic sulfide (As_2S_3).

In Cave 2, arsenic was mainly detected by portable X-ray fluorescence analysers (p-XRF), though the yellow colour was not observed. Examination of cross-sections with a secondary electron microscope equipped with energy dispersive X-ray spectroscopy (SEM-EDS) showed no clear particles, but arsenic was detected in a characteristic linear shape (Fig. 8.13). In the case of unaltered orpiment, sulfur is also detected. Its absence from the Cave 2 samples indicates the presence of altered orpiment.

Findings of the ISCR also suggest the use of the red-orange pigment, realgar, which is another type of arsenic sulfide. Compared with orpiment, the colour difference is due to its different crystal structure. The ISCR report describes a number of possibilities for obtaining a pinkish or an orange-yellow tone for skin colour, by mixing pigments such as different types of white with red ochre and/or orpiment/realgar. However, it must be noted that realgar is an unstable compound, and our studies have not so far indicated its presence as a painting material.

(a) Sampling location

(b) Photomicrograph of cross-section

(c) SEM back-scattered image, in which arsenic was detected linear shape (arrowed)

Fig. 8.13 Detection of arsenic in areas of white paint, indicating the presence of altered orpiment ©ASI/NRICPT

8.2.4.3 Green

In the Ajanta paintings, the use of green earth is reported from the earliest technical studies onwards. Indeed, geological deposits of green earth are observable around the Ajanta Caves (Fig. 8.14). Two types of mineral have been widely identified as green earth pigments, glauconite and celadonite. Lal thought that the green earth at Ajanta was derived from glauconite, but the ISCR's investigation confirmed the use of celadonite.

Distinguishing the two minerals by scientific methods is not easy. However, the geological occurrence of the two minerals is distinct, and thus they cannot be found in the same location at the same geological time. The origin of the green earth around Ajanta is from basalt, which is a volcanic rock formed by the rapid cooling of lava. Basalt is classified as a mafic type of igneous rock that contains relatively lower amounts of silica, and is rich in iron and magnesium. Celadonite is typically found in vesicular basalt [18] (Fig. 8.15). In contrast, glauconite is found in sedimentary

Fig. 8.14 Deposits of green earth in the terrain around the Ajanta Caves

(a) Ajanta

(b) Pitalkhora

Fig. 8.15 Celadonite in vesicular basalt at the Ajanta and Pitalkhora Buddhist cave sites, Maharashtra

rocks—especially those formed under marine environments of relatively low temperature. It can therefore be stated that the green earth found in the paintings at Ajanta is celadonite, as reported by Mariottini [19].

The colour green can also be created by mixing blue and yellow. In its investigations of Cave 17, the ISCR carried out XRF analyses of green painting in 7 locations. Results showed a high detection for arsenic. It was therefore suggested that orpiment was mixed with blue pigments such as indigo or ultramarine blue, or organic black, to produce a green colour. Since our investigation was limited to Cave 2, which may not be directly comparable with Cave 17, nevertheless no evidence was found to indicate that secondary colours were produced by combining pigments/colourants, such as blue and yellow to make green, red and blue to make purple, or red and yellow to make orange.

8.2.4.4 Blue

The number of naturally occurring blue pigments available for use when the Ajanta Caves were painted was limited. Among pigments that have been reported, the precious blue pigment, ultramarine blue, made from lapis lazuli, is most notable. Considering possible trade links with Afghanistan, the source of lapis lazuli, it is not surprising to find ultramarine blue in the paintings. Use of the copper-containing blue pigment, azurite, is also possible but it has not been reported so far. The organic blue colourant, indigo, is mentioned in many articles and reports, but our investigation in Cave 2 did not indicate its use there.

In the wider context, technical studies of the Ajanta paintings are quite limited, and references to blue pigments tend to be confusing. In the Japanese translation of Lal's study [3], ultramarine blue is called '*gunjo*' and '*lapis lazuli*' [6]. In Japan, '*gunjo*' meant only azurite until the import of synthetic ultramarine blue, which the earliest known records indicate occurred in the middle of nineteenth century. After the Meiji period (which began in 1868), synthetic ultramarine blue was gradually introduced as a pigment, and was referred to as '*jinzo gunjo*', which literally means artificial ultramarine blue (in colour); later this was abbreviated to '*gunjo*'. In the twentieth century, production of azurite as a pigment declined, and it only continued to be made in small amounts for use by certain people such as artists. Therefore, it is quite possible that when this book was published in 1971, 'ultramarine blue' could have been translated into '*gunjo*'.

8.3 Use of Secondary Colours

As mentioned above, from the investigations carried out in Cave 2, there is no clear evidence that colours were produced by mixing together two or more pigments and/or colourants. Previously, Sadakane [8] also noted that the superimposition or mixing of colours was rarely done in the Ajanta paintings. However, he suggested that white or black pigments were combined with other colours to achieve a range of lighter or darker tones: for example, combining red ochre and green earth respectively with kaolin or lamp black.

In Cave 2, some areas are painted in lighter tones of red or blue, which may indicate that kaolin was added to modify red ochre or ultramarine blue pigments. It is also possible that red paint was diluted with more binder or water and thinly applied over the white ground. To achieve gradations of colour to a lighter tone, such as can be observed in the use of red in the Main shrine (Fig. 8.16), this method of dilution would have been practicable. In contrast, to achieve darker tones, a very small amount of black could be added to the paint.

Fig. 8.16 Use of gradations of red colour on the painting inside the Main shrine of Cave 2 ©ASI/NRICPT

Another method of producing a secondary colour is by superimposing one paint layer on top of another, allowing the lower layer to modify the optical properties of the upper one. Some of the examined cross-sections from Cave 2 showed two or three layers of superimposed paint (Fig. 8.17). However, considering the nature of these layers, the primary intention does not seem to have been to obtain a secondary colour.

In summary, the mixing or superimposing of paints to obtain a secondary colour, or to produce a variety of colour shades, does not seem to have been common practice. In Cave 2, colours appear to have been achieved by using paints composed of single pigments or colourants.

(a) Painting on the right wall, sampling location

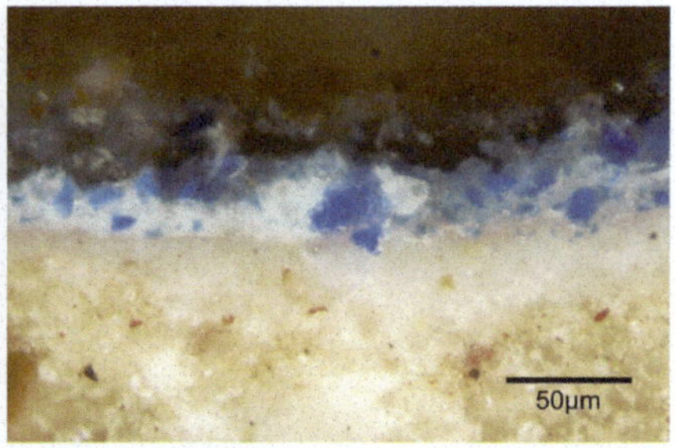

(b) Photomicrograph of paint sample showing a black layer applied over a blue layer, which is composed of natural ultramarine blue

Fig. 8.17 Evidence for the use of superimposed paint layers in Cave 2 ©ASI/NRICPT

8.4 Summary Conclusions

Although publications on the Ajanta Caves are quite numerous, descriptions of the materials and techniques used in the wall paintings are limited. This paper summarizes these publications and more recent research.

Technical findings can be summarized briefly. The main plasters are earthen, originating from local soils formed from basalt rocks. White ground-layers are used, which comprise lime, occasionally superimposed by kaolin. White paint can also be similarly identified as lime- or kaolin-based. White ground-layers sometimes function as white paint. Red colours include red ochre, red lead, red lac resin, and, in Cave 17, use of vermilion and realgar has been suggested. Yellows include yellow ochre and orpiment, the latter typically existing in an altered state. Green is celadonite. The principal blue used is ultramarine blue, derived from lapis lazuli, but other blues may be present, too. The black is carbon black.

The earth pigments—red and yellow ochres, and green earth—were probably available locally, while ultramarine blue was imported from outside of the region. Lac was probably also sourced locally. Other organic colours, including indigo, have not been definitely identified. The sources of red lead and orpiment, and—assuming they were also used—of vermilion and realgar, are unclear as yet, and these questions should be investigated further in future. No gold or silver has been found.

Regarding painting application techniques, the limited investigations did not indicate that mixing or superimposition of paints was a common practice. Instead, there seems to have been a preference for applying paint in single colours. However, colours are gradated, for instance from deep to light red. Thus, although the number of colours used seems to be limited, these gradation techniques would make the paintings appear much richer.

The conclusions of this paper are mostly based on findings made in Cave 2 alone. To increase understanding of the painting materials and techniques used in the other caves, more scientific investigations are required. This would facilitate comparative assessment between the caves, revealing new insights of how the Ajanta paintings were created.

Acknowledgements This paper presents the investigative findings obtained through the following grant-aided projects:

– Indo-Japanese Joint Project for the Conservation of Ajanta Paintings, facilitated by the Networking Core Centres for International Cooperation on Conservation of Cultural Heritage by NRICPT and ASI (funded by the Agency for Cultural Affairs);

– MEXT KAKENHI (Grants-in-Aid for Scientific Research), Grant Number JP24101014

– JSPS KAKENHI, Grant Number JP24501261.

Thanks are due to the institutes and organizations involved, and to the project members. The FTIR analysis was performed with a BL43IR of SPring-8 with the approval of the Japan Synchrotron Radiation Research Institute (JASRI) (Proposal No. 2013B1232, 2014B1243, and 2015A1896).

References

1. Lal, B.B.: Ajanta Paintings Their Composition, Technique, Deterioration and Preservation, pp. 145. Islamic Wonders Bureau (2007)
2. Set, S.C.: Indian Wall Paintings: Analysis of Materials and Techniques. In: Agnew, N., (ed.) Conservation of Ancient Sites on the Silk Road: Proceedings of the Second International Conference on the Conservation of Grotto Sites, Mogao Grottoes, Dunhuang, People's Republic of China, June 28-July 3, 2004. Los Angeles, CA: Getty Conservation Institute. pp.336-342 (2010) http://hdl.handle.net/10020/gci_pubs/2nd_silkroad
3. Lal, B.B.: The Murals: Their Composition & Technique. In: Ghosh, A. (ed.) *Ajanta Murals*. Archaeological Survey of India, pp. 53–55 (1967)
4. Singh, M., Arbad, B.: Chemistry of Preservation of the Ajanta Murals. *International Journal of Conservation Science*. **4**(2), 161–176 (2013)
5. Dabhade, S.B.: The technique of wall painting in ancient India, pp. 48. Accessible through Google books (1895 [actually in 1973]) (http://books.google.com). Date last accessed on 4th January 2020
6. Takada, O.: Indo Kodai Chusei no Hekiga [The wall paintings in ancient and middle age of India]. Bukkyou bijutsu-shi ronkou. Chuokouronbijyutsu, p. 153 (in Japanese) (1971)
7. Yamazaki, K.: Ajanta Hekiga no Zaishitsu [Materials for Ajanta wall paintings]. Ajanta: Sekkutsu-jiin to Hekiga [Ajanta: Stone cave temples and wall paintings], pp. 76–78 (in Japanese) (1971)
8. Sadakane, K.: Ajanta Hekiga no Kenkyu: Kenkyu-hen [Research on Ajanta wall paintings: volume for research], Chuokouronbijyutsu, pp. 235–236 (in Japanese) (2009)
9. Sharma, R. K.: Painting techniques and materials of cave mural paintings in India and their conservation problems. In: Mural paintings of the Silk Road: Cultural exchange between East

and West, Proceedings of the 29th Annual International Symposium on the Conservation and Restoration of Cultural Property. NRICPT, pp.101–106. Archetype publications (2007)
10. ASI/NRICPT.: Indo-Japanese Research Project for the Conservation of Ajanta Paintings—Conservation and Scientific Investigation of the Paintings of Ajanta Caves 2 and 9 (2009–2011). Rana, K.S., Yamauchi, K. (eds.), ASI/NRICPT, pp. 174 (2015)
11. ISCR.: Painted Ajanta: Vol 1 & 2: Studies on the Techniques and the Conservation of the Indian Rock Art Site. Bon, C., Ioele, M., Capanna, F. (eds.) pp. 424, Gangemi; Bilingual edition (2014)
12. Taniguchi, Y.: Issues of conservation for the Bamiyan Buddhist mural paintings. In: Mural paintings of the Silk Road: Cultural exchange between East and West, Proceedings of the 29th Annual International Symposium on the Conservation and Restoration of Cultural Property. NRICPT, pp.144–151. Archetype publications (2007)
13. ASI/NRICPT.: Indo-Japanese Project for the Conservation of Ajanta Paintings, 2008. Rana, K.S., Yamauchi, K. (eds.) ASI/NRICPT, pp. 115 (2010)
14. Ghose, N.C.: Composition and origin of Deccan basalts. *Lithos* **9**, 65–73 (1976)
15. Shimadzu, Y., Taniguchi, Y., Yamauchi, K.: A red organic colorant observed in the Ajanta Cave 2 of a Buddhism temple site. *36th Annual Meeting of the Japan Society for the Conservation of Cultural Property*. 238–239 (in Japanese) (2014)
16. Kumar, R.: Paintings and lifestyles of Jammu region: from 17th to 19th century A.D, Kalpaz Publications, p. 514 (2006)
17. Taniguch, Y., Kitagawa, M., Murofushi, M., Yang J., Shimadzu, Y., Ichiro, S.: Lac resin as an organic red colourant in ancient and Medieval Asia. *31st Annual Meeting of the Japan Society for Scientific Studies on Cultural Property.* 60–61 (in Japanese) (2014)
18. Hendricks, S.B., Ross, C.S.: Chemical composition and genesis of glauconite and celadonite. *Am. Mineral.* **26**(12), 683–708 (1941)
19. Mariottini, M.: The caves of Ajanta: Geology and History. In: DVD attached to ISCR 2014 (2014)

Chapter 9
Materials and Techniques of the Polychromy of the Giant Buddha Statues in Bāmiyān

Catharina Blaensdorf

Abstract Despite the world-famous status of the two Giant Buddha statues of Bāmiyān, their painting techniques had not been investigated in detail before the Taliban destroyed the monuments in 2001. Small fragments of the clay modelling rescued from the debris showed that they were originally painted in bright colors. Examination carried out by the Technische Universität Munich between 2007 and 2013 (with an extension until 2016) focused on characterizing the materials and stratigraphy of the paint layers. Analyses also demonstrated that the Eastern Buddha is slightly older than the Western Buddha. The examined fragments represent a small proportion of the entire sculptural polychromy. Findings are partial but instructive. Phases of painting are best preserved on the outer garments. Among the identified pigments is the precious blue, ultramarine. Different organic binders were identified; egg may have used for the original paint layers. Despite their enormous dimensions, the statues were repainted at least twice between their creation in the sixth or seventh century and 977, when the region converted to Islam. While the sequence of paint layers can be reconstructed for probably large parts of the outer garments, this is not possible for other areas such as the undergarments, flesh tones and hair. These appear to have been completely destroyed in 2001.

Keywords Bamiyan · Buddha statues · Clay sculptures · Polychromy · Pigment analysis

9.1 Introduction

The two Giant Buddha statues, measuring 35 m (Eastern Buddha) and 53 m (Western Buddha) in height, were the most prominent works of art in the extremely rich context of Buddhist temple sites in Bāmiyān. When they were destroyed in March 2001 by the Taliban, they had never been examined regarding their polychromy. Although colors had been mentioned in connection with the Buddha statues and traces of paint

C. Blaensdorf (✉)
Archäologische Staatssammlung, Himbselstraße 1, 80538 Munich, Germany

S. Aoki et al. (eds.), *Conservation and Painting Techniques of Wall Paintings on the Ancient Silk Road*, Cultural Heritage Science,
https://doi.org/10.1007/978-981-33-4161-6_9

layers had been observed, the perception of the Giant Buddha statues is still shaped by their appearance and the color photographs taken before the destruction. At that time, the Buddha statues appeared light brown, like the stone of the cliff around them. The knowledge that they were not cut directly from the rock and were not meant to look like a part of it had been forgotten.

In 2004 a campaign funded by ICOMOS and UNESCO began, aiming to secure the remains of the Giant Buddha statues. While removing the enormous piles of rubble left by the explosion, fragments of the surfaces of the statues came to light among rock fragments, powdered stone and ammunition. By 2007, about 6100 fragments had been catalogued. Clay fragments showing paint layers, some still brightly colored, motivated Edmund Melzl,[1] the conservator working on site, to send several hundred small samples to Munich for examination.

The aim of research carried out between 2007 and 2016 was to gain information about the materials and techniques used, as well as the history of the statues, including all intermediary steps from sculpting the rock to modern restorations. In addition, research on literary and pictorial sources was carried out to investigate the historical context of the Bāmiyān Buddhas. This paper summarizes the most important results regarding the polychromy.

9.2 Creation and History of the Buddha Statues

The production process and the history of the Bāmiyān statues have been described in detail elsewhere [1, pp. 197–280]. The important facts for the interpretation of the paint layers are briefly summarized here.

9.2.1 Production Technique

In terms of their fabrication technology, the Buddhas must be considered within the tradition of painted clay statues, even though their main bulk consists of rock. Clay statues have a long history in large parts of Asia, from the Neolithic Age to modern times. In Central Asia (Turkmenistan, Uzbekistan, Tadzhikistan, Afghanistan, northern India, and China) there is evidence for the highly skilled production of clay statues since the first century BC, which became an integral part of the newly developing Buddhist art [2]. It can be assumed that in general Buddhist clay statues were painted.

There are technical variations in the process of making clay statues, depending on the size and use of the statues, their integration into architectural contexts, as well as local traditions and resources. In general, however, a clay statue is made starting with a rigid core that defines the rough design and carries the main load. This core can consist of wood, brick or stone. Modelling in clay starts with a coarse mixture of clay, often containing straw, which is applied and shaped by hand. One or two layers of fine clay are then applied to reach the definitive shape. They often contain fine sand and plant fibers or animal hair to prevent shrinkage cracks. Besides bare hands, wooden tools, cloths or chamois leathers are used to smooth the surface. After drying, the statues are painted, usually on a white preparation layer. In architectural contexts such as temple halls or caves, statues and surrounding walls are often designed using the same (or similar) materials and techniques.

The Giant Buddhas were roughly cut from conglomerate rock. Since this rock contains coarse inclusions of variable hardness, it does not permit the sculpting of smooth or finely structured surfaces. All rock surfaces were therefore covered with applications of clay, which also completed the three-dimensional shape: smaller details thus are only modelled in clay. To increase the adhesion of the clay layers to the rock, a variety of keying methods were employed. On the smaller Eastern Buddha, numerous holes (about 7 cm wide and deep) were cut into the surface. Before applying the clay, pebbles were pressed into the holes to provide an anchor (Fig. 9.1). On the larger Western Buddha, details such as the garment folds were not carved from the rock. Instead, rows of holes (3.5–4 cm wide, 7 cm deep) were made along the fold ridges with a pointed chisel. Wooden pegs were inserted into the holes, connected with ropes made of an endemic plant (*astragulus coneifolius*)

Fig. 9.1 Eastern Buddha: right arm and right leg in the 1950s: the clay modelling of the *sangati* is preserved in large areas Taken by Edmund Melzl in 1955. By courtesy of Erwin Emmering

and then covered with coarse clay. The protruding forearms of both statues were modeled over massive wooden beams, which were inserted into the rock and then covered with clay; in the case of the Western Buddha, adobe bricks were also used. On the Eastern Buddha, a sacred relic was placed into the opening cut for inserting the beam. The coarse first clay layer contains wheat and barley straw as well as goat hair. The application of this plaster varies in thickness from 2 to 3.5 cm in the flat parts, to about 20 cm in the fold ridges, which were roughly modelled by hand.

The top layer is only 0.5–5 mm thick and contains sand, charcoal and fine sheep wool. The surfaces are smooth and show impressions of wooden modelling tools. A leather rag in which a round seed or nut had been tied up was found in the rubble, which may have been used as smoothing tool. The walls of the niches were also coated with clay plaster. Finally, the walls and statues were painted.

9.2.2 *Important Events in the History of Bāmiyān*

Since the examination allowed us to conclude that the clay modelling is original and thus dates to the same time as the rock-cut sculptures and niches,[2] radiocarbon (14^C AMS) dating was done on 22 samples of organic components of the clay layers and anchoring pegs. It resulted in the following dating [1, p. 235]:

- Eastern Buddha: 544–595 AD (2 σ)
- Western Buddha: 591–644 AD (2 σ)

This indicates that the Eastern Buddha was the first one to be created and is slightly older than the Western one.

The first scholar to mention the Giant Buddhas, who arrived in Bamiyan around 630, perhaps even before the Western Buddha was finished, was the Chinese monk Xuanzang.[3] From 770 Islam became predominant in the region of Bāmiyān for a period of about a hundred years, which was followed by a second Buddhist phase. Archaeological evidence indicates that Islamic and Buddhist traditions co-existed peacefully. Since 977, Bāmiyān has been Muslim. Maintenance and repainting of the Buddha statues was therefore only undertaken during the four previous centuries [1, pp. 231–236].

In medieval times, Islamic scholars mentioned the giant “idols”,[4] and not knowing their Buddhist origins, locals associated them with kings or heroes.[5] Even today, they are regarded as a couple, the smaller Eastern Buddha representing the “female”.[6] In the eighteenth century legends about the idols reached Europe.[7] Around 1830, Europeans who came to Bāmiyān as adventurers or soldiers sketched and described the statues, such as William Moorcroft and George Trebeck [23], followed by Alexander Burnes [3], Charles Masson [24], and Vincent Eyre [4]; and in 1895, the first photograph, showing the Eastern Buddha, was published.[8] The first descriptions of a more scientific nature were made in 1885 by Talbot and Simpson, with sketches by P. J. Maitland [5, pp. 303–350].

In the twentieth century there were two important campaigns that focused on the study and the conservation of the Bāmiyān Buddhas:

- In the 1920s and 1930s, the DAFA (Délégation Archéologique Française en Afghanistan[9] carried out an archaeological survey and made structural interventions, building a buttress on the left side of the niche of the Eastern Buddha [6–8]. At the same time, the first analyses on colorants and binders were performed [9, pp. 168–193].
- Between 1969 and 1978, the Indo-Afghan project organized by the ASI (Archeological Survey of India) restored the Buddha statues [10, 11]. This included partial reconstructions of the clay modelling. As a final intervention, a clay suspension was applied to all surfaces, probably to homogenize the overall appearance. Perhaps it was also assumed that this would reconstruct the original design, even though extensive traces of polychromy were found and mentioned. At the same time, Tarzi, described the technique of the clay modelling of the statues [12; vol. 2, p. 177]. Between 1978 and 2001, when most of the color photographs of the Buddha statues were taken, they consequently looked very much like the surrounding stone cliffs.

9.3 Examination of the Paint Layers

The main question, and thus the intended aim of the examination, was to understand what the statues looked like originally. This included research and interpretation on several levels:

1. Identification of the materials and
2. understanding the paint stratigraphy.
   Step 2 proved to be the most difficult and was done before and in parallel with step 1.
3. Understanding how the individual parts of the statues were painted in order to reconstruct the painting scheme.
4. Reconstruction of changes in the painting schemes over time.

At present, step 3 and 4 can only be answered partially.

All fragments were examined using a stereomicroscope (max. 40 times magnification). Cross sections were imbedded in epoxy resin (Araldite® 2020) and observed using VIS and UV (bandpass 355–425 nm), at max. 500 times magnification. The identification of pigments was mainly done with polarised light microscopy (PLM); for a few farther-reaching questions micro x-ray fluorescence (μ-XRF), x-ray diffraction (XRD) and energy dispersive X-ray spectroscopy (EDX) could be applied.[10]

9.3.1 Preserved Paint Layers

After the destruction in 2001, almost no paint layers were preserved in situ: on the Western Buddha, no original intact surfaces remain; on the Eastern Buddha, small areas of painted clay layers are preserved below the right arm (Fig. 9.2) and next to the left leg, but only to a marginal extent compared with the dimensions of the statue.

Fragments found in the rubble were catalogued separately from the stone fragments, using the initials GBL for the Western and KBL for the Eastern Buddha.[11] Mostly each entry corresponds to one fragment, but tiny fragments were listed in groups of up to 40 pieces. For the Western Buddha, 3517 painted clay fragments were recorded, for the Eastern Buddha, 2574 clay fragments.[12] Despite the large number of fragments it is difficult to allocate them unambiguously to specific areas of the statues. First of all, they only represent a small portion of the clay layers that were still preserved before the destruction, at most approximately 2.5% for the Western Buddha, and 4% for the Eastern Buddha.[13] Most of them are too small to draw conclusions based on their shape, but some larger fragments can be identified as fold ridges (Fig. 9.3). Due to the enormous height of the statues, fragments of the lower parts had a better chance of surviving the fall after the explosion. But large parts of the modelled surface, especially in the regions below the knees, were already lost before the destruction. Thus, no clearly discernible fragments from the heads of the statues and the murals around them have been found. Often the surfaces of the clay fragments are soiled with dirt or soot, or the paint layers are altered, so notes in the find list characterizing the surfaces as "dark," "brown," "yellowish" or "reddish" can refer to soiling, the clay support, or discolored pigments, as well as to the actual color of the paint layer. This means that a reconstruction of the polychromy of the statues based on the examination of preserved paint layers is only possible to a limited extent.

Fig. 9.2 Preserved clay modelling with blue paint layer under the right arm after conservation, October 2008 [© Bert Praxenthaler 2008. All Rights Reserved]

Fig. 9.3 Part of a fold-ridge from the Western Buddha (GBL 852). The paint layer originally was in red hues (fragment ID 48: group 2–4; ID 169: group 2) Taken by Edmund Melzl in 2005. By courtesy of Erwin Emmerling

The material sent to Munich for examination contained 276 painted clay fragments, 173 coming from the Western Buddha and 103 from the Eastern Buddha (Fig. 9.4). Most of the tiny fragments measure less than 3 cm × 3 cm. Fragments were selected with the goal of representing as large a range of different paint layers as possible, and thus do not correspond to the quantitative distribution of colors in the retrieved fragments. Furthermore, only tiny fragments found in isolation were

Fig. 9.4 Overview of the painted clay layer fragments examined in Munich [© Catharina Blaensdorf, 2007. All Rights Reserved]

taken in order not to damage preserved larger fragments by sampling. For practical reasons, the fragments received a numbering system independent from the find list (marked with ID + sequential number).

9.3.2 *Classification of Paint Layers*

During the recovery of the fragments, the presence of several paint layers was already noted. Underneath the clay suspension applied during the restoration by the ASI team (1969–1978) there are up to four layers. Mostly they have the same color though in different hues. This indicates that the original color scheme was repeated in the later overpaintings. Despite similarities, there are also differences that show that the history of (re)painting was not the same for both statues.

The fragments were classified into groups 1–6 according to the main colors. In a later stage, subgroups, e.g. 2–3, were introduced for stratigraphies with characteristics of two groups. The main colors are red [group 1 and 2, Figs. 9.5, 9.6, and 9.7] and blue [group 5, Fig. 9.8]. Numerous fragments show brown layers (group 3 and 4), sometimes covering remains of red and blue layers (groups 2–3, 2–4 and 5–1). The polychromy of a preserved fragment of ridge fold [GBL 852, Fig. 9.3] belongs to group 2 (ID 169, red without a white preparation layer), but in another sample, a brown layer is present over traces of red (ID 48, group 2–4). Thus brown layers seem to be a later modification, not belonging to the original color scheme. Fragments of group 6 have a white preparation layer and paint hues that do not appear elsewhere (dark red, blue, yellow ochre, white, pale blue, partly in translucent layers). Some of them preserve several colours applied adjacent to each other, reminiscent of mural painting techniques.

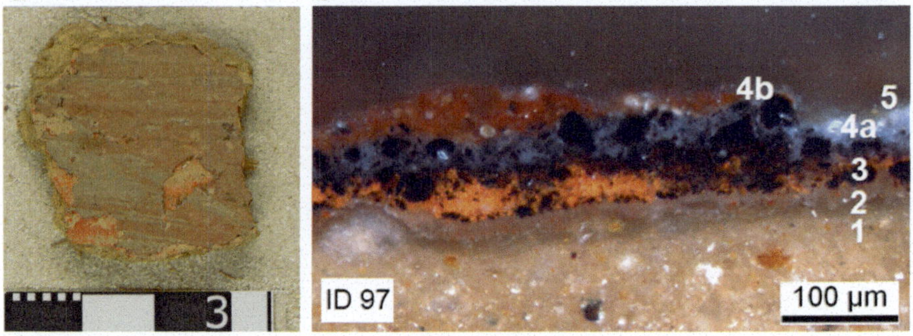

Fig. 9.5 **a** and **b**. Western Buddha, Fragment ID 97, paint layers in red hues (group 2): (1) clay, (2) original pink paint layer, (3) one or two layers of red lead, partly discoloured to black, (4a) white underpainting for (4b) red layer, (5) traces of clay suspension from 1969–1978 [© Stephanie Pfeffer //Catharina Blaensdorf 2007. All Rights Reserved]

Fig. 9.6 Western Buddha, Fragment ID 59, group 2, with visible black discoloration of red lead layers [© Catharina Blaensdorf 2007. All Rights Reserved]

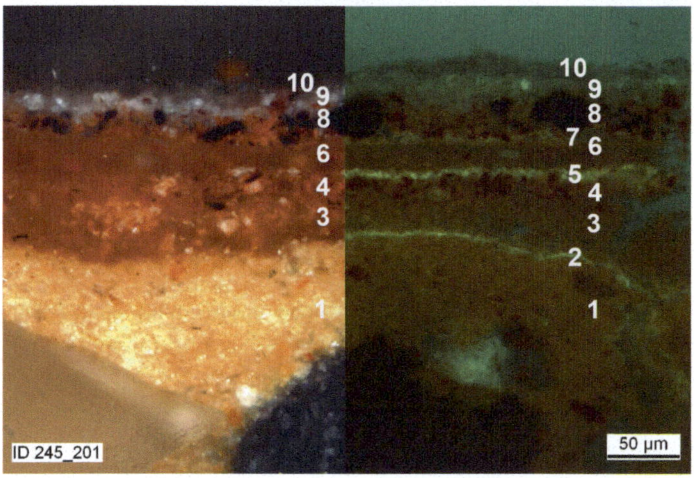

Fig. 9.7 Eastern Buddha, Fragment ID 245, group 2, VIS and UV: (1) clay, (2) sealant, preparation, (3) original pink layer, (4) partial red lead layer, maybe original (5) sealant, preparation, (6) pink overpainting, (7) sealant, preparation, (8) orange overpainting with partly discoloured red lead, (9) white overpainting, (10) clay suspension from 1969–1978 [© Catharina Blaensdorf 2013. All Rights Reserved]

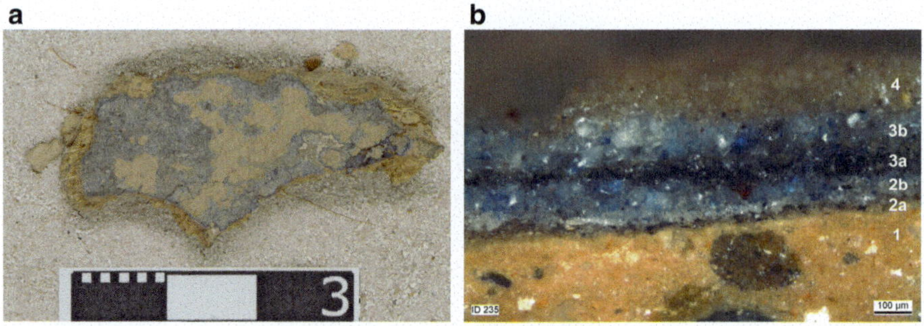

Fig. 9.8 **a** and **b**. Eastern Buddha, Fragment ID 235, blue paint layers (group 5): (1) clay, (2a) original grey underpainting, (2b) blue layer, (3a) grey underpainting, (3b) blue overpainting, (4) clay suspension from 1969–1978 [© Catharina Blaensdorf 2013. All Rights Reserved]

9.3.3 Preparatory Layers

A sealant layer underneath the first paint layer can be distinguished in most of the cross sections and also on some of the fragments. It has a strong white to yellowish UV fluorescence (Fig. 9.9, layer 2).

One puzzling observation is that on the majority of the fragments there is no white preparation layer underneath the paint layers. Only on a few fragments is there a continuous white layer, while others show only small traces or tiny splashes of white material. In wall paintings in the region, a white—or light-colored—preparation layer was usually applied before painting clay surfaces, containing gypsum, chalk or white clay minerals [1, p. 260]. This type of preparation layer is also present in the caves in Bāmiyān, and can be assumed from photographs for the murals of the niche vaults. On the niche walls it is not discernible, but their clay layers were already in a poor state of conservation before 2001, and little photographic evidence exists. The absence of a white preparation layer can either mean that the original polychromy was almost totally lost or removed at an early time, only leaving small traces of the white preparation layer; or that the original paint layers are still preserved, but were executed without a preparation layer.

Considering the enormous dimensions of the statues and the challenge craftsmen faced in accessing every surface, it is difficult to imagine that the polychromy would have been removed completely before repainting. Thus, the most recent conclusion is that only the niche walls, or at least the vaults, received a white preparation layer, but not the statues. Splashes of the white preparation may have dripped down on the still unpainted statues while working on the walls or vaults. The fragments with a preparation layer presumably come from these locations, or from adjacent surfaces that were accidentally covered. The fragments from group 6, which have a thick continuous white preparation layer superimposed by multiple colours, may belong to the murals themselves. This assumption is also supported by the fact that the clay layers of group 6 contain a very low amount of hair. The presence or absence

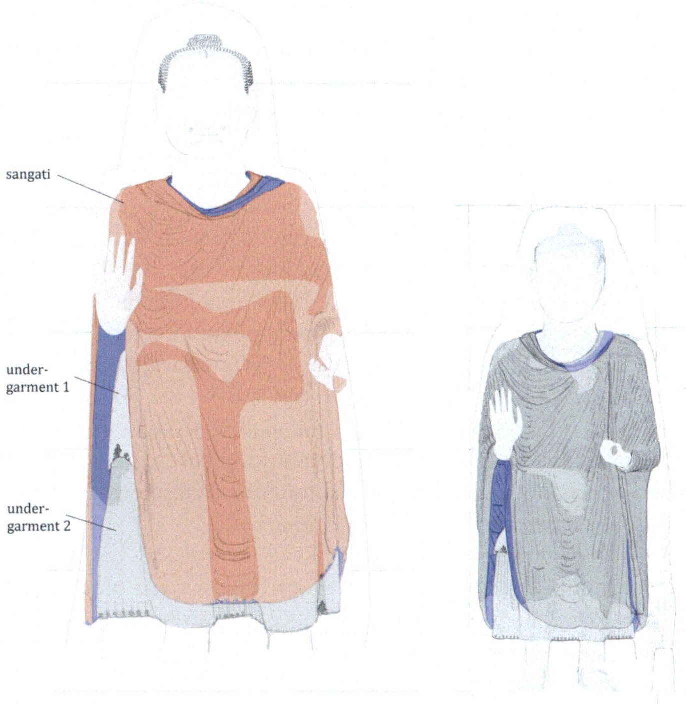

Fig. 9.9 Allocation of colors for the original paint layers to parts of the statues. In the light-colored areas, the clay layers were already lost before 2001. The outside of the *sangati* of the Eastern Buddha was also probably mainly pink. [© Catharina Blaensdorf 2019. All Rights Reserved]

of hair in the plaster was a characteristic used to distinguish paint fragments originating from the statues (plaster with hair) from those belonging to the murals of the surrounding caves (plaster layers without hair). On fragments from both statues, the white preparation layer contains white clay minerals, gypsum (probably natural calcium disulphate) and some calcite. On fragments from group 6, it consists of very finely ground gypsum with some calcite.

9.3.4 Paint Layers—Structure and Pigments

9.3.4.1 Oldest Preserved (Original?) Paint Layers

The first preserved layers in the red and blue groups (1, 2 and 5) are currently interpreted as remains of the original polychromy. They appear rather matt, with a soft and porous structure and no discernible craquelure. The thickness of the layers is about 20–50μ for the reddish layers and up to 100μ for the blue layers.

The first "red" was in fact a pink, composed of red iron oxide and a white containing white earth and/or gypsum, comparable to the composition of the white preparation layer mentioned above. The pink was of a slightly darker hue on the Eastern Buddha than on the Western Buddha. In cross sections of two fragments from the Eastern Buddha, there is a partial orange layer of red lead on top of the pink (Fig. 9.7, layers 3 and 4). This may indicate a two-layered structure of which the top layer is partly lost, or it may originate from a subsequent decoration, even though orange on pink seems a strange color combination. A third possibility is that an overpainting penetrated between separated older layers.

The blue layers were applied on a dark grey underpainting that was composed of black and white. The black is charcoal black, partly containing rather coarse particles with discernible cell structures (Western Buddha). White earth minerals and gypsum were identified as the white component in samples from the Western Buddha, and gypsum in those of the Eastern Buddha. A sealant layer, discernible by its strong UV fluorescence, was applied onto the grey underpainting before the application of the blue paint layer—probably to impede a contamination of the bright blue by the grey. The blue consists of rather pure and dark ultramarine.

One fragment from the Eastern Buddha (ID 256) probably comes from an area where painting in pink (paint layers as in group 1) bordered blue painting (paint layers as in group 5). A thin red line (based on its color, probably iron oxide) applied before the colours were applied is probably an underdrawing indicating the border between the pink and blue areas of painting.

The fragments of group 6 again differ from the ones of group 1, 2 and 5. The pigments are very finely ground. Ultramarine, red iron oxide and a very pure yellow iron oxide hydroxide were used. On one fragment (Western Buddha, ID 65) a mixture of red oxide and red lead was used.

9.3.4.2 Overpainting Layers

The red parts of both statues show two to three overpainting layers. On the Eastern Buddha the first overpainting was a pink of the same hue (Fig. 9.7, layer 6) as the oldest preserved paint layer. There is a sealant layer with a strong yellowish UV fluorescence between the two pink layers, probably applied as a preparatory layer for the overpainting (Fig. 9.7, layers 5 and 7).

The second overpainting is orange, containing red lead. On the Western Buddha the first overpainting is orange (red lead, Fig. 9.5b, layer 3). In some fragments there are two orange layers, indicating a total or partial repainting. On both statues the red lead has discolored to a considerable degree (Fig. 9.6). Especially on the Eastern Buddha many fragments have turned completely black, probably enhanced by the fact that the red lead layer is thinner than the ones on the Western Buddha. Using XRD, white lead sulfate (anglesite, $PbSO_4$) was found next to the red lead, which can also be seen as white inclusions in the orange layers. The brown to black transformation products could be identified as plattnerite (PbO_2, black) and scrutinyite (PbO_2, greyish to brown).

The discoloration of the orange red paint layers may have been the reason for the last overpainting. A white layer consisting of lead white was applied on both statues, perhaps with the aim of covering the partially blackened surfaces. On the Western Buddha a red layer containing iron oxide was applied over the white layer, which therefore can be interpreted as a preparation layer (Fig. 9.5b, layers 4a and b). On the Eastern Buddha no red layer was found and may never have been applied here (Fig. 9.7, layer 9).

The blue parts were overpainted once on both statues, following the same system as on the original polychromy: a sealant layer served as preparation for the new paint layers. They consist of a grey underpainting (a lighter hue than the original grey), a sealant layer, and a layer of ultramarine. The particles of the ultramarine are finer and the blue has a slightly lighter tone than the original paint layer (Fig. 9.8).

According to the interpretation of fragment ID 256, which shows a transition from red to a blue painting, the blue overpainting was done after the red lead layer had been applied, and thus belongs to the second (orange) or the third (white) overpainting. The reason why the red parts were overpainted more often than the blue ones may be that the red paint was more exposed, or showed more damages or changes that demanded repair. It may also be related to the cost of the pigments, with ultramarine being precious even in Afghanistan.

It is not clear to which parts the brown fragments may have belonged, and if the layers are really paint layers or soil/clay accidentally or intentionally soaked with binding material. Most fragments show two brown layers on top of each other, some with particles in between that may be soil or soot. The layers are translucent and show fine craquelure patterns, suggesting a higher amount of binding medium. The surfaces vary from glossy to matt, scaly to smooth. As some of the red layers preserved underneath the brown layers contain iron oxide, the brown layers may have been either applied rather late or come from parts that deviate from the main system of the red layers of the statues.

The clay suspension applied on all painted surfaces of the statues in 1969–1978 contains diluted earth, gypsum and an organic binder.

9.3.5 Binding Media, Sealants and Mordants

Very little is known about painting techniques of clay statues in Asia. Technical literature or treatises from the region and the time of the Bāmiyān Buddhas dealing with painting clay supports are not known. Studies on artistic traditions regarding painting technique and scientific investigations are still rare, and analyses have mainly concentrated on use of inorganic materials and wall paintings, not sculptures. Indian treatises on painting, written between the early Middle Ages (*Visnudharmottara*, dated between the fourth and the seventh century AD) and the sixteenth century mention organic binders and associated materials in all steps of painting on clay plasters, from the preparation of the walls to the final steps of decoration with gold leaf. The named materials comprise gums and saps of plants (probably mainly saccharides), legume broths (containing proteins and saccharides), the addition of beeswax, and in the later texts, animal glue [13, pp. 11–54; 22, p. 43].

Analyses carried out on fragments of wall paintings from the caves in the Bāmiyān valley revealed the use of different types of binders—proteins (animal glue or egg white), plant gums, natural resins and drying oil—with a different binder or material mixture identified in each layer in the overall painting stratigraphy (i.e., surface of the clay wall, preparation layers and paint layers).[14] This shows that a wide range of organic materials may be present in the samples, including a number that may not have been previously documented or identified through analysis (such as proteins from plants), and are thus more difficult to identify. Furthermore, different materials may have been used for overpainting layers and as restoration materials. Due to the porous structure of the layers and to losses that have occurred over the time, the more recently applied materials penetrated into the lower layers. Thus, it is likely that there has been some contamination of the sampled layers by materials applied later.

Several different analytical approaches were used in studies undertaken between 2007 and 2017. 19 fragments (14 from the Western Buddha and 5 from the Eastern Buddha) were selected as typical representatives of the groups 1–5. From each fragment up to 8 sub-samples were taken following the stratigraphy of the paint layers. CS/MS analyses were performed on all selected sub-samples [1, pp. 265–276], in order to identify polysaccharides, proteinaceous and glycerolipid materials, as well as waxes and terpenoid materials [14, p. 7]. A sub-set of 30 samples from 11 fragments was analysed in order to identify the source of the proteinaceous materials. Synchrotron Radiation micro-imaging techniques (SR-μ ATR-FTIR) were applied to a few cross sections to allocate organic materials to single layers [22, pp. 44–45].

GC/MS analyses proved evidence for the presence of proteins and polysaccharides, while natural resins and drying oils were not detected in any of the sub-samples. Proteins derived from egg, milk (casein) and animal glue were all identified, but in some samples the amino acidic profiles remained unidentified. Protein identification (proteomics) showed that the milk of cows, as well as goats or sheep was used [14, pp. 8–12]. Polysaccharides were found in a third of the sub-samples and always together with proteins. Tragacanth gum was indicated in most cases.

Interpretation of the results with regard to the painting materials and techniques used originally and in later phases is extremely complex, and this work has not yet been completed. Each additional tranche of analysis brought more results, often apparently contradicting earlier analyzes or those of comparable fragments. This can be explained by factors such as contamination by later applications, which penetrated into the lower layers, and probably not in the same way in all parts of the giant statues. The surface of the clay layers and the binding medium layer applied as preparation contains proteinaceous materials. It is possible that egg was used as a binder of the original paint layers, while milk and egg were used in the overpainting phases, although milk was also detected in original layers. Animal glue is never present alone, but is found combined with egg, and only in the top layers of the fragments from group 3 and 4, and thus may be interpreted as a material belonging to an overpainting or restoration. Polysaccharides were also detected in the clay suspension of the restoration of 1967–1978. Thus the last restoration may have been the source, or a supplementary source, for the saccharide material analyzed in so many samples.

9.3.6 The Colouration of the Statues

To understand the principal color scheme of the Buddha statues, it is necessary to assign fragments or paint layers to specific compositional areas. Contemporaneous Buddha statues from Central Asia which are similar in style (although much smaller)[15] show that different areas can be distinguished: the exposed parts of the body and three layers of garments, of which also the lining (inside) may be visible. The flesh tone can be "natural" (i.e. some hue of pink) or gilded. The main garment is the *sangati*, a large rectangular piece of cloth draped around the shoulders, which reaches to the lower legs. Underneath the *sangati*, portions of two undergarments may be visible, which usually have a color different from the *sangati*. Historical photographs of the Bāmiyān Buddhas show that the *sangati* covered the main part of the body. The lining was visible underneath the arms and, judging by the modelling, also at the left side of the neckline. The undergarments were visible below the right arm and at the hemline across the legs, both ending in undulating folds, where small areas of the lining could be seen. Of the shorter garment, ending above the knees, only a very small area was visible. This means that eight areas can be distinguished where different colors were probably used: 1. flesh tone, 2. hair, 3–5. three layers of garments, and 6–8. small parts of their linings (Fig. 9.9).

Historical descriptions of the colors include medieval accounts that were based on the rather well-preserved state of the statues at the time, but these do not provide any details. The medieval sources—as for example that of Yaqut al-Hamawi, who described Bāmiyān in 1218, just a few years before the city of Bāmiyān was destroyed by Genghis Khan in 1221[16]—mention a "red idol" (*surkhbud*) and a "moonwhite idol" (*khinkbud*). It can be assumed that these characterizations were based on an overall impression or the dominant colors of the statues. In descriptions from the nineteenth century, paint layers are rarely mentioned, indicating that they were probably not predominant anymore. In the twentieth century, colors are mentioned in the reports by the DAFA and the ASI, but the DAFA expedition did not focus on polychromy, and in the restoration by the ASI, it only played a minor role.

The Western Buddha can be identified as the one described as the "red idol" in the medieval sources. The DAFA and the ASI reports both mention red as the color of the outside of the *sangati*, but no other colors. The Eastern Buddha corresponds to the one described as the "white idol." If we deduce that this also refers to the color of the largest part of the statue, the outside of the *sangati* would have been white or light grey, a very unusual color for the robe of a Buddha. The DAFA (1934) and the ASI reports (1965 and 1984) mention gold on the flesh tones, of which only the face was preserved (the forearms were already lost and the feet had no clay layers). Red and blue traces are mentioned for the garments: in 1934 "*traces of red and blue*" were described "*on the sleeve.*" The ASI report of 1965 gives contradictory information, describing the *sangati* as "*probably blue as opposed to the red robe*" of the Western Buddha, but also as "*reddish to pink*" in a later paragraph [10]. In 1984 Sengupta only mentions blue [11, pp. 31–46].

Xuanzang (writing between 629–645 AD) mentions golden hues and precious ornaments. This may be reflected in a later description published in 1801 in London that mentions "*embroidery and figured work*" on the robes, and also says that one statue was painted in red and the other either painted in grey or the color of the stone [15, p. 464]. No traces of any decorations of the robes, neither plastically applied nor painted, could be found. There is also no evidence in drawings or photographs, and no mention in descriptions of the nineteenth and twentieth century.

Interpreting these descriptions in combination with the results of the examination led to the conclusion that most of the clay and paint layers of the undergarments and the flesh tones had already been lost before the destruction of the statues. Therefore, most of the preserved fragments probably belonged to the *sangati*. For the Western Buddha, fragments of fold ridges with reddish paint layers are preserved (Fig. 9.3). This supports the assumption that the outside of the *sangati* was of a red tone. Fragments with blue paint layers, which are less frequent, are classified as the lining of the *sangati*. On the Eastern Buddha, the lining of the *sangati* was definitely blue, as parts are still preserved on-site (Fig. 9.2). There are few fragments described as red in the find list, but this may be due to the strong discoloration of the red lead. As red is also mentioned as the color of the robe, the most recent interpretation is that the outside of the *sangati* was painted in red tones, either completely or partially. In the last overpainting, only the white layer, (probably a preparation layer) was applied, not the subsequent red layer. The reason for this is unknown: the question if it was intentional, or whether the work was interrupted by some kind of difficulties, cannot now be answered. However, it would explain the description of the Eastern Buddha as the "white idol."

The pigments used to paint the statues are widely used all over Asia and were also found on the wall paintings at Bāmiyān [1, p. 260; 16, pp. 37–47]. What seems remarkable is that the original coloration of the outside of the *sangati* was not red, but *pink*. While cheaper materials—iron oxide, rather than cinnabar—were used for painting the largest area (outside of the *sangati*), expensive ultramarine was applied on smaller, but nevertheless still extensive parts. The only identifiable colors are pink and blue. Green, yellow, red, black, or white, as well as metal foils, which are present in the murals of the caves, were not found. This very limited range of colors can probably be explained by the fact that the polychromy of the body and the undergarments was already severely reduced before the destruction of the statues.

9.4 Conclusion

Thousands of fragments of the painted clay surfaces of the two giant Buddha statues of Bāmiyān were rescued from the piles of debris resulting from the destruction in 2001. They provide valuable material for study, allowing a technical examination of the manufacture and polychromy of the Buddha statues for the first time in their history. Though abundant as material for a scientific examination, the fragments however only comprise a very small part of the original painted surfaces. Questions therefore remain over the complex processes involved in the production of the statues, their original appearance, as well as the physical history.

The most recent interpretation is that most of the identifiable fragments come from the largest area of the statues: the main garment (*sangati*). For the Western Buddha, the outside of the *sangati* was originally pink and was later repainted in orange and finally in red, while the lining of the *sangati* was blue. For the slightly older Eastern Buddha, a similar situation can be assumed. But the outside of the *sangati* may originally also have contained blue areas and partial decorations of which few, if any, traces remain. In the last overpainting, the outside of the *sangati* probably remained white. Two to three overpainting phases on both statues can be dated prior to 977: despite their enormous dimensions, on average the statues were repainted at intervals of about 100 years. The severe discoloration of the red lead may have been one reason for the overpainting of reddish areas. In the restoration of the 1969–1978, most of the colors were covered by a clay suspension ranging in color from light brown to grey.

The analyzed pigments—iron oxides, red lead, ultramarine, gypsum, white clay minerals, and charcoal black—are widespread in Asia. The use of precious ultramarine even for large areas is remarkable despite the extensive deposits of lapis lazuli in Afghanistan. Different kinds of organic binding media could be identified, comprising polysaccharides and proteins from egg, milk of cows and goats or sheep, as well as animal glue. Mostly there is more than one organic material in each layer.

The results achieved to date have answered many questions regarding the polychromy of the Buddha statues and their history, but understanding of the painting techniques used and a reconstruction of their original appearance remain at a very preliminary stage.

Notes

1. Edmund Melzl (1937–2015) lived in Afghanistan between 1956 and 1963. In 1958 he visited Bāmiyān. In 2004 he joined the team working in Bāmiyān to help with the conservation work on-site.
2. Larger repairs seem to be restricted to the twentieth century; they can be distinguished from the original clay layers. ICOMOS XIX [1, p. 225].
3. In 627 Xuanzang 玄奘 (ca. 602–664), travelled from Xi'an (China) to India where he arrived around 630. He described his journey after he returned to China in 645. For his description of Bāmyiān see Beal, S. [17, pp. 50–51].
4. For example, Yaqut al-Hamawi (see Note 16, below).
5. These legends are reflected in descriptions made by Europeans in the 18th and nineteenth centuries: Burnes reported that in the history of Tamourlane (or Tamerlane, 1336–1405) Sherif o deen, Tamerlane's historian, called the statues *Lat* and *Munat*, while in his own day they were called *Silsal* and *Shahmama*, representing king Silsal and his wife [3, pp. 185, 187, 188]. In 1798 Wilford

reported that "*the Musulmans insist that they are the statues of Key-Umursh and his consort, that is to say, Adam and Eve*" [18, p. 464, cited in 6, p. 85].

6. In 1843 Vincent Eyre described the eastern statue as male (Eyre 1843, cited in [6, pp. 87–88]; Talbot and Simpson identified the smaller statue as female [5, pp. 303–350]).
7. Thomas Hyde was the first European to mention the Bamiyan buddhas, based on Arab literary sources (Hyde 1769, pp. 129–130, cited in [6, p. 3, reference [19]]). Even in the early nineteenth century, descriptions still based on hearsay were published, for example by Captain Francis Wilford [18, pp. 462–468; 15, p. 464] and Montsyuart Elphinstone [19].
8. For historical descriptions and records of the Buddha statues, see ICOMOS XIX [24, pp. 18–26], compiled by E. Melzl and M. Petzet, with the assistance of C. Blaensdorf.
9. The DAFA was founded in 1922 under the protection of King Amanullah of Afghanistan.
10. Equipment used for the examination: at the Technical University Munich, Chair of Conservation: stereo microscope: Stemi 2000 (Zeiss) and M50 (Leica); cross sections: DMLM (Leica) with camera EC3; polarised light microscope DMLP (Leica) with camera EC3; SEM-EDX (since 2013): Phenom Pro X (Phenom). Roman Germanic Central Museum Mainz: μ-XRF: MiniPal 4025/00 (Philips); Bayerisches Landesamt für Denkmalpflege, Central Laboratory: XRD: PW 1800 (Philips); SEM-EDX: DSM 960 (Zeiss), EDX (until 2012): X-Flash SDD detector (Bruker).
11. The initials GBL = *Großer Buddha Lehm* (Western Buddha clay) and KBL = *Kleiner Buddha Lehm* (Eastern Buddha clay).
12. The find list for the Western Buddha comprises 3672 entries, including clay layers, pegs, ropes and a few other materials. The find list for the Eastern Buddha has 1939 entries, of which 1924 are clay fragments, but the presence of paint layers was not always recorded, due to omissions of a system that had not been fully developed at the start of the project.
13. These numbers are rough calculations, based on estimates of the modelled surfaces preserved before destruction and the maximum dimensions of the fragments in the find list: Western Buddha, surface c. 900 m^2, fragments max. 23.048 m^2; Eastern Buddha: surface c. 700 m^2, fragments max. 33.793 m^2.
14. See, for example, analysis of a paint sample from Cave 4 at Foladi [20, pp. 130–131].
15. Observations mainly of clay sculptures from the Museum für Asiatische Kunst, Staatliche Museen zu Berlin.
16. Yaqut ibn-Abdullah al-Roumi al-Hamawi (1179–1229) included a brief account of the Bāmiyān Buddhas in his *Geographic Dictionary.* For a translation, see Barbier de Meynard, C. [21, p. 80].

Acknowledgements The work on-site was funded by UNESCO and ICOMOS, under the overall direction of Michael Petzet. Erwin Emmerling was the head of the project at the Technical University Munich, Chair of Conservation-Restoration, Art Technology and Conservation Science. Edmund

Melzl and Bert Praxenthaler worked as conservators on-site and collected the fragments. The investigation included many scientists from different institutions and would not have been possible without their work. At the Technical University, Chair of Conservation, Stephanie Pfeffer, started the examination of the fragments and prepared the cross sections with the help of Vladimir Ruttner, Department of Engineering Geology; Christina Elsässer analysed selected cross sections. Maximilian Knidlberger identified many of the pigments with PLM. The analyses of binders involved many scientists, of whom only the responsible persons can be named here. GC/MS and FTIR: Ilaria Bonaduce and Anna Lluveras-Tenorio, Dipartimento di Chimica e Chimica Industriale, Università di Pisa; Proteomics: Leila Birolo, Dipartimento di Scienze Chimiche, Università di Napoli Federico II, Napoli; Marine Cotte, European Synchroton Radiation, Grenoble: Synchrotron Radiation micro imaging techniques (SR-μ ATR-FTIR). Ursula Baumer, Doerner Institut Munich: FT-IR of the preparation layer on two samples. Analyses of inorganic components were done by Susanne Greiff, Roman Germanic Central Museum: μ-XRF analyses, Christian Gruber, Bayerisches Landesamt fuer Denkmalpflege, Munich: SEM-EDX, Vojislav Tucic, Bayerisches Landesamt fuer Denkmalpflege, Munich, and Klaus Rapp, Munich: XRD. Radiocarbon dating was performed by Marie-Joseé Nadeau, Pieter M. Grootes and Matthias Hüls, Leibnitz Laboratory for Radiometric Dating and Isotope Research, Christian-Albrechts-Universität Kiel. Many more scientists were also involved, for analyses of clay, hair, and plant material (see ICOMOS XIX). Lilla Russel Smith, Staatliche Museen zu Berlin, Stiftung Preußischer Kulturbesitz, Süd, Südost und Zentralasien, Curator of Central Asian Art, kindly allowed close examination of the clay statues from Xinjiang in their collection.

References

1. ICOMOS XIX: Petzet, M. (ed.): *The Giant Buddhas of Bamiyan, Safeguarding the Remains.* ICOMOS Monuments and Sites of the International Council on Monuments and Sites XIX (2009)
2. Blaensdorf, C.: Zur Technik polychromer Lehmfiguren in China und Afghanistan. In: Beiträge des 9. Konservierungswissenschaftlichen Kolloquiums in Berlin/Brandenburg am 20. November 2015 in Berlin-Dahlem: Historische Techniken und Rezepte, Teil 2 – vergessen und wiederentdeckt. Arbeitshefte des Brandenburgischen Landesamts für Denkmalpflege, Berlin, pp. 124–141 (2015)
3. Burnes, A.: Travels into Bokhara, Together with a Narrative of a Voyage on the Indus, London (1834). Reprint: Oxford University Press (1973)
4. Eyre, V.: The Military Operations at Cabul (...): with a Journal of Imprisonment in Afghanistan, London (1843)
5. Talbot, M.G., Simpson, W.: The rockcut caves and statues of Bamian. J. Roy. Asiatic Soc. **XVIII**, 303–350, London (1886)
6. Godard, A., Godard, Y., Hackin, J. Les Antiquités Bouddhiques de Bāmiyān. Mémoires de la délégation archéologique, Française en Afghanistan, vol. II, Paris, Brussels (1928)
7. Hackin, J., Carl, J.: Nouvelles recherches archéologiques à Bâmiyân. Les Mémoires de la DAFA, vol. III (1933)
8. Hackin, J., Hackin, R.: Le site archéologique de Bâmiyân. Guide du visiteur, Paris (1934)
9. Gettens, R.: The materials in the wall paintings of Bāmiyān, Afghanistan. Technical Studies in the Field of the Fine Arts, vol. VI, pp. 168–193 (1937–1938)
10. Lal, B.B., Sengupta, R.: A Report of the Preservation of Buddhist Monuments at Bamiyan in Afghanistan, Archaeological Survey of India, New Delhi, unpublished report (1965)
11. Sengupta, R.: The Restoration of the Small Buddha at Bamyian. ICOMOS, York, pp. 31–46 (Momentum 27.1) (1984)

12. Tarzi, Z.: L'architecture et le décor rupestre des grottes de Bamiyan, Paris, 2 vols (vol. 1: texts, vol. 2: sketches, photographs) (1977)
13. Gunasinghe, S.: La technique de la peinture Indienne d'après les Texts du Silpa, Paris (1957)
14. Lluveras-Tenorio, A., Vinciguerra, R., Galano, E., Blaensdorf, C., Emmerling, E., Perla, Colombini M., et al.: GC/MS and proteomics to unravel the painting history of the lost Giant Buddhas of Bāmiyān (Afghanistan). PLoS ONE **12**(4), e0172990 (2017). https://doi.org/10.1371/journal.pone.0172990
15. Wilford, F.: Asiatick Researches, or Transaction of the Society Instituted in Bengal, vol. 6, XII, London (1801)
16. Japan Center for International Cooperation in Conservation/National Research Institute for Cultural Properties, Japan, Study of the Afghanistan's Displaced Cultural Properties. Material and Techniques of the Bamiyan Mural Paintings. Recent Cultural Heritage Issues in Afghanistan, vol. 3. Tokyo (2006)
17. Beal, S.: SI-YU-KI, Buddhist Records of the Western World, translated from the Chinese of Hsiuen-Tsiang, London (1884)
18. Wilford, F.: On mount caucasus. Asiatick Researches, vol. VI (1798)
19. Elphinstone, M.: An Account of the Kingdom of Caubul, and Its Dependencies in Persia, Tartary, and India. Comprising a View of the Afghaun Nation, and a History of the Dooraunee Monarchy, London (1814)
20. Taniguchi, Y.: Conserving the Buddhist wall paintings of Bamiyan in Afghanistan: practical issues and dilemmas. In: Park, D., Wangmo, K., Cather, S. (eds.) Art of Merit: Studies in Buddhist Art and its Conservation, pp. 124–139. Archetype, London (2013)
21. Barbier de Meynard, C.: Dictionnaire Géographique, Historique et Littéraire de la Perse et des contrées adjacentes: Extrait du Mu´ǧam al-Buldān de Yāqūt, Paris (1861). Reprint: Sezegin, F. ed. Publications of the Institute for the History of Arabic-Islamic Science, Islamic Geography, vol. 221. First Part Ābaǧ – Sirkān, Frankfurt am Main (1994)
22. ICOMOS XIX: Emmerling, E., Petzet, M. (eds.): The Giant Buddhas of Bamiyan, Safeguarding the Remains 2010–2014. ICOMOS Monuments and Sites of the International Council on Monuments and Sites XXI (2016)
23. Moorcroft, W. and Trebeck, G.: Travels in the Himalayan provinces of Hindustan and the Panjab, in Ladakh and Kashmir, in Peshawar, Kabul, Kunduz and Bokhara from 1819 to 1825. Prepared for the press (...) by H. H. Wilson, London (1841)
24. Charles Masson: Sketches by Charles Masson were published in: Journal of the Asiatic Society of Bombay, vol. 6 (1836)

Chapter 10
Materials and Technologies of the Bamiyan Wall Paintings

Yoko Taniguchi

Abstract Bamiyan is located at a cultural cross-roads between Iran and the Mediterranean to the west, China to the east, India to the south, and nomadic cultures to the north. Approximately 1,000 caves were modified to serve as temples, residences, and storage chambers. As is well known, the site's two colossal Buddha sculptures were shattered in March 2001, and wall paintings were either destroyed or looted for sale in international art markets. Over 80% of the paintings were lost. In 2005 a UNESCO project entitled "Safeguarding the Bamiyan Valley" was launched. Prior to conservation work, minute samples were taken from 31 caves (of a total 279). These were used to investigate original painting technologies and their mechanics of deterioration. Analysis confirmed the Buddhist paintings of mid-seventh century Bamiyan as the world's earliest examples to use drying oil as binding media. In addition, the creation of the paintings is distinguished by other complex technologies, including the skilful use of earthen materials; and the sophisticated application of multiple colour layers and finishing techniques.

Keywords Bamiyan · Wall paintings · Sculpture · Conservation · Analysis · Drying oils · Multi-layered oil painting

10.1 Introduction

The Bamiyan site is located in the highlands of Afghanistan, approximately 2500 m above sea level, between the Hindu Kush and Koh-i-Baba mountains. Locally, it is situated within an elongated basin that consists of 3 valleys: Bamiyan, Foladi and Kakrak. Bamiyan is known as the cultural cross-roads to Iran and the Mediterranean to the west, China to the east, India to the south, and nomadic culture to the north (Fig. 10.1).

Y. Taniguchi (✉)
Conservation Science and Archaeological Science, Faculty of Humanities and Social Sciences, University of Tsukuba, 1-1-1 Tennoudai, Ibaraki 305-8577, Japan
e-mail: Taniguchi.yoko.fu@u.tsukuba.ac.jp

S. Aoki et al. (eds.), *Conservation and Painting Techniques of Wall Paintings on the Ancient Silk Road*, Cultural Heritage Science,
https://doi.org/10.1007/978-981-33-4161-6_10

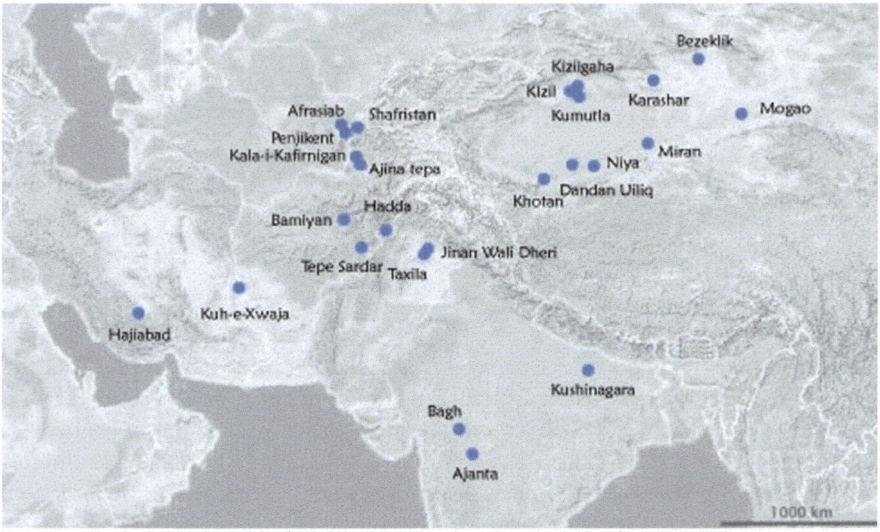

Fig. 10.1 Location of the major sites with wall paintings in Asia (earthen render with *secco* technique) (© Y. Taniguchi)

At the Bamiyan site, approximately 1,000 caves were modified to serve as temples, residences, and storage chambers. Most people are familiar with the two colossal Buddha sculptures carved into the eastern and western ends of a high cliff, which were shattered in March 2001. At that time, wall paintings that decorated the cave temples were either destroyed or looted for sale in international art markets. Over 80% of the paintings were tragically lost [1].

In 2005 a UNESCO project entitled "Safeguarding the Bamiyan Valley" was launched, although it was later terminated due to security concerns [2]. Prior to conservation work, minute samples were taken in Bamiyan, Foladi and Kakrak from 31 caves (of a total 279) with intact wall paintings. These samples were used for examination of the mechanics of deterioration and to investigate painting technologies. Because most samples were collected from among fragments on the floor, it has not yet been possible to identify exactly where in each cave they came from. Samples were selected after an inventory of all the fragments had been created. The fragments were classified according to colour and condition, and then recorded on inventory sheets. Where large areas of wall painting remained in situ, samples were selected from these. The samples were then brought to Japan with the permission of the Afghan Ministry of Information and Culture and examined using a PLM (polarised light microscope) (Olympus BX51) under normal and UV light sources. Each of the selected samples was then prepared as a polished cross-section by mounting in polyester resin (Struers No. 105) to enable examination of the paint-layer stratigraphy. The cross-sections were then dry polished with 4,000–12,000 Micromesh® to avoid unnecessary dissolution of water-soluble binding media and colourants.

The polished sections were then observed and documented using the polarised light microscope under both normal and UV emission modes following guidelines provided by Plesters [3]. Lastly, the remaining minute samples were used for analyses via the synchrotron-radiation-based μFTIR, μXRF and μXRD, and LC/MS, nanoLC/ESI-MS/MS, GC/MS and ELISA.

10.2 Classification of Wall Painting Technologies

Over the site's history, painting materials such as pigments and binding media were brought from various locations by way of the Silk Road. How these materials and technologies were adapted may be interpreted through knowledge of each site in its geological setting, and by exploring the provenance of materials and ancient trading routes.

Available literature, including primary source material, does not yet offer a clear picture of ancient painting techniques. Therefore, direct analysis of the paintings is crucial. The words *fresco* and *tempera* are commonly used to describe painting technologies, yet these terminologies are often misused or misunderstood. It would therefore be useful, before proceeding further, to provide more precise definitions for these terms.

Although often misused as a synonym for 'wall painting', the term *fresco* (meaning 'fresh' in Italian) refers to a specific technique in which pigments suspended in water are applied directly to fresh lime plaster. The term *tempera* is also often misused. It is erroneously applied to the use of pigments mixed with egg only, while it does, in fact, apply to painting using a range of different water-soluble binding media such as animal glue, plant gums, and egg (whole egg, yolk or white). However, other, non-water-soluble organic binders can also be used, including various drying oils and resins [4]. Painting in which any type of organic binding medium is used is defined as *secco* (meaning 'dry' in Italian). All paintings from medieval Central Asia are applied *a secco*, so, in order to avoid mixing terminologies, I will refer to the paintings as *secco* paintings on earthen renders. These are observed across a wide area of eastern Eurasia.

Historically, many pigments could not be sourced locally. They would have been transported long distances with travelling artisans, making them expensive commodities. The quantities required would, however, have been relatively low. In contrast, the large amounts of bulky, heavy materials required for construction precluded sourcing from far afield, and construction technologies tend to be largely circumscribed by conditions of local geology. For example, creating a lime mortar requires copious quantities of raw limestone, fuel and water, so lime technology tends to dominate in geologically calcareous areas like West Asia or the Mediterranean. In the region of Afghanistan and further east along the Silk Road, where these materials are geologically scarce, earthen mortar was naturally popular.

In research on wall paintings, it must be stressed that wall paintings are immovable properties. They are applied onto rock surfaces or onto the walls of architectural structures, and usually also onto secondary supports of for example lime, earth or other materials that reflect local geological features. The coloured pigments and binding media used may have been transported from afar, depending on the political and economic conditions of the time. Not only technologies and painting materials, but also artistic concepts were mobile, reflecting the political and religious milieux of the artisans who were brought into create them at the invitation of local elites and donors.

Creating paint requires combining coloured pigments with a binding medium that fixes pigment particles to the wall surface. There are many techniques to achieve this. In *fresco* technique, painting is applied onto a lime plaster, and setting depends on the transformation through carbonation of calcium hydroxide in the lime plaster [$Ca(OH)_2$] to calcium carbonate [$CaCO_3$] on contact with carbon dioxide in the air. Watercolours use gum Arabic as a binder, while wax is the binder in encaustic painting. Oil paintings are bound using drying oils, which create a film through a process of oxidative polymerisation.

Although there are multiple publications that offer elemental analysis of the inorganic pigments used in wall paintings, few exist which provide analytical identification of their organic components, such as colourants, binding media, organo-metallic compounds, glazes, and sizing materials. This is likely to be due to the technical difficulty of analysis of aged organic materials. Therefore, publications that examine the comprehensive range of materials used by ancient artisans and provide a broad interpretation of their painting technologies are limited.

While analyses of both the organic and inorganic components of wall paintings is essential, a full understanding of historical painting technologies requires more than this. It is equally important to shed light on the relationship between support and paint layers, between colourant particles and binding media, the optical effects sought through stratigraphic sequencing of paints, the role of painting tools in application methodology, the causes and effects of deterioration, and more [5].

Although the accuracy and variety of non-invasive analytical instruments has considerably increased in recent years, these techniques produce limited results. For instance, surface analysis cannot provide information on the layering of paints or on the physical properties and deterioration phenomena occurring in individual particles. This information can only be derived through interrogation of stratigraphic samples. Because all such information is crucial, we apply both invasive and non-invasive methodologies in our research.

10.3 Characteristics of the Bamiyan Wall Paintings

Wall paintings in Central Asia, Xinjiang and India were applied onto earthen plasters using colourants mixed with organic binders. They have a multi-layered structure. First, earthen renders were applied onto the surface of the rock. Next, an organic sizing layer was applied, followed by the application of ground and paint layers. The use of earth for construction and plastering can be found in almost all parts of the world, but it is a particular characteristic of eastern Eurasia, which lacks the calcareous geological features that enable the use of lime elsewhere.

Bamiyan's *secco* painting technique is characterised by multiple layering to produce various skilful optical effects, for example, a deep green colour is achieved by layering green over white and red layers: the white layer between the green and red layers acts as a reflective interface, while the red layer, which is the complementary of green, works to deepen the uppermost green colour. A deep blue is produced by applying coarsely ground lapis lazuli over carbon black.

Analysis has been able to help further the understanding of the methods and materials used. In one example from Cave N(a), where the National Research Institute of Cultural Properties, Tokyo (NRICPT) has been working on conservation projects since 2005, the green was analysed with SR-μXRF, SR-μFTIR.

Synchrotron radiation can enhance microbeams with a high S/N ratio.[1] Therefore, it is useful for analysing thin paint layers (of a few μm) at high resolution [6]. In a collaborative research project with the European Synchrotron Radiation Facility (ESRF), micro analyses of both inorganic and organic materials were carried out to identify constituent materials and deterioration mechanisms. ESRF is a synchrotron beam facility of about 6 GeV with a large 844-metre circumference ring. At ID21 and ID18F, SR-μFTIR and SR-μXRD were carried out (Figs. 10. 2 and 10.3).

From the analyses, the green was identified as a copper-based green, and lead carboxylate was identified in the layers beneath [7]. In some areas of painting on the ceiling of Cave N(a), which shows a pattern of palmettes and animals (Figs. 10.4 and 10.5), a yellow resin was found over a lead/tin alloy, clearly in imitation of gold

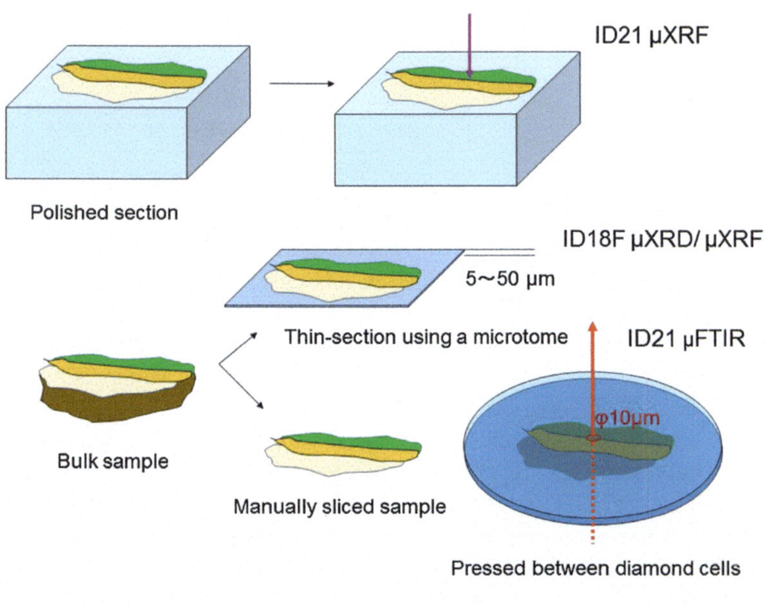

Fig. 10.2 Sample preparations for SR-μFTIR and SR-μXRF/XRD (© Y. Taniguchi)

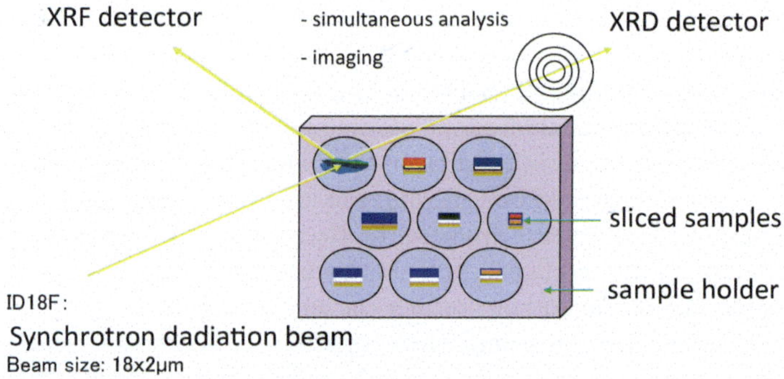

Fig. 10.3 Sample setting for SR-μXRF/XRD (© Y. Taniguchi)

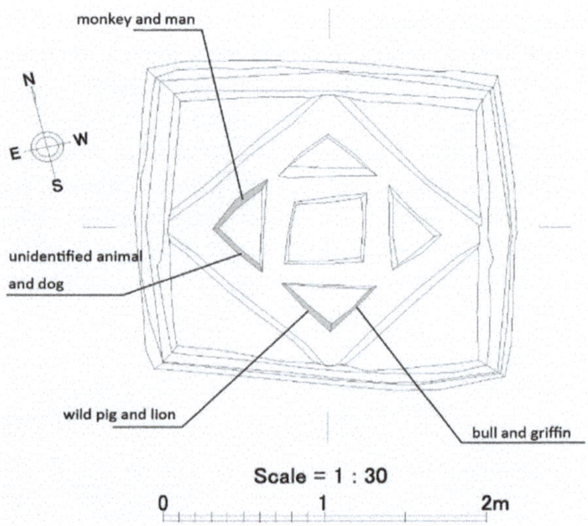

Fig. 10.4 Drawing of the ceiling of Cave N(a) (© Y. Taniguchi, courtesy of the National Research Institute for Cultural Properties, Tokyo [NRICPT])

(Figs. 10.6, 10.7, 10.8, 10.9, and 10.10). This is similar to the *mecca* technique, which employs golden yellow resin on tin leaf. In this case, a tin-containing lead-based alloy was used, since the technology did not exist at the time to allow the tin to be extracted from the lead (Figs. 10.6–10.10). Yuki Watanuki created a reconstruction of the ceiling of Cave N(a) using the original vivid palette, which is now in the collection at Teikyo University (Fig. 10.11).

A sample of paint with a green surface from Foladi Cave 4 was also analysed with Synchrotron radiation (Figs. 10.12 and 10.13). Particular points of absorption within the fingerprint region of FTIR spectra were selected and mapped. Additionally, maps combining SR-μXRF and SR-μXRD were created. From this analysis the following stratigraphy was established (from the lower layers): two sizing layers (the first, a natural resin; the second, a proteinous material with polysaccharide content); a white ground layer (lead white), a red layer (minium with red ochre), a further white

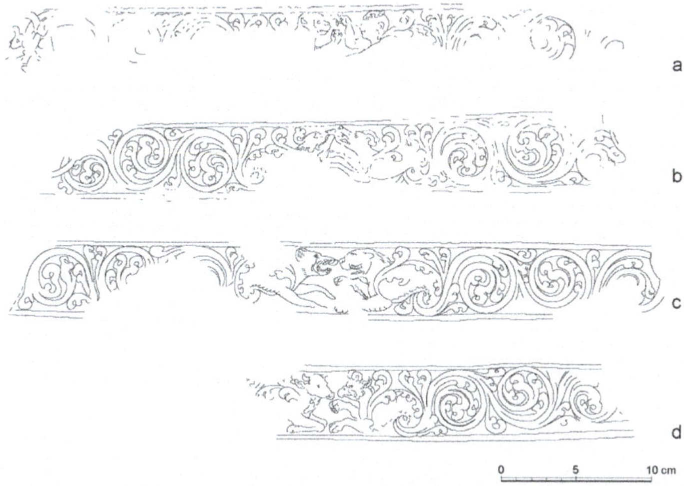

Fig. 10.5 Patterns of palmettes and animals on the ceiling of Cave N(a) (© Y. Taniguchi, courtesy of NRICPT)

Fig. 10.6 Microphotograph of sample BMM186: lead/tin alloy coated with a yellow resin (© Y. Taniguchi)

layer (lead white), and finally a layer of green (atacamite), which had altered to copper oxalate. From the top layer of green to the white ground, drying oils were used as binding media (Figs. 10.14, 10.15, 10.16, 10.17, 10.18, 10.19, 10.20, 10.21, Table 10.1) [4, 8].

The collagens present in the proteinaceous layers of samples taken from four different caves were analysed by Nippi (Co.) and Nara Women's University. Nippi employed a marker peptide method using LC/MS, and Nara Women's University applied nanoLC/ESI-MS/MS. They both identified horse glue from Cave N(a). Proteins present in layers from Cave A lower sale, Cave M, and Foladi Cave 5 were identified as cow milk casein, probably present as sizing or binding media.

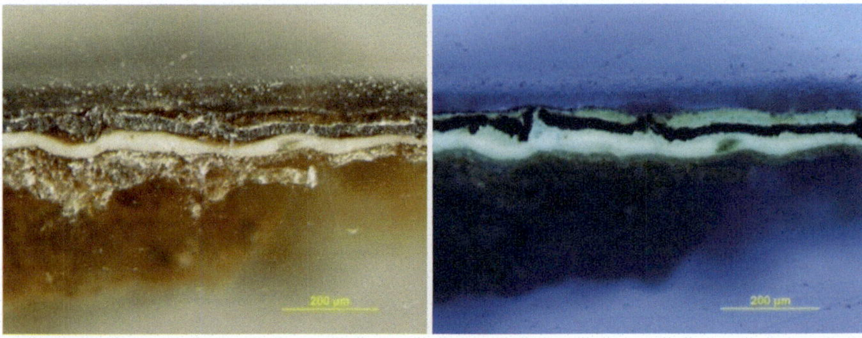

Fig. 10.7 Microphotographs of BMM186 under PLM (left: normal light; right: UV fluorescence) (© Y. Taniguchi)

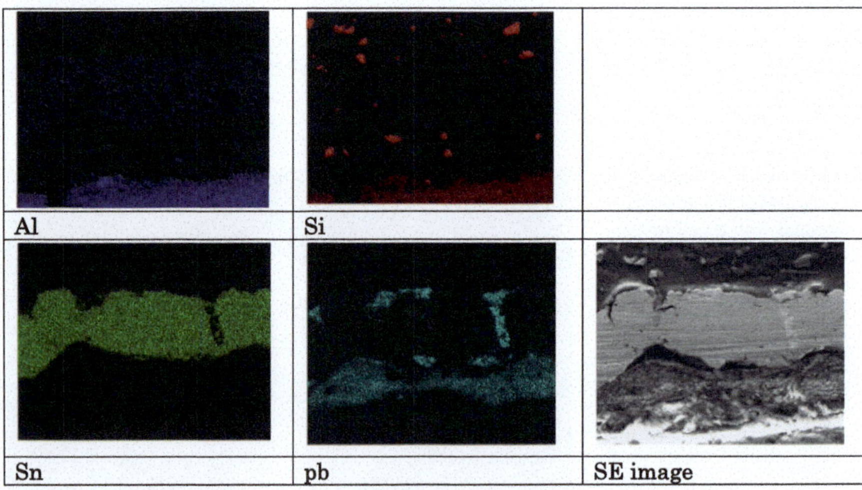

Fig. 10.8 SEM-EDS imaging of sample BMM186 (© Y. Taniguchi)

Cow glue and cow milk casein were both detected from the polychromy both of the East and West Giant Buddhas (BMM191 and BMM201, respectively), and it is assumed that the casein was applied as a size, and that the cow hide glue was used as a binding medium [9].

Further analyses on drying oils were conducted at the Getty Conservation Institute using GC/MS. The P/S ratios (palmitic acid/stearic acid) of obtained fatty acids were about 3:1, which indicates that fatty acids originating from either walnut or poppy seed oils were used [10].

Similar fatty acids were detected in samples from Bamiyan Caves F(c), S(a), N(a), B(d), and L; Foladi Caves 2, 3, 4, and 6; Kakrak Caves 43 and 44 and Col-e-Jalal [11]. A tentative connection may be made between the materials used and the styles and iconography employed. Each of the caves where evidence for the use of drying

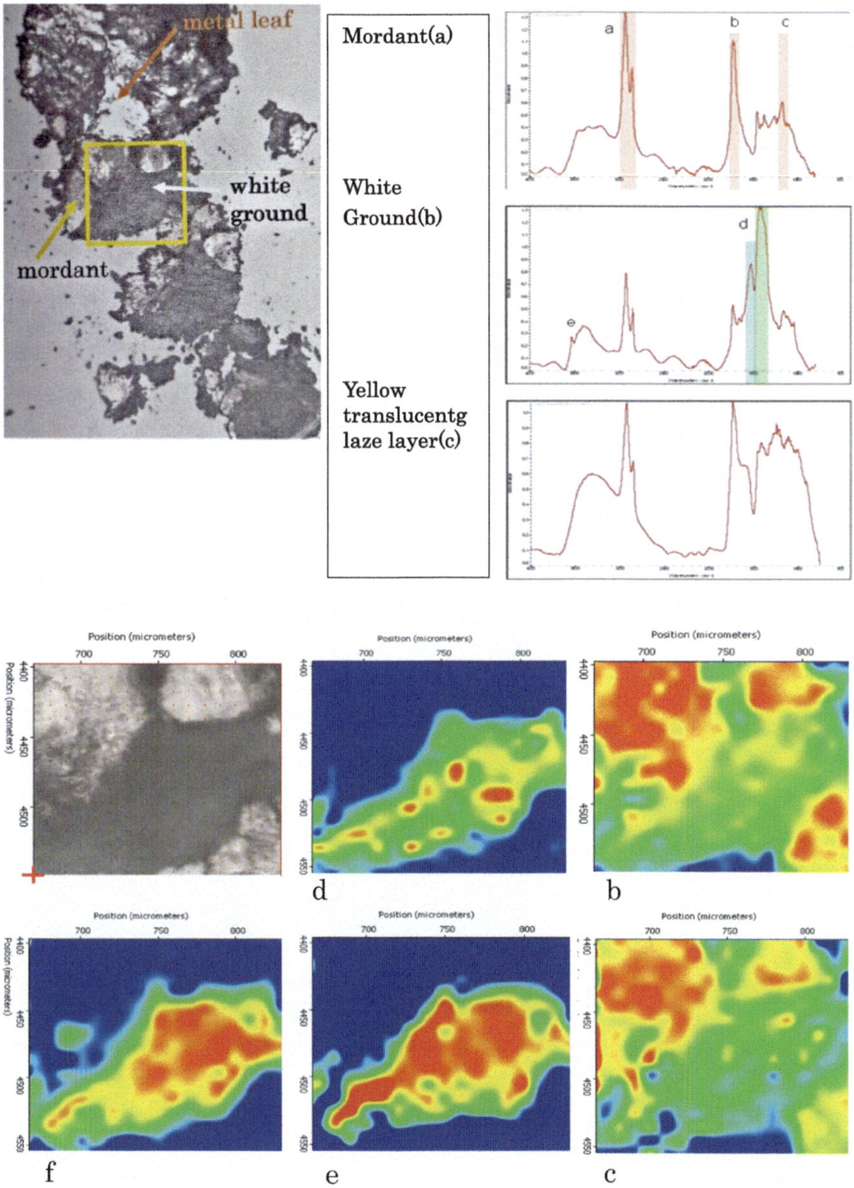

Fig. 10.9 SR-μFTIR imaging of sample BMM178, also of the lead/tin allow coated with a yellow resin (© Y. Taniguchi)

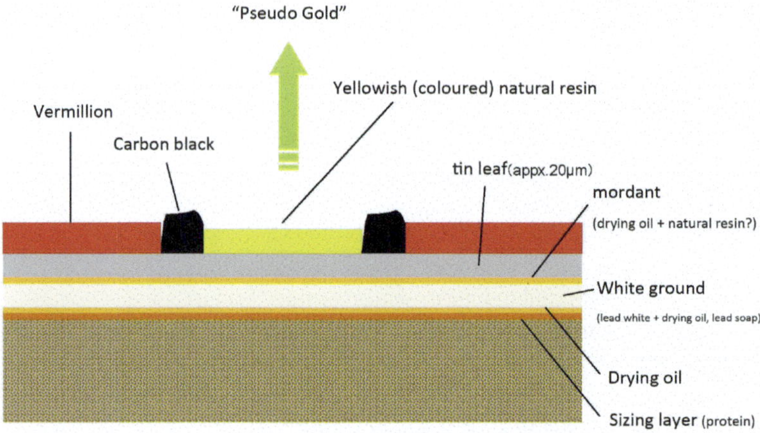

Fig. 10.10 Painting stratigraphy of the animal pattern found on the ceiling of Cave N(a) (© Y. Taniguchi)

Fig. 10.11 Reconstruction of the ceiling of Cave N(a) by Watanuki Yuki (© Y. Watanuki)

oils has been established has *laternendecke*-style ceilings (with four bars placed diagonally across the corners of the room and additional beams arranged diagonally to each other). Similar designs are seen widely across the Eurasian continent, from the Korean Peninsula in the east to the East Mediterranean Basin in the west, and their origin may be associated with ceiling beams in wooden architectural structures in Kashmir, Nisa, and Funza, or from buildings from Armenia to Gandhara [12]. Further study is needed to clarify this question.

Wall painting traditions in Bamiyan are classified as belonging to one of three phases. Phase I spans the third to early sixth century and includes Caves M and J and the East Giant Buddha, which employs either gypsum or lime as a white ground

Fig. 10.12 The ceiling of Foladi Cave 4 (© Y. Taniguchi, courtesy of NRICPT)

Fig. 10.13 A fragment from Foladi Cave 4 (© T. Kijima courtesy of the Tokyo University of the Arts)

fixed with a water-soluble binder. Phase II spans the sixth and mid-Seventh centuries and includes Caves C and D and the West Giant Buddha, which employs a gypsum ground only with a water-soluble binder. Phase III spans the mid-seventh and eighth centuries and includes Caves N(a), F(c), and B(d) and the Foladi/Kakrak caves, which use either a gypsum ground with a water-soluble binder or a lead white ground with a drying oil binder [11, 13].

Besides the use of the *laternendecke* ceiling style, the cluster of caves where drying oils have been identified also reveal other common technical features, including the

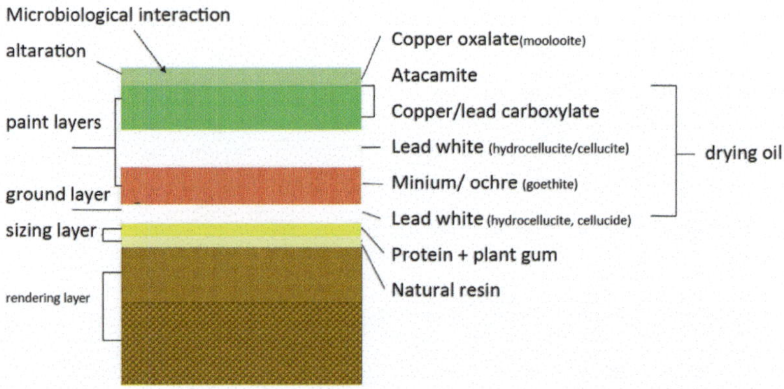

Fig. 10.14 Painting stratigraphy of a paint sample with a green surface layer from Foladi Cave 4 (© Y. Taniguchi)

Fig. 10.15 Flattened sample from Foladi Cave 4 between diamond cells (© M. Cotte)

use of lead white grounds, metal leaf gilding (*mecca* method), and a clear multiple stratigraphy. Tentative connections can be made with painting techniques in the Chitora Sutra in India, where stylistic and iconographical parallels can also be found, including the use of palmetto patterns, thousand-Buddha themes, and Mandala-like compositions. These observations would benefit from further research.

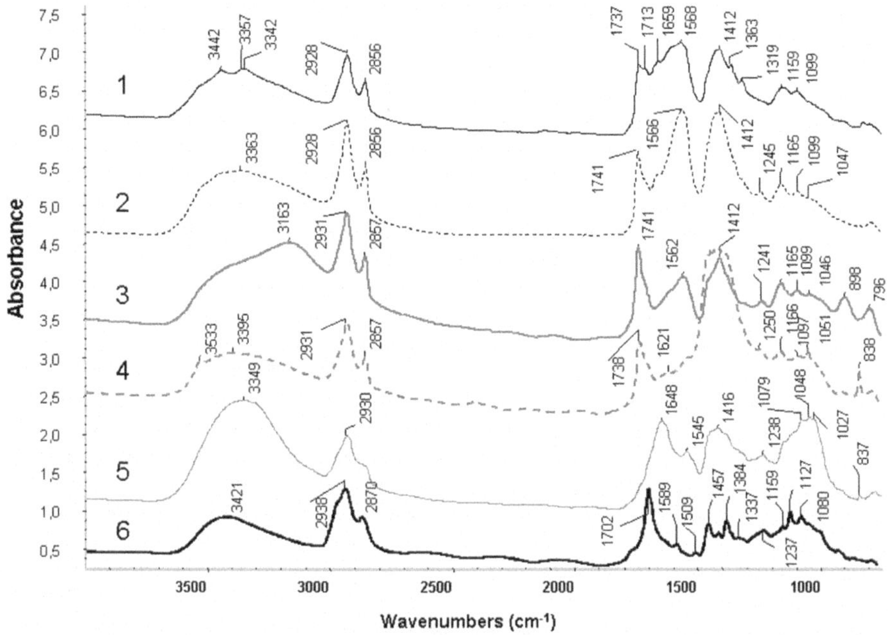

Fig. 10.16 SR-μFTIR spectra for each layer of the paint sample from Foladi Cave 4 (© M. Cotte)

10.4 Wall Paintings as an Oil Painting Technique

This analytical work has established the paintings of mid-seventh century Bamiyan as the world's earliest examples of the use of drying oil as binding media. However, drying oils have a long history of use for medicinal or craft applications. Roman references describe this use, including the text 'De Materia Medica' by Roman physician and botanist Pedanius Dioscorides (c.40–90 AD), which mentions poppy seed oil and walnut oil as medicines [14]. The Greek doctor, Aetius (late-5–6th century), mentions the application of walnut oil as a protective coating for gold leaf and for use in encaustic paintings [15], which may be the origin of *mecca* techniques.

Texts post-dating the use of oil at Bamiyan include the 'Codex Lucensis', written around the 8th century, which mentions the use of linseed oil, galbanum, balsam, and saffron to create a bright varnish on paintings [16]. In addition, Eraclius (10–11th century), the pseudonymous Theophilus Presbyter (12th century), and the painter Cennini (14th century) describe the application of drying oils to create a golden varnish, for use as paint binding media, and as oil glazes on *tempera* paintings [17–20].

There is no mention of drying oils in ancient Asian manuscripts. However, there was knowledge that oils could be extracted by squeezing walnuts, as evidenced by a mention of 'walnut oil' in *Sino-Iranica* [21], and walnut oil was used both as a lamp oil or an edible oil. Walnuts were domesticated and cultivated from an early period in

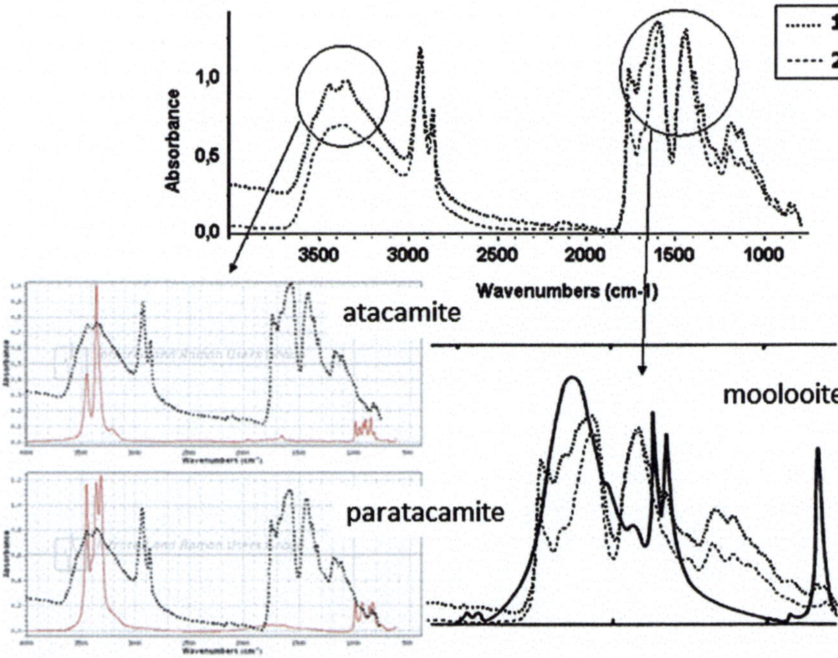

Fig. 10.17 Comparison of SR-μFTIR spectra of the topmost layer and the green paint layer of the sample from Foladi Cave 4 (see Fig. 10.14 for full stratigraphy). The reference spectra are for moolooite and atacamite[2] (© M. Cotte)

Iran and Central Asia (the Chinese characters for walnut are 胡桃, the character 胡 referring to an Iranian ethnic group), while linseed and poppy seeds are also grown and readily available in the region.

Previously, painting using drying oils was believed to have originated in the Scandinavian Peninsula, and the oldest example of polychromy applied using drying oils was from the 12th century wooden crucifix of Hemse church in Godland Island, Sweden [22]. It is then thought to have been adopted by Early Netherlandish artists such as Van Eyke.

Although drying oils may have been used as paint media in the occident at an earlier date than currently thought, their use in the Bamiyan wall paintings is unprecedented. Not only is there no geographical, stylistic, iconographical or technical connection between the painting of Northern Europe and Bamiyan, but there is a 400-year gap in time between Bamiyan and the first known instance in Europe. The use of drying oil technology in Buddhist art therefore, more specifically, in Bamiyan wall paintings, remains unique in world history.

In Bamiyan the innovation of using oil as a binder is accompanied by the use of lead white as the prevalent pigment used in the ground. While oil mixed with gypsum or lime becomes translucent owing to the closeness of their refractive indices, oil mixed with artificial lead-based pigments such as lead white and massicot produces

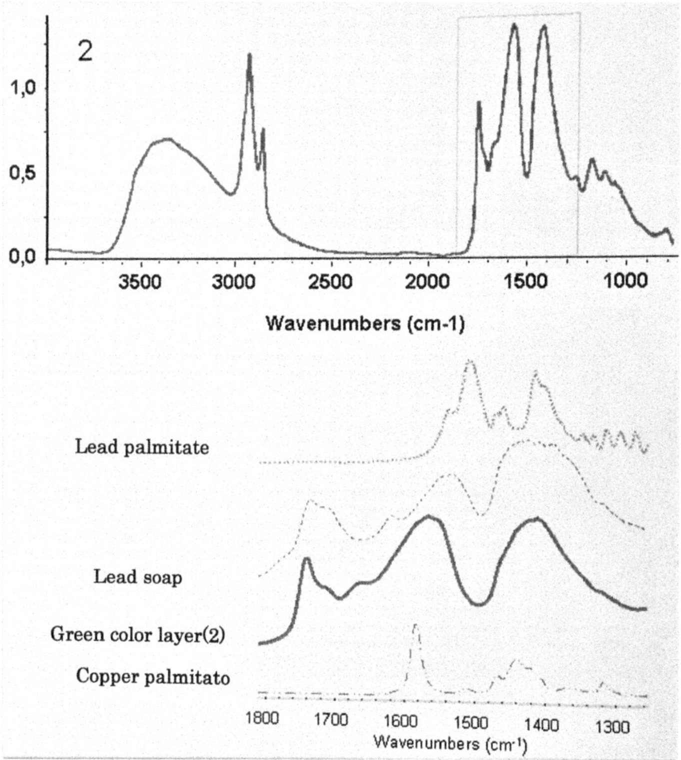

Fig. 10.18 SR-μFTIR spectra with the green layer (2) and references to copper palmitate, lead palmitate and lead soap (FDM055) (© M. Cotte)

paint layers with excellent covering power. The use of drying oils as paint media is rooted in this discovery. Analytical research on binding media in Central and East Asia is rare as yet, and no definitive links have so far been made between the use of drying oils and the use of the yellow pigment, massicot, which can be found in Iran.

10.5 Final Remarks

Technically complex, multi-layered wall paintings at Bamiyan dating to the mid-7th century were established through analysis as having been created using drying oils. Sadly now only surviving in a fragmentary condition, these exceptional Buddhist wall paintings are currently the earliest known examples of *secco* painting using drying oils in the world, meriting further research, particularly into the painting materials at other sites in Central Asia.

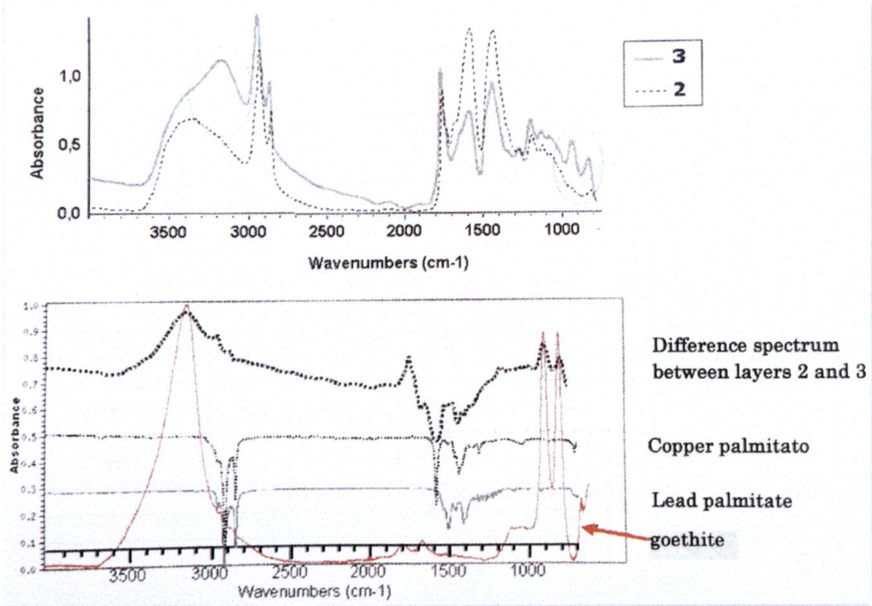

Fig. 10.19 Comparison of SR-μFTIR showing difference in the spectra of layers 2 and 3, with spectra of copper palmitate, lead palmitate and goethite (FDM055)[2] (© M. Cotte)

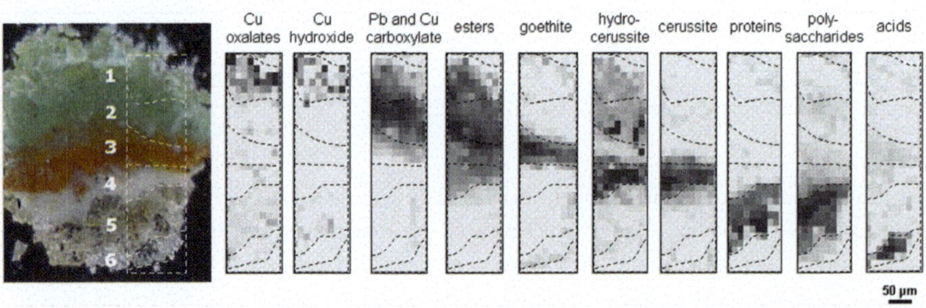

Fig. 10.20 Imaging of selected absorption of SR-μFTIR spectra (FDM055)[2] (© M. Cotte)

Besides this evidence for the early use of drying oils, other aspects of technical advancement are also demonstrated at Bamiyan, including the creation of earthen supports, sophisticated colour layering, and the use of metal leaf and glazes. Each material was carefully chosen, and painting techniques were highly skilful. However, it is likely that many of these technological features did not originate in Bamiyan, but were established elsewhere, meriting further exploration, analysis and research.

Although local geological conditions and the availability of construction materials inevitably enforced constraints on some aspects of wall painting technology, other materials, particularly those used for the painting itself could much more easily

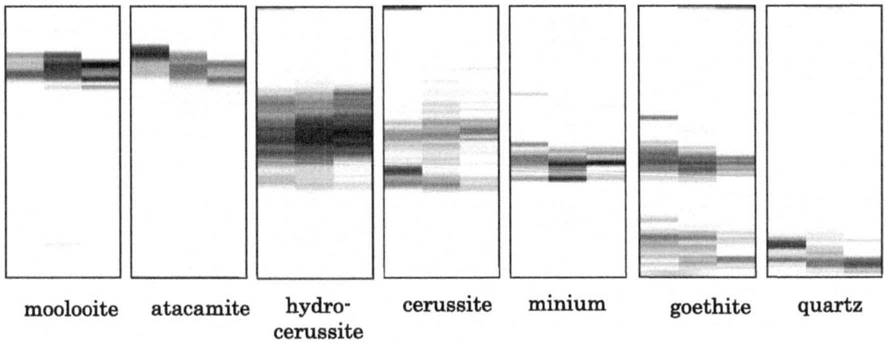

Fig. 10.21 Imaging of SR-μXRD analyses (FDM055) (© M. Cotte)

Table 10.1 Pigments, alteration products, organic materials and binding media, and other inclusions in the layer structure of the green paint sample from Foladi Cave 4 (FDM055)

Layer	Location	Colour	Major elements	Estimated pigments, alteration products, inclusions
1	Surface	Pale green	Cu, Cl, Ag, (K)	Copper oxalate (moolooite)
2	Paint layers	Green		Atacamite, copper carboxylate, lead carboxylate, drying oil
3		White	Pb, Cl, Ca, Mn, As	Lead white (hydrocerussite > cerussite), orpiment, lead carboxylate, drying oil
4		Red	Pb, Ca, Fe, Co, As	Lead red, red ochre (goethite), lead carboxylate, drying oil
5	Ground	White		Lead white (hydrocerussite > cerussite), Lead carboxylate, drying oil
6	Sizing	Translucent yellow		Protein, polysaccharide (plant gum)
7		Translucent pale yellow		Resin
8	Render	Brown	K, Mn, Ti, Fe	Earthen material, quartz

be transported along the trading routes. These would probably have been carried long distances by artisans from different traditions within Buddhist Central Asia, who were invited to create temple paintings in great Buddhist centres like Bamiyan, Kizil and Mogao. These artisans brought not only the materials of their trade, but also a range of painting technologies and artistic conceptions formed in widely differing artistic, political and religious milieux. The evidence of these circumstances of cultural exchange and innovation is embedded in the material remains of the wall paintings of Bamiyan, and it is to be hoped that future research may continue to yield further remarkable discoveries.

Notes

1. S/N ratio: Signal-to-noise ratio is a measure used to compare the level of a desired signal to the level of background noise.
2. Reference FTIR spectra were cited from the IRUG database [23].

   1) Spectrum of atacamite: Philadelphia Museum of Art, 2000. "IMP00198 Atacamite, natural". Spectrum of paratacamite: "IMP00166 Paratacamite".
   2) Spectrum of goethite: Philadelphia Museum of Art, 2000. "IMP00188 Goethite".

Acknowledgements I would like to express my gratitude to the Ministry of Culture and Information in Afghanistan, as well as Kazuya Yamauchi and his colleagues at the ex-National Research Institute of Cultural Properties in Tokyo and Masanori Nagaoka at the UNESCO Kabul office. Analyses at ESRF (European Synchrotron Radiation Facility) were carried out as part of the "Combination of Micro-X-ray Diffraction and Micro-Infrared Spectroscopy for the Study of Multi-Layered Buddhist Mural Paintings from Bamiyan" (EC101 Y. Taniguchi, M. Cotte, E. Checroun). Reference FTIR spectra were cited from the IRUG database (www.irug.org). I would like to thank the following colleagues who contributed important data and advice: Kosaku Maeda, Shigeo Aoki, Hidemi Otake, Takeshi Nakazawa, Miho Takashima, Yuki Kumazawa, Hiromi Yasui, Yuki Watanuki, Fabio Colombo, Georgios Toubekis, Catharina Blaensdorf, Omara Khan Masoudi, Abbas Kawari, Joy Mazurek, Stephen Rickerby, Rayna Rusenko and, finally, Michael Petzet, who sadly passed away in 2019. This work was supported by JSPS KAKENHI Grant Number JP18700680.

References

1. Taniguchi, Y., Otake, H., Maeda, K.: Constituent material analysis of the Bamiyan Buddhist mural paintings (I): examination of painting techniques from polished sections. Sci. Conserv. **46**, 1–8 (2006). NRICP, Japan (in Japanese)
2. Yamauchi, K. (ed.): Preliminary report on the safeguarding the Bamiyan site 2006—6th and 7th missions, Recent Cultural Heritage Issues in Afghanistan, Preliminary report series, 2, Ministry of Culture and Information, National Research Institutes for Cultural Properties, Tokyo and Nara (2006a)
3. Plesters, J.: Cross-sections and chemical analysis of paint samples. Stud. Conserv **2**, 110–131 (1956)
4. Taniguchi, Y.: Constituent material analysis of the Bamiyan Buddhist wall paintings in Central Asia (1): using synchrotron-based μFTIR, SR-μXRF/SR-μXRD analyses. Bull. Natl. Mus. Jpn. Hist. **81**, 29–79 (2012a)
5. Taniguchi, Y.: Issues of conservation for the Bamiyan Buddhist mural paintings. In Mural Paintings of the Silk Road: Cultural Exchanges Between East and West, Proceedings of the 29th Annual International Symposium on the Conservation and Restoration of Cultural Property, National Research Institute for Cultural Properties, 144–151. Archetype Publications, London (2007)

6. Cotte, M., Dumas, P., Taniguchi, Y., Checroun, E., Walter, P., Susini, J.: Recent applications and current trends in cultural heritage science using synchrotron-based fourier transform infrared micro-spectroscopy. C.R. Phys. **41**, 590–600 (2010)
7. Taniguchi, Y., Cotte, M., Checroun, E.: Constituent material analysis of the Bamiyan Buddhist mural paintings (II): a study of pseudo-gold leaf technique discovered at cave N(a) using synchrotron-based μFTIR. Sci. Conserv. **46**, 181–188 (2007). NRICP, Japan (in Japanese)
8. Taniguchi, Y.: Conserving the Buddhist wall paintings of Bamiyan in Afghanistan: practical issues and dilemmas. In: Park, D., Wangmo, K., Cather, S. (eds.) Art of Merit: Studies in Buddhist Art and Its Conservation, pp. 125–140. Archetype Publications, London (2013)
9. Nakazawa, T., Kawahara, K., Fukakusa, S., Karino, M., Takashima, M., Taniguchi, Y.: (2016) Mass spectrometry of collagen and casein in the remains of the 5th to 7th century Bamiyan Buddhas. 64th ASMS Conference on Mass Spectrometry and Allied Topics
10. Taniguchi, Y.: Do archaeological and conservation sciences save cultural heritage?: cultural identity and reviving values after demolishment. In: Tsuneki, A., Yamada, S., Hisada, K. (eds.) Ancient West Asian Civilization: Geoenvironment and Society in the Pre-Islamic Middle East, pp. 179–197. Springer, New York (2016)
11. Taniguchi, Y.: Constituent material analysis of the Bamiyan Buddhist wall paintings in Central Asia (2): organic analysis using GC/MS and ELISA. Bull. Natl. Mus. Jpn. Histo. **81**, 81–106 (in Japanese)
12. Godard, A., Godard, Y., Hackin, J.: Les antiquities bouddhique de Bâmiyân (Mémories de la Délégation Archaéologique Française en Afghanistan, tome II). Les Editions G. Van Oest, Paris and Brussels (1928)
13. Taniguchi, Y., Mazurek, J.: Constituent material analysis of the Bamiyan Buddhist mural paintings (III): analysis of organic materials in mural paintings from Cave B(d) by gas chromatography/mass spectrometry. Sci. Conserv. **48**, 79–88 (2008b). NRICPT, Japan (in Japanese)
14. Dioscorides: De materia medica, Ibidis press, Johannesburg (2000)
15. Eastlake, C. L.: Methods and Materials of Painting of the Great Schools and Masters, vol. 1. Dover Publications, New York (2001)
16. Burnam, J.M.: A Classical Technology Edited from Codex Lucensis 490. R.G. Badger-the Gorham Press, Boston (1920)
17. Merrifield, M.: Eraclius De Cooribus et Artibus Romanorum (XLIV), Medieval and Renaissance Treatises on the Arts of Painting. Dover Publications, New York (1967)
18. Cennini, C.D.A.: The Craftsman's Handbook: Il Libro dell' Arte. Daniel V. Thompson (Translation) (1954)
19. Thompson, D.V.: The Materials and Techniques of Medieval Painting. Dover Publications, London (1956)
20. Theophilus: On Divers Arts (Hawthorne, J. G., Smith, C. S. trans.). Dover Publications, New York (1979)
21. Laufer, B.: Sino-Iranica: Chinese Contributions to the History of Civilization in Ancient Iran. Ch'eng-wen Publishing Company, Taipei (1967)
22. Tångeberg, P.: The crucifix from Hemse. In: Nadolny, J., Kollandsrud, K, Sauerberg, L., Frøysaker, T. (eds.) Medieval painting in Northern Europe: Techniques, Analysis, Art History, pp. 1–10. Archetype Publications, London (2006)
23. IRUG: Infrared and Raman Users Group Spectral Database. In: Price, B., Pretzel, B. (eds.) IRUG, Philadelphia (2007)

Chapter 11
Conservation of Detached Wall Paintings from the Bamiyan, Foladi and Mes Aynak Sites

Takayasu Kijima

Abstract This paper reports on the conservation and restoration of wall paintings carried out at the Tokyo National University of the Arts. The examples are of fragments of excavated Buddhist wall paintings from the Bamiyan and Foladi caves, and the Mes Aynak archaeological site in Afghanistan. The restoration of 42 fragments from the Bamiyan and Foladi caves was carried out from 2007 to 2009 [1, 2]. Painted in *secco* techniques on plasters made of earth and straw, the fragments were in such fragile condition that they were on the verge of disintegration. Inappropriate previous treatments were removed where safe to do so, and treatments were carried out to consolidate both the paint layers and the supporting plasters. For reinforcement, the reverse sides of the fragments were lined with triaxial woven rayon fabric. The frames were designed to enable the fragments to be displayed in museum exhibitions. All the fragments have returned to Afghanistan by 2016. The Mes Aynak archaeological site is an ancient Buddhist city that prospered from the third to the seventh centuries [3]. The Tokyo National University of the Arts restored a fragment of wall painting from the site, which was unearthed in a rescue excavation. Its previous treatment included the application of protective facings on the painting surface with Paraloid B-72, and the reinforcement of the reverse side with mortar, aluminum bars and glass fiber adhered with epoxy resin. These problematic treatments were removed where possible. The fragment was then provided with a frame of similar design to those made for the Bamiyan and Foladi fragments.

Keywords Conservation · Detached wall painting · Bamiyan · Foladi · Mes aynak · Archaeological site · Remounting · Exhibition

11.1 Introduction

The Tokyo National University of the Arts has twice engaged in the task of restoring detached Buddhist wall paintings from Afghanistan. The first was from 2007 to

T. Kijima (✉)
Tokyo National University of Arts, 12-8, Ueno-Park, 110-8714 Taito, Japan
e-mail: t-kjm@fa.geidai.ac.jp

S. Aoki et al. (eds.), *Conservation and Painting Techniques of Wall Paintings on the Ancient Silk Road*, Cultural Heritage Science,
https://doi.org/10.1007/978-981-33-4161-6_11

Fig. 11.1 Wall painting fragments from Bamiyan Archaeological site (Seventh–eighth century)

2009, which involved 42 wall painting fragments from the seventh to the eighth century that had been removed from archeological sites in Bamiyan (Fig. 11.1) and Foladi (Fig. 11.2). The second involved one wall painting fragment, dating from between the third–seventh century, that had been unearthed from remains in Mes Aynak (Fig. 11.3). The shared characteristic in both cases was the fact that the wall painting fragments had been stripped from ancient sites.

In addition to the obvious task of strengthening the wall paintings, the concept of the restoration procedure was focused on respecting their originality, as well as creating mounts and frames that would be easy to handle during future public exhibitions and associated transportation.

11.2 Restoration of the Fragments from Bamiyan and Foladi

The Buddhist wall painting fragments that were selected for restoration during the first project comprise 18 fragments from the Bamiyan site, and 24 from the Foladi site. In both cases, the painting fragments had been detached from the walls of rock-cut caves, illegally exported from Afghanistan, and shipped to Japan. The National Commission for Protection of Exported Cultural Properties then took on the task of

Fig. 11.2 Wall painting fragments from Foladi archeological site (Seventh–eighth century)

Fig. 11.3 Wall painting fragment from Mes Aynak archaeological site (third-seventh century)

protecting the fragments after identifying them as items of displaced cultural property [4].

The wall painting fragments were of numerous sizes, with diameters ranging from just a few centimeters up to 30 cm. While some of these stood alone, others appeared to belong to groups of fragments that could be conjoined. Our project involved piecing together these latter fragments, from which we were able to frame a total of 28 paintings. The restored wall painting fragments were then returned to

Afghanistan in 2016. In this paper, we will describe the restoration process from its experimental stages to the methods that were eventually adopted for the successful preservation and presentation of the paintings.

11.2.1 Condition Prior to Restoration

The fragments originated from paintings on earthen supports, mixed with straw. Their painting may be said to have been executed largely using water-based paints mixed with a binder, although there is a possibility that on some oil-based paints were used. The earthen plaster layers that had been removed from their sturdy bedrock base survived in a fragile, brittle state. Amateur restorative procedures had already been carried out on the fragments so that they could be sold by antique art dealers, and the pieces had been carefully framed.

Careful observation of the wall painting fragments revealed that some of them had been intentionally broken, and then reassembled and readhered unevenly using epoxy resin. The use of epoxy resin presented particular difficulties in terms of restoring the fragments a second time.

Around half of the wall painting fragments had been reinforced from behind, and gypsum plaster thickly applied to attach them to their frames (Fig. 11.4). This type of framing prioritized their aesthetic appearance for sales-related purposes, however, with no consideration given to preservation. As a result, the wall painting fragments were not securely affixed within their frames, making them unstable. In some cases, they had been broken inside the frames and then repaired.

Flaking paint had been previously treated, probably with an animal glue. Around the edges of the fragments, the original plaster had weakened and collapsed, exposing the straw additives. Attempts had been made to repair the edges with an earthen filler

Fig. 11.4 The reverse of the wall painting fragment was fixed with gypsum plaster

of similar color to the original plaster. Joins between the fragments had also been repaired and colored to disguise them.

The thickness of the earthen plaster layers differs between the Bamiyan and Foladi sites. The bedrock in Bamiyan is a conglomerate rock, the cutting of which leaves a highly irregular surface. This required thick plaster layers to be applied in order to produce an even surface. In contrast, the primary support at Foladi is sandstone, which could be easily cut back to an even level, and consequently only thin earthen plaster layers were applied here. Despite these differences, two plaster layers were applied at both sites. The lower layers were composed of earth and straw, while the upper layers were mostly comprised of sand and earth, mixed with a small amount of straw.

The original paint materials of the fragments were examined using a variety of optical methods and contact-free investigative techniques, which were conducted jointly by the Tokyo National Research Institute for Cultural Properties and the Tokyo National University of the Arts between 2004 and 2005. In addition, projection radiographic images were taken to examine the structure of the earthen plaster layers.

11.2.2 Rationale of the Restoration Procedures

(1) **Removal of previous treatments**

The process of removing the previous mounts and frames and replacing them with a new system was conducted with the primary purpose of strengthening the wall painting fragments. In addition to changing the frames, a major component of the treatment was to consolidate the fragments to provide them with adequate reinforcement. In preparation for the new strengthening procedures, we removed as much as possible of the gypsum plaster that had been applied to the back of the earthen plasters during the previous restoration work. However, the gypsum was removed only from areas where it was determined that it presented a problem to the preservation of the wall painting fragments: for example, where it was unevenly applied across broken parts; or where it had been additionally compromised by other non-original materials or coloring agents. Even in these cases, the substances were removed in a manner so as not to damage the fragments.

In this project, within our objects, we have framed together for those fragments which can be conjoined to each other. However, there is a possibility of new findings, which can be other pieces of the same painting group. In such a case, it is necessary to take off our frame and restore it again. It was necessary that the process of dismantling the frames and removing the wall painting fragments be done in a safe and simple manner. While our restoration materials must be appropriate for the treatment of cultural assets, we also opted for easily obtainable materials, including with respect to the factor of cost. It is our hope that the methods of this project will become a model-case to be applied not only to wall paintings, but also to the treatment of other cultural objects, or in Japan, or in Afghanistan.

(2) **Stable and adaptable framing**

Since it was the intention that the wall paining fragments would be returned to Afghanistan for exhibition purposes, it was necessary for the frames to be stable and safe enough to allow for transportation and display.

In carrying out the restoration of the wall painting fragments in our care, we also had to anticipate that, in future, other fragments could be located that might need to be incorporated into the new framing system. Therefore, an adaptable system was required, whereby the frames could be easily and safely dismantled and reassembled, to allow access to the fragments when necessary. While restoration materials were used that are considered appropriate for the treatment of cultural assets, we also opted for easily obtainable materials, including with respect to the factor of cost. It is our hope, then, that these procedures will become a model to be used not only for wall paintings, but also in the treatment of other cultural objects.

(3) **Framing method that displays the sides of the wall painting fragments**

Previous methods of presenting wall painting fragments from Central Asia and China tended to concentrate on displaying only the painted surface. In this approach, the earthen plasters were removed, and the thin upper layer of painting was then lined and mounted on an artificial support, such as a honeycomb board, and any exposed edges concealed by fillers.

Removing the earthen plaster layers, however, means that the structure of the wall painting itself is lost—along with accompanying information there may have been regarding its fabrication. When considering the fact that wall paintings are a multi-layered structure comprising painting, preparatory and plaster layers, it is important that this entire stratigraphy is kept intact.

In order to preserve the original nature of the wall painting, framing should be carried out in a manner that allows the sides to be viewed. In addition, when considering their history of dislocation and removal from Afghanistan, it becomes clear that there is also significance in displaying the detached wall painting fragments with their sides exposed.

Exposing the full stratigraphy of the fragments, however, results in less available surface area for securing them, presenting the problem that further strengthening is required. There is a fundamental inconsistency, then, between the need to use a mounting system that ensures security and stability, and also exhibiting the wall painting fragments in a manner that reveals their structure. To address these conflicting requirements of both preserving and exhibiting the wall painting fragments, a mounting system is needed that provides appropriate reinforcement, and additionally secures them in a manner fitting the specific requirements of each individual fragment.

11.2.3 Treatment Processes

Because the state of damage and past treatment history of each wall painting fragment differed, retreatment materials and approaches were varied to match these individual circumstances. Here, we describe the major treatment methods and materials that were used.

(1) **Removing wall painting fragments from their previous frames**

The wall painting fragments were first taken out of their frames with the mounting brackets left intact; the mounting brackets were subsequently removed from the fragments.

(2) **Reinforcing the paint layers**

The paint layers, which had disintegrated and become fragile, were re-adhered and strengthened using a sprayed solution of HPC (H) 0.25~0.5wt% in ethanol. Special care was taken with respect to the level of adhesive reinforcement and also to avoid darkening the colors.

(3) **Surface facing (Fig.** 11.5)

Before treating the reverse side of the fragments, it was first necessary to protect the surface of the painting. A protective facing system was applied, comprising rayon paper saturated with naturally sublimated cyclododecane. This facing system was able to last from one two months.

(4) **Removing the gypsum plaster from the reverse side**

This was carried out mechanically, using an ultrasonic scalpel.

(5) **Removing the old filler**:

In order to maintain the original fragments in an undamaged and intact state, mud that had been used as a filler material in joins between the fragments was removed only where absolutely necessary.

Fig. 11.5 Covering the fragments with Rayon paper and cyclododecane

(6) **Removing non-original coloring**

Filling agents used to bridge joins between fragments had also been coloured as part of this treatment. The old fills were removed using an ultrasonic scalpel. The added colors that also extended onto the painting were removed only insofar as this did not cause damage to the original.

(7) **Reinforcing the earthen layers**

This was done using a 2~5wt% Primal E-330S ethanol dispersion (10wt% in particularly fragile areas). The penetration of this acrylic emulsion is improved by the inclusion of ethanol, making it ideal for reinforcing earthen layers.

(8) **Consolidation and repair of the reverse side**

The purpose of the consolidation and repair procedures was to strengthen the wall painting fragments, rather than to reconstruct them. The preliminary treatment involved an infusion of 10 wt% Primal E-330S ethanol fluid dispersion to reinforce the earthen plaster, which, after drying, was followed by repair using a filling agent. The filling agent comprised a mixture of phenolic microballoons with HPC(M) and Primal E-330S0, which was injected into indentations with a syringe; or was applied with a spatula to fill cracks. These procedures were also used on the sides of the fragments that seemed to be in danger of crumbling.

(9) **Additional strengthening of the reverse side (Figs.** 11.6, 11.7)

Despite the consolidation of the earthen plaster layers with Primal E-330S, they remained in a fragile state. To provide additional strengthening, triaxial fabric was applied as a facing layer. The triaxial fabric is a textile made from weaving yarn on three axes at an angle of 60°, which disperses its strength and gives it superior protection against rupturing and tearing. With its ability to retain a balanced level of strength in all directions, the fabric does not tend to become twisted in its application, allowing it to retain a stable form and shape for small loads. This fabric was highly useful for the purpose of reinforcing the reverse side of the wall painting fragments, with all of their surface irregularities.

Fig. 11.6 Lining with traxial woven rayon fabric impregnated with Paraloid B72 to the reverse side

Fig. 11.7 The reverse of the fragment after completing the lining

The fabric used was a 32-gauge triaxial woven rayon fiber, which was dyed a brownish-yellow color resembling that of the earthen plasters. Woven sheets of around 15 × 15 cm in size were impregnated with Paraloid B-72 (15% in acetone), and then allowed to dry. Pieces cut to 3 cm square were subsequently applied using a hot press to the sides and reverse of the earthen plasters, to which 10 wt% Paraloid B-72 in acetone had been previously applied. This lining system provided ample reinforcement of the wall painting fragments.

(10) **Surface fills and repairs**

The composition of the filler used on the surface of the wall paintings was the same as that used on their reverse side. It was not the purpose of the repairs to replicate the exact nature and appearance of the materials of the wall paintings. For uneven joins exposed by the removal of the previous fills, the new filler was applied using tools such as a syringe or spatula. The filler was applied to a level just below that of the painted surface, allowing a clear visual differentiation with the original painting.

(11) **Added colors (Retouch)**

Colors were not added to restore the iconography of the fragments. However, because the fillers used in the repair of the fragments had a distracting appearance, these were colored so that they did not negatively affect the aesthetic quality of the wall painting. Natural pigments combined with an 0.5wt% HPC (M) ethanol solution were used to create the added coloring. Its reversibility was exceptionally good, allowing for easy erasure if further repairs are required.

(12) **New mounts**

To ensure the long-term security and preservation of the wall painting fragments for their exhibition, it was necessary that they were fixed to stable mounts following their repair, in addition to framing them. Since the earthen plasters were left intact, the fragments were heavy. As a consequence, the mounts and frames had to be kept as light as possible, while also providing adequate reinforcement.

The mounts were made by adding well-blended methylcellulose (viscosity, 400cps) and Primal AC-2235 to a repair mixture of diatomaceous earth, clay, finely cut acrylic fibers and glass micro balloons, and applying this formulation in layers to the desired thickness. The triaxial fabric was sandwiched between the layers to provide reinforcement. Finally, a cosmetic mortar was applied to the mounts, which were fixed to the wall painting fragments using silicone resin.

(13) **Framing (Figs.** 11.8, 11.9, 11.10)

The frames have a dual-layered structure, with inner and outer sections. The outer frame incorporated an acrylic sheet to filter out ultraviolet rays, in order to protect the surface of the paintings during exhibition. To ensure ease of viewing when the wall paintings are exhibited in flat display cases, non-acrylic sheets were utilized for the inner frames. The frames can be opened and closed exclusively by means of screws affixed to the sides, which increases the safety of handling due to the ability to

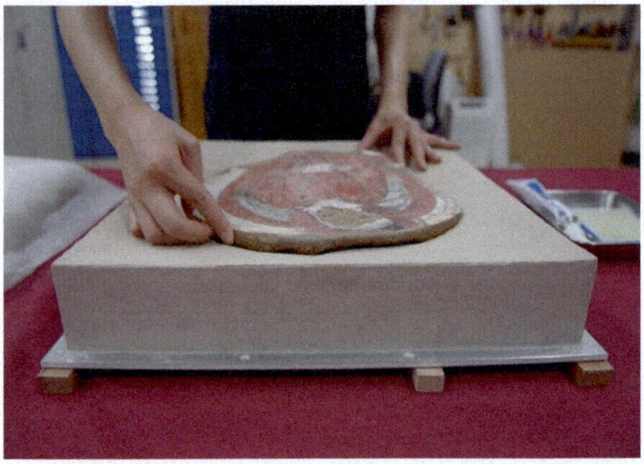

Fig. 11.8 Placing the fragment on the mount as the sides remain visible

Fig. 11.9 After mounting and framing

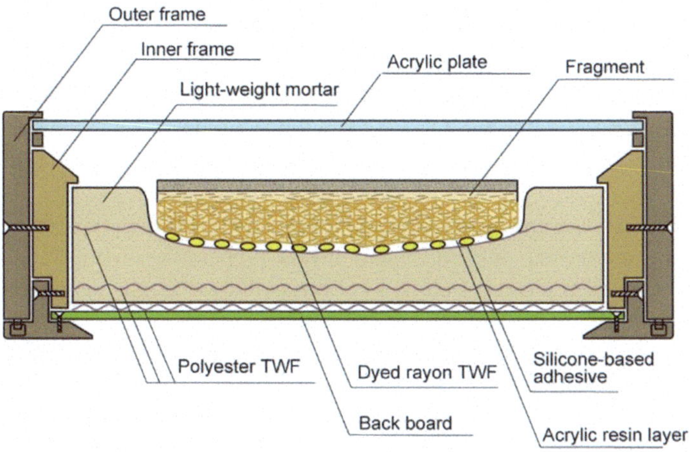

Fig. 11.10 Frame structure diagram

manage them while they are in a flat position. In addition, this permits investigation and photography of the inner frames for research-related purposes.

11.2.4 General Project Overview

The 42 wall painting fragments from archaeological sites in Bamiyan and Foladi were studied and restored over a three-year period that began in 2007.

In addition to photographs taken in normal light, the investigation also involved detailed examination using ultraviolet fluorescent photography, infrared reflectography, X-ray transparent photography, and imaging with digital microscopes. With these techniques, we gained tremendous understanding of the wall painting fragments in terms of their original materials and techniques, and layer structures, as well as of aspects of their condition, such as the presence of fissures and gaps, and the state of their previous mounting systems.

Findings of the investigations helped to define the full range of preservation and restoration procedures, such as how best to dismantle the old frames and mounts; the procedures that were necessary to preserve the wall painting fragments' outer appearance and condition; and how to devise new framing and mounting methods for public display-related activities such as exhibition and transportation of the wall painting fragments. Indeed, the new mounts and frames were designed to fulfill stringent display requirements, incorporating custom-made features. In addition, the restoration process allowed us to obtain valuable insights into the behavior and properties of the treatment materials used, such as certain solvents and adhesives. Key aspects of the remounting process can be summarized as follows:

(1) Preserving the integrity of the wall painting fragments

The wall paintings were executed on earthen supports, but have been dislocated from their original contexts by looting on the one hand, and by emergency archeological excavation on the other. Given these circumstances, the fundamental conservation-related question remains: In what way should these wall paintings best be preserved?

As described above, most approaches to restoring detached wall paintings involve preserving and mounting only the upper layers of painting, while removing and discarding the earthen plasters below. In our view, however, the entire stratigraphy of the wall painting, including the earthen support layers, should be preserved. If the earthen layers are removed, this erases contextual information about the original setting of the wall painting fragments, such as the time period when they were created, which can be revealed through the preservation of certain application techniques or by features such as remaining accumulations of straw, etc.

In the process of removing the previous mounting materials, interesting original features were revealed. We discovered that cow manure was additionally added to the earthen plasters, presumably to increase their adhesiveness to the rock layer [5]. The wall painting fragments from the Foladi site also incorporated different ratios of broken rock in the two plaster layers. This was in contrast to the wall painting fragments from the Bamiyan site, resulting in plasters of quite different color and performance, for example in terms of the way in which they had separated from their rock supports.

Precisely in order to preserve this type of valuable information, we did not remove the earthen layers from the wall painting fragments. In addition, we aimed to research and develop the use of minimum restoration materials and techniques, so as not to alter the fragments' authentic appearance and shape. By analyzing the structure of the wall painting fragments, we were able to develop new methods for restoration and mounting that enable them to be displayed, and, if required, also studied. During the treatment process, new application procedures were developed, such as improving adhesive performance by the addition of specific solvents.

(2) Mounting and framing for exhibition purposes

After being returned to the National Museum of Afghanistan, the wall painting fragments were intended to be put on display. This placed several requirements on their mounting and framing, as follows:

- The mounts and frames must preserve the wall paintings, and be reversible and able to be dismantled.
- Because the wall painting fragments are research documents, the framing must be done in a way that permits in situ study, including photography and non-contact analysis.
- In order to preserve the wall paintings, the framing must not permit contact with harmful agents such as ultraviolet rays.
- Both wall exhibitions and flat-display exhibitions must be permitted.

As the frame structure and the methods for mounting and strengthening the wall paintings were interrelated, we had to consider all these factors in terms of one composite whole. The mounts were affixed to the back covers as one unit, which were designed to be removed from the frames altogether. With the structures laid out flat, we then affixed detachable back covers to the frames from the top using screws.

If it proves necessary to remove the wall painting fragments from their mounts, the system allows this to be done easily and safely. First, mortar used in their fixing is removed, enabling each fragment to be taken out as a whole, cushioned and strengthened by the polyester fiber-based triaxial woven fabric. For small and fragile wall painting fragments, such as those from the Foladi archaeological site, we added reversible layers between the lining on the reverse of the fragments and the surface of the mount. These layers, could be dissolved by a solvent that had no effect on the lining on the reverse of the fragments, thereby enabling the fragments to be safely taken off of their mounts while leaving the reverse lining intact.

Few examples exist in Japan of wall paintings on earthen plaster layers, and there have been no cases to date involving their comprehensive preservation and preparation for exhibition. Our restoration project demonstrated a completely new set of methods and procedures for preserving detached wall paintings, which respected their physical integrity, from their painted surface to their underlying plaster layers.

11.3 Restoration of the Fragments from Mes Aynak

The Mes Aynak archaeological site is found in the location of a Buddhist city, which flourished between the third and seventh centuries A.D. It is known that the city was influenced by the spread of Greek civilization via the expeditions of King Alexander, and it is an immensely important site for researching the cultural exchanges that occurred between the east and west in this period. Mes Aynak sits above untapped copper deposits, and plans are being made to carry out extensive open-pit mining at the site, leading to fears of the imminent extinction of its archaeological ruins. Consequently, the Mes Aynak archaeological site is currently undergoing emergency excavation, but due to the extraordinary large-scale nature of its ruins, completing this rescue work before the mining development project commences is an impossibility.

UNESCO is therefore planning to relocate the Mes Aynak site, for which the conservation of the excavated artifacts is an urgent task [6]. The archaeological objects unearthed at Mes Aynak were transported to the National Museum of Afghanistan, located in the country's capital, Kabul, but its conservation infrastructure is not adequate. Consequently, the Tokyo National University of the Arts began assisting with the task of preservation in 2016, and a research team of painting conservators was put in charge of investigating and restoring one wall painting fragment and two Buddha figure heads, beginning in November 2017.

To support sustainable conservation initiatives at the site, we helped with efforts at the museum to procure conservation-related materials, and to train human resource professionals in the field of conservation [7]. This report, however, focuses on the wall painting fragment restoration.

Through scientific-based inquiry we gained significant knowledge of the structure and materials of the wall painting fragment. In addition, the project was greatly informed by the approaches to restoration, mounting and framing involved in the previous case study.

11.3.1 Condition Prior to Restoration

The dimensions of the wall painting were 400 mm at the top, 1170 mm at the bottom, 790 mm on the right side, and 430 mm on the left side; its thickness was 21 to 35 mm (including the painted layers). Numerous layers of the wall painting fragment had been stripped away in its removal, and the entire underside had been covered with a thick layer of supplementary mortar. The painted surface and underlying plaster were badly damaged in five different places, and the edges of the damaged areas were extremely brittle and in danger of disintegrating.

The sides of the wall painting fragment, and the earthen plaster below the upper paint layers, had also disintegrated, and numerous internal gaps had formed. A number of previous procedures had been undertaken to strengthen the underside of the fragment, using materials including mortar, gauze and triaxial woven fabric, together with several adhesives including acrylic resin emulsions, Paraloid B-72 (35% in acetone), and epoxy resin. Aluminum bars of 5 mm thickness had also been randomly placed across the underside, adhered in place with substantial quantities of epoxy resin.

Information on these materials and procedures was provided by staff at the National Museum of Afghanistan in Kabul, who had undertaken the work. Much of the previous treatment was concentrated on the underside of the fragment (Figs. 11.11, 11.12), but the edges were in an exceedingly fragile state and were on the verge of disintegration, making transportation of the fragment difficult and risky. In addition, fibers from the gauze and the triaxial woven fabric that had been applied to support the underside were protruding in numerous places, which were worrying signs of failure. It was also feared that the epoxy resin used to adhere the aluminum bars to the underside was also liable to failure. In addition to reinforcing the weakened areas, it was therefore necessary to consider preservation methods to prevent damage to the earthen layers during transport.

For its protection, Japanese paper and gauze had been adhered to the surface of the wall painting fragment using Paraloid B72. The acrylic resin had penetrated into the painting and earthen layers, and it seemed likely that the facing materials would be difficult to remove.

Local attempts to remove the gauze and Japanese paper from the left half of the painting had resulted in cracking, peeling and powdering of the paint layers. Some paper that had been partly peeled away had also taken paint with it. Reportedly, these

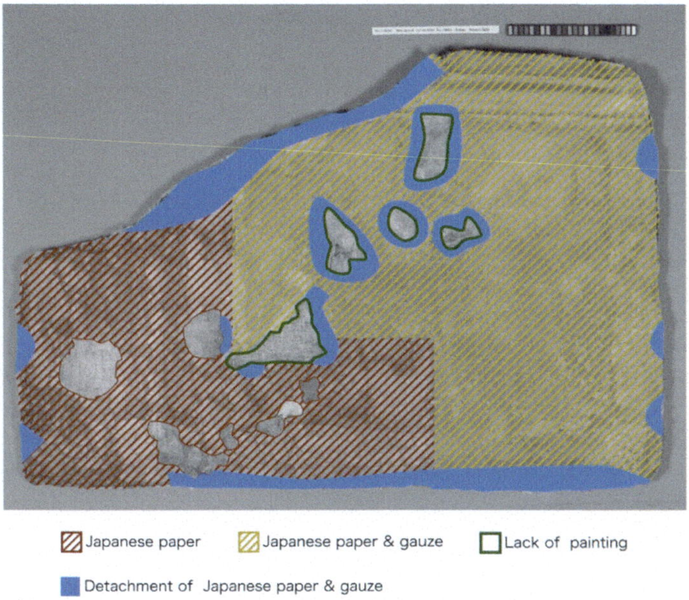

Fig. 11.11 Showing condition and damage prior to restoration

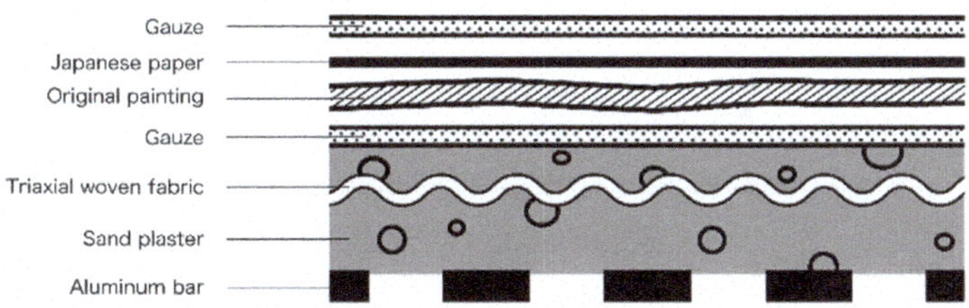

Fig. 11.12 Diagram showing the structure prior to restoration

mishaps occurred after absorbent cotton soaked in acetone was placed on the facing materials for 20 min in order to try to remove them.

Considering that the use of acetone to remove the facing materials adhered with Paraloid B72 was probably unavoidable, it was necessary to consider a procedure that could be carried out with great precision and care, and thereby cause minimal damage. Although it was difficult to observe the colors and the condition of the wall painting due to the facing materials that covered it, fluorescent X-ray analysis allowed us to deduce some of the pigments that had been originally used. Since Fe (iron) was detected in areas of red painting, and Cu (copper) in the green painting, it may be speculated that iron oxide-based and copper mineral pigments were both used.

11.3.2 Rationale of the Restoration Procedures

Based on a preliminary inquiry and consultation with responsible officials at the National Museum of Afghanistan in Kabul, it was decided that the restoration procedures would involve the following:

(1) **Use of restoration materials that could be procured in Afghanistan**

It is difficult to source restoration materials in Afghanistan that are appropriate for use on cultural objects. Moreover, because conservation education and reliable human resources are both lacking in the country, there has been no research of technologies for treating wall paintings.

Since an aim of our project was to enable local museum staff to undertake similar restoration procedures unilaterally in the future, we selected materials that could be locally procured wherever possible. In addition, we invited relevant professionals from the National Museum of Afghanistan to attend courses, where they could participate in exchanges regarding the technologies and materials of restoration.

(2) **Safe removal of facings and other previous treatments**

As mentioned above, protective facing materials were applied to the wall painting fragment when it was sent from the site to the National Museum of Afghanistan. Removing these was a priority, but careful consideration had to be given to methods that could be implemented with precision and safety, so as to avoid damaging the painting. In addition, it was necessary to remove previous fills that had been made in the fragment, while taking care not to damage it; and to replace these with fills that were more appropriately colored and textured.

(3) **Devising a frame for secure transport, exhibition, and storage**

Following its restoration, it was intended to exhibit the wall painting fragment at the National Museum of Afghanistan in Kabul. For this reason, framing had to be carried out in a manner that facilitated ease of handling, as well as providing stability for its exhibition. Because the wall painting fragment was already quite heavy due to the mortar and aluminum bars that had been previously applied to its reverse side, it was important that the materials selected for the framing should be as light as possible. A removable frame was also necessary, moreover, to accommodate the possibility that changes might be required in the future.

11.3.3 Treatment Processes

(1) **Removing the gauze (Figs.** 11.13, 11.14)

This was accomplished by dabbing acetone-soaked absorbent cotton onto the surface of the gauze, and then slowly dissolving the Paraloid B72 that had penetrated into the wall painting fragment. After the Paraloid had softened, the gauze was

Fig. 11.13 During removal of the gauze

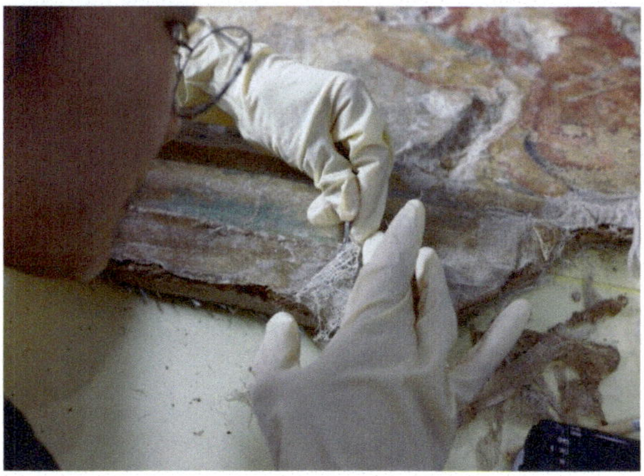

Fig. 11.14 Peeling off the gauze carefully with tweezers

carefully peeled away. To avoid the gauze re-adhering as the solvent evaporated, the best course of action was to cut it away with scissors gradually during the removal operation. Fortunately, since the layer of Japanese paper protected the painting, the gauze did not come into direct contact with the painting layers. The process of gauze removal was therefore relatively easy.

By using solvent-resistant gloves, masks with activated charcoal, tweezers to handle the absorbent cotton, and exhaust air ducts, we were able to maintain a safe environment for both the restoration work and human health.

(2) **Strengthening of the painting and earthen materials (Fig.** 11.15)

After removing the strengthening gauze from the surface of the fragment, a Primal AC-2235 ethanol dispersion (2.5wt%) was injected by syringe to consolidate the painting and earthen layers. This use of ethanol to enhance the penetration of the Primal AC-2235 was based on the experience gained in the treatment of the Bamiyan wall painting fragments, described above.

Fig. 11.15 Inserting Primal AC2234 by syringe

As well as consolidating the painting and earthen layers, treatment with Primal AC-2235 was also intended to address problems of separated layers within the fragment. Injections were therefore made under the edges of delaminated paint layers. Because the earthen plaster was particularly fragile in these areas, and was also in danger of being lost, the density of the Primal AC-2235 solution was increased from 2 to 5%, and introduced repeatedly but gradually.

(3) **Removing the Japanese paper** (**Fig.** 11.16)

Absorbent cotton soaked with acetone was applied to dissolve the Paraloid B72 that had been used to adhere the Japanese paper to the surface of the painting. A solution of 5wt% Primal AC-2235 ethanol dispersion was used for consolidation during the process of carefully peeling away the Japanese paper using tweezers.

In areas where attempts had already been made to remove the paper, less acetone application was required.

(4) **Modifying chromatic alterations**

Paraloid B72 forms a coating on the surface of a wall painting, and when it is applied in large amounts, a glossy sheen is created. This happened in the treatment

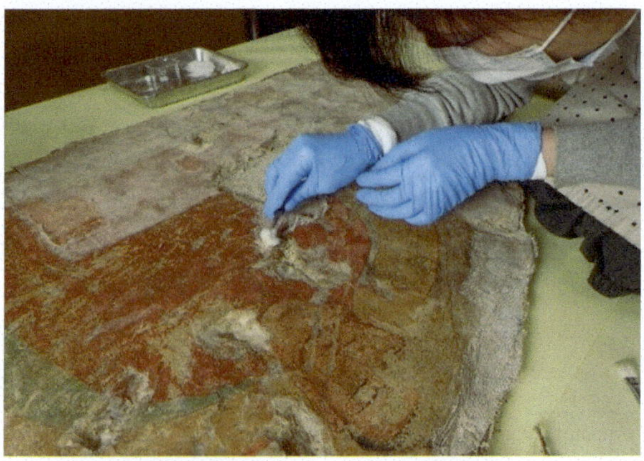

Fig. 11.16 Dissolving Paraloid B72 using absorbent cotton soaked with acetone

of the wall painting fragment after its emergency excavation. Previous efforts to remove the facing materials using large amounts of acetone had resulted in chromatic alterations of the paint surface, especially on the left side of the fragment. When restoration work was initiated, substantial flaking of the paint layer also occurred. To even out the chromatic irregularities in the paint surface and provide further reinforcement, a solution of 5wt% Paraloid B-72 was applied three or four times.

(5) **Restoring the underside and removing the old filler**

Previous treatment of the underside of the fragment had included adhering triaxial fabric with large amounts of epoxy resin. Much of the hardened fabric was protruding from the fragment. Excess materials that were obscuring viewing of the fragment were carefully cut away. In addition, a fill used to repair the surface of the fragment, which appears to have been made from gypsum and clay, and which had spread onto the paint layer, was gently removed using an ultrasonic scalpel.

(6) **Filling and shaping**

Past treatment efforts had resulted in damages and fissures of various types and sizes that undermined the earthen plasters. Because these damages included areas of separation that appeared to present structural dangers, as well as impeding appreciation of the wall painting, these were treated with a filling agent.

Based on trials, the filler was applied in successive layers, as follows: a foundation layer (filler A); and the insulation and cosmetic layers (filler B). To ensure that the filler reached inside small crevices and gaps, Primal AC-2235 or HPC(M) (hydroxypropylcellulose) was dispersed or dissolved using ethanol for filler A.

Glass micro-balloons were selected as a filler component due to their light weight, small size, and ability to contribute to good fluidity. For the large gaps inside and around the edges of the wall painting fragment, filler A was inserted using a syringe. In order to provide a good key for filler B, the surface of filler A was lightly scored with a scalpel. A 15wt% Primal AC-2235 aqueous dispersion was applied to provide insulation at the interface between the two fillers.

Juraku and Kanuma soils were used as components in the filler B mixture. The Juraku soil, which originates in the Jurakutei area of Kyoto, was used in Japan during ancient times as a preferred finishing material for the walls of tea-houses. Kanuma soil, which comes from the city of Kanuma in Tochigi prefecture, is a yellow-tinged pumice with excellent aeration qualities. This was used after grinding with a pestle and mortar. Filler B was then applied as a thin cosmetic layer on top of the insulation filler.

Trials were carried out to determine the best color tone and texture of the surface coating. When the filler was too fluid, this resulted in an inappropriate surface appearance. Combining small amounts of methylcellulose with the Primal AC-2235 aqueous dispersion formed a more viscous mixture, which could be easily spread by brush. A small *tataki-bake* brush was used to achieve a uniform surface finish. The upper filler was applied slightly lower that the level of the original painting (Table 11.1).

Table 11.1 Filler mixtures

Materials		Filler A (parts by volume)	Filler B (parts by volume)
Fillers	Glass micro-balloons	9	
	Juraku soil		3
	Kanuma soil		1
Binders	15wt% Primal AC 2235 ethanol dispersion	3	
	15wt% Primal AC 2235 aqueous dispersion		optimal quantity
	0.2wt% HPC (M) ethanol solution	1	
	5wt% methylcellulose aqueous solution		small quantity

(7) **The mounting system** (**Figs.** 11.17, 11.18)

For mounting the fragment, a sturdy 20 mm-thick polymethacrylimide plastic closed-cell, foam-body structure (ROHACELL®) was selected, due to its lightweight and high-strength performance. For additional reinforcement, a 3 mm-thick aluminum plywood segment was fixed with epoxy resin to the underside of the fragment. Aluminum plywood consists of closed-cell foam resin sandwiched between aluminum plates. In addition to its lightweight qualities (with a specific gravity of

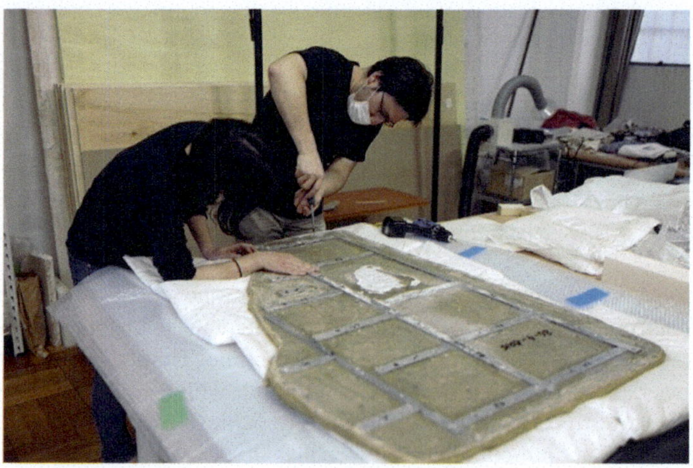

Fig. 11.17 The firmly adhered aluminum bars of previous treatment were utilized for fixing the fragment with mount

Fig. 11.18 Adhering ROHACELL on to the aluminum board

0.8), it is also smooth, and has superior heat-resistance and durability. The ROHACELL® was lightly shaped using sandpaper to accommodate the uneven underside of the wall painting fragment. After fixing the mount, exposed parts of the fragment were then finished with a cosmetic mortar. This was designed to be close in configuration and texture as the original plaster, but also easily distinguishable from it. A mixture was created using gypsum dihydrate with Kanuma and Juraku soils. A small amount of methylcellulose was added as a finishing layer to the surface of both the mortar and the filler.

As part of the stabilization measures, 49 holes for screws were drilled in the mount and the bottom of the wall painting fragment. In addition, holes were drilled into the foam-based ROHACELL®. To minimize the impact of these interventions, holes were drilled close to the supplementary aluminum bars. Drilling was done just deep enough to pass through the bars, as well as through the mounting, after marks were made on the back of the wall painting fragment to ensure their correct alignment. The radiuses of the holes were slightly smaller than the screws (Table 11.2).

(8) **Framing (Stabilizing the Frame) (Figs.** 11.19, 11.20, 11.21)

A wooden (oak) frame was chosen in order to enhance the aesthetic appearance of the wall painting fragment. Its thickness was set at 85 mm, a little greater than the fragment placed inside it. To ensure the safety of the painting during international transport and exhibitions, as well as to enable its easy removal and replacement, the fragment was secured to the mount, and the mount to the frame, with screws. To avoid misalignment of the screw-holes during framing, the wall painting fragment was positioned on its mount inside the frame, and all of the screw-holes were drilled together at the same time. To fix the wall painting fragment to the mount, stainless

Table 11.2 Composition of the cosmetic mortar mixture

Ingredients		Parts (volume)
Fillers and binders	Gypsum dehydrate	4
	Juraku soil	3
	Kanuma soil	1
Adhesives	5wt% Primal AC 2235 aqueous solution	as required
	5wt% methylcellulose aqueous solution	small quantity

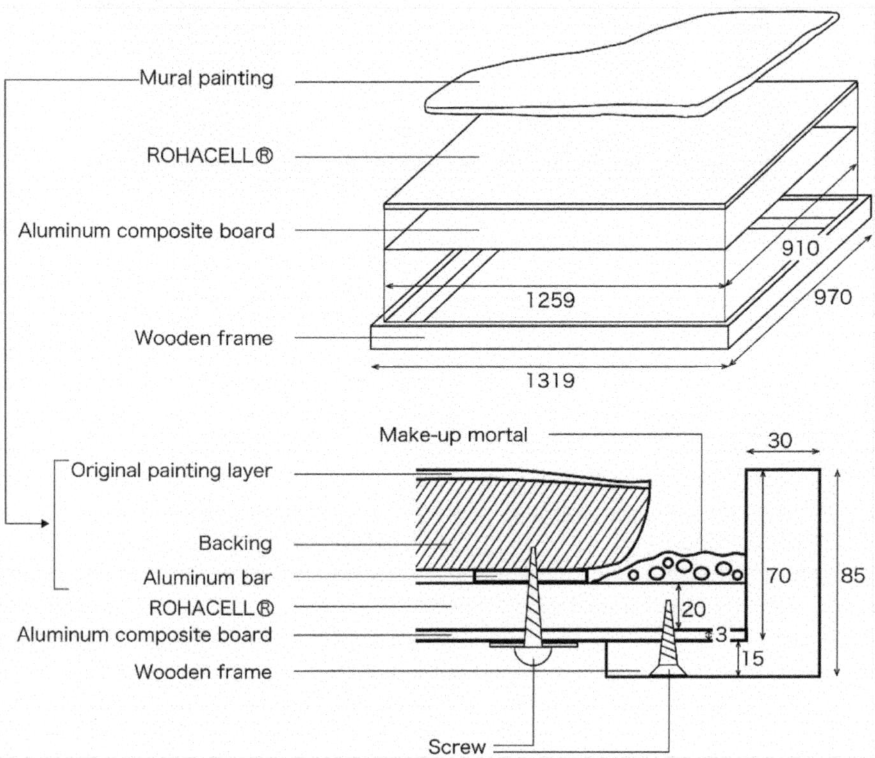

Fig. 11.19 Diagram of the framing structure (unit: mm)

steel pan-head screws (sizes: 25 mm, 30 mm and 35 mm) were adjusted to accommodate the surface irregularities of the reverse side of the fragment. To ensure that the framed wall painting fragment could be laid down flat, stainless steel flat-head screws (size: 25 mm) were used to fix the mount to the frame

Fig. 11.20 Positioning the fragment on the mount

Fig. 11.21 After mounting and framing

11.3.4 Successes and Challenges

This initiative to restore a wall painting fragment that had been displaced from an archaeological site in Afghanistan, and which had been neglected without proper treatment, was without precedent.

The National Museum of Afghanistan houses numerous wall painting fragments that, like the case study reported here, have only been partly restored. Anticipating that museum staff will need to carry out further restoration work in future, the project was planned on the basis of using equipment and materials that could be procured in Afghanistan. As a result of the approaches adopted for this project, we were able to restore the wall painting fragment to a state of preservation and stability that will enable its transportation and exhibition.

In November 2017 and March 2018, while the work was underway, two staff-members from the National Museum of Afghanistan traveled to Japan, and we were able to hold very meaningful exchanges of opinion. A central topic of discussion was the nature of the previous restoration work, which had been carried out under extremely constrained conditions using very limited materials, and how to deal with these. The two main areas of concern were: (1) the surface of the wall painting fragment, which had been faced with gauze and Japanese paper; and (2), the reinforcement of the reverse side of the fragment using gauze, mortar, triaxial woven fabric, and aluminum bars. Different treatment approaches were adopted to address these concerns. For the surface of the painting, complete removal of the facing materials was necessary, since these obscured the painting and prevented its appreciation. Regarding the strengthening materials applied to the reverse of the fragment, it was too risky to remove these and instead the security of the wall painting fragment was prioritized.

11.3.5 Summary Conclusions

The restoration work carried out on this project involved a comparatively large wall painting fragment (around 1.2 m). This, and the fact that it had been excavated under emergency conditions and then poorly treated, made our work considerably more challenging than any we had engaged in previously.

That we were able to restore the wall painting fragment to a state where it may be safely exhibited, and the substantial contribution to international cooperation in cultural preservation that this represents, are the resounding achievements of this project. The experiences that we gained in the treatment of the fragments from Bamiyan and Foladi were extremely useful in this process, allowing us to experiment with various materials and application methods over the course of the restoration work to reach a successful outcome.

References

1. Kijima, T.: Displaced cultural treasures: Preserving and restoring Bamyan Buddhist mural painting fragments (displaced cultural assets, Silk Road, Foundation for Preserving and Revitalizing Cultural Treasures, No. 56) (2008)

2. Kijima, T., Sato, I., Kudo, H., et. al.: Study and restoration of Buddhist mural paintings displaced from Afghanistan, Tokyo University of the Arts. Art Dept. Bulletin. 47, 113–136 (2010)
3. Iwai, S., Taniguchi, Y., Aoki, S.: Conserving wall paintings, protecting the World Heritage site of Bamiyan: Key issues for the establishment of a comprehensive management plan 2004, NRICP, Japan (2004)
4. Yamauchi, K. (ed.): Preliminary report on safeguarding the Bamiyan site, Recent cultural heritage issues in Afghanistan, preliminary report series, 3, Ministry of Culture and Information, National Research Institute for Cultural Properties, Tokyo (2006)
5. Taniguchi, Y.: Era of mural painting creation, and basecoat layer including straw, Afghanistan Cultural Property report No. 2, pp. 29–31 (2006)
6. Curves, H.H.: Mes Aynak (Afghanistan), global standards and local practices. Post-conflict Archaeology and Cultural Heritage, pp. 263–281 (2018)
7. Aoki, S.: Preservation of archeological artifacts and training personnel at Mes Aynak, Afghanistan, Tokyo University of the Arts Public Collaboration Center Research Report, pp. 33–418 (2018)

Chapter 12
Deterioration and Conservation of Wall Paintings in Uzbekistan

Otabek Aripdjanov

Abstract Uzbekistan lies at the heart of Asia where the western flanks of the great mountain massifs of the Tien Shan and the Pamirs meet the vast steppes and deserts. Watered by the lengthy inland river systems of the Amu Darya, the Syr Darya (the Oxus and Jaxartes of antiquity) and the Zarafshan, all fed by the melt from mountain snows and glaciers, its croplands, orchards and pastures nurtured the development of important centres of civilization. The nation's extremely rich and ancient history provides valuable insights regarding the general processes of mankind's social and cultural development, as well as those specific to the unique cultures of the region with multiple archaeological sites. The most important issue confronting conservation professionals in Uzbekistan is the preservation of wall paintings found at archaeological sites. Their conservation will allow us not only to preserve these unique samples of ancient material culture and art, but also to study the working methods of ancient artists and craftsmen, and the composition of their original materials, This will help to determine the unique features that characterize artistic development in antiquity in the territory of Uzbekistan. This paper provides an overview of general developments in conservation and restoration practices of wall paintings, and proposes future research directions. It also describes developments in training in the conservation of wall paintings being carried out in the State Museum of the History of Uzbekistan with colleagues from Japan and the USA.

Keywords Uzbekistan · Wall paintings · Conservation · Restoration · Cultural heritage · Training

12.1 Introduction

The preservation of wall paintings at archaeological sites in Uzbekistan has been carried out by specialists from various fields, such as artists, archaeologists and sometimes chemists. However, most of this work was episodic in nature, since it was

O. Aripdjanov (✉)
State Museum of History of Uzbekistan, 3, Sharov Rashidov Street,
Tashkent 100029, Uzbekistan

S. Aoki et al. (eds.), *Conservation and Painting Techniques of Wall Paintings on the Ancient Silk Road*, Cultural Heritage Science,
https://doi.org/10.1007/978-981-33-4161-6_12

done in connection with certain unique archaeological discoveries. Remedial treatments have mainly relied on the use of the widely accepted consolidation polymer, polybutyl methacrylate (PBMA), which was introduced into treatment practice in the 1940s (Fig. 12.1).

In the second half of the twentieth century, PBMA found wide application in the post-Soviet Union States, in particular in Central Asian countries, in the conservation of archeological paintings on earthen supports, as well as of other types of murals [1]. PBMA was chosen because of its good solubility parameters in a wide range of solvents, its workability in different concentrations, and its perceived ability to be used without causing colour changes in original materials. It has therefore been used to perform deep impregnation and consolidation of wall paintings so that they could be detached and removed from excavation sites, or, alternatively, it facilitated their conservation in situ. PBMA was introduced into the region in 1949 by P. I. Kostrov, an employee of the State Hermitage Museum [2]. Over the following decades, it became the standard material used in archaeological conservation.

After Uzbekistan became an independent state in 1991, exchanges intensified in the protection and preservation of the country's unique tangible and intangible cultural heritage. In the study and preservation of wall paintings, the restoration department of the Institute of Archeology of the Academy of Science of the Republic of Uzbekistan plays an important role. For many years, in collaboration with foreign colleagues, it has been responsible for carrying out major conservation-restoration

Fig. 12.1 Major archeological sites located in the territory of Uzbekistan

works. Special mention should be made of the head of the department, M. A. Reutova, who was principally responsible for the preservation of wall paintings from Afrasiab, Balaliktep, Karatepa and other sites. Advances in equipment and techniques, and the sharing of knowledge and experience, have been promoted by international cooperation.

12.2 Discovery of the Wall Paintings

12.2.1 Balalyktepa

Archaeological research in different territories of the Republic of Uzbekistan resulted in the discovery of many monuments that preserved painting on earthen supports. A majority of these were found in the region of Surkhandarya, in the southern part of Uzbekistan. Among the discoveries, the wall painting found at the site of Balalyktepa (Angor district) by L. I. Al'baum in 1953–1954, which dates from the end of the fifth and the beginning of the sixth centuries AD, is extremely interesting in terms of its artistic conception, content and technical proficiency [3].

The painting depicts a feast scene in which men and women sit on rugs (Fig. 12.2). The composition can be divided into separate figure-groups. In the foreground are large figures dressed in richly ornamented clothes, who delicately hold cups and bowls by their fingertips. In the background, servants are shown in a smaller scale. Judging by their clothes, both men and women servants are present, and most of

Fig. 12.2. Fragment of wall painting from Babayktepa in the Surkhandarya region

them hold a fan. Behind the figures, a wall is depicted on which are hung rectangular, square and round objects in white, some decorated with traces of drawings. At the top of the composition, under the eaves and at the level of the servants' heads, multi-colored circles are depicted in different sizes and distances from each other. Undulating coloured ribbons with bells are attached to the center of these circles. This description provides only a general overview of the painting, which belies the detail and interest of a composition that preserves some 47 figures, arranged in distinct groupings.

12.2.2 Afrasiab

The early medieval paintings at Afrasiab at the ancient site of Samarkand are well known to specialists, but as is often the case, general knowledge of them is lacking. The first wall painting at the site was discovered by the historian and regional expert, N. Viatkin, in 1913 [4], but the main body of paintings was found only in 1965–1967, when they were investigated by L. Al'baum [5] (Fig. 12.3). Discovered by chance in the course of road building, they are unique in many respects. The paintings were found on the walls of a domestic building, probably a kind of palace, and their subject matter is secular. The themes include hunting, crossing a river and a solemn procession, but the principal scene is the arrival of emissaries of various countries at the court of Varkhuman, the ruler of Sogdia (Fig. 12.4). In it, the inhabitants of various Asian regions, namely Sogdians, Turks, Iranians, Chinese, Koreans, and envoys of mountain tribes wearing their ceremonial dresses, are shown side by side. This true-to-life scene has immense documentary and ethnographic importance, providing unique knowledge of socio-political interactions in early-medieval Central Asia. Moreover, the scheme is the only figurative one known with an inscription, which enables it to be dated to the second third of the seventh century AD. One of the chief features of this Sogdian inscription is that it contains the address of the Chaganian

Fig. 12.3 The scholar L. Al'baum explains the early medieval wall paintings discovered at Afrasiab, Samarkand

Fig. 12.4 Detail of the wall painting from Afrasiab showing part of a great procession

ambassador to Varkhuman, the ruler of Samarkand, in 655, which is confirmed by Chinese sources. These wall paintings are not the only type of their kind. The early-medieval paintings at Balalyktepa, Varakhsha and Penjikent were already known and are similar. However, the Afrasiab paintings are nonetheless unique [6].

12.2.3 Karatepa

Excavations of a large religious centre at Karatepa, near Termez, a "re-transmitter" of Buddhism to Central Asia, China, Korea and Japan, revealed a site that can be regarded as having made a major contribution to artistic and spiritual culture. Archeologists of Uzbekistan together with Japanese researchers studied the monumental complex, which can be dated to the first–fourth centuries AD. Its cultic buildings are remarkable not only for their size, but also for their innovative architectural and construction concepts, and the richness and elegance of their interior decoration. The walls of the monastic rooms were decorated with colourful paintings and furnished with clay-based sculptures. The paintings contain scenes and characters from the Buddhist world, including depictions of the Buddha, bodhisattvas, ascetics, monks and mythical birds and animals. Paintings of secular persons, who are represented in dynamic and very expressive styles, are no less interesting. Discovered objects from the site testify to the highly developed nature of the original culture of ancient Bactria, which had much in common with Gandhara-Buddhist and Greek-Roman art. Each object discovered at Karatepa has unique value, demonstrating that in the first–fourth centuries AD, the highest class of artists lived and worked in the territory of Termez. Their artistry delights viewers even after the passing of thousands of years.

In order to clarify the architectural development of the western half of the monastic complex, archaeological research was begun in 2016. During excavations in one of the rooms, a colourful wall painting scheme was discovered [7] (Fig. 12.5). The plan of the complex has not yet been precisely determined, but based on investigations carried out to date, there was a principal hall of at least 25 m in length. Painting was found on the east and south walls of this hall, preserved to a height of more than 250 cm. Unfortunately, some areas were spoiled by the burrowing of termites and rodents, which means that certain details and some whole images cannot be accurately identified. Nevertheless, in comparison with other ancient wall paintings in Central Asia, those at Karatepa can be considered to survive in reasonably good size and condition. Prior to this discovery, the presence of wall paintings at Karatepa was only known from fragmentary remains. Only one painting, from the ground-level cave "B" in the area of the Southern Hill, showing an image of the Buddha and monks, was distinguished by its somewhat better preservation, but it was also small and fragmentary.

The hall to which the paintings belong is dated to the second–third centuries A.D. In this range, a later date for the painting is confirmed by coin finds. When the hall was destroyed, debris from its ceiling and upper walls was used in the construction of a nearby dam, in which was discovered two minted coins, from the period of Kanishka III. According to most researchers, the circulation time of these coins belongs to the second half of the third century AD.

We hope that further investigation of the great hall at Karatepa will reveal more wall paintings that help expand our knowledge of the rich and diverse artistic culture of Bactria in the Kushan era, which certainly holds a worthy place in the history of the culture of world civilization.

Fig. 12.5 Fragments of wall painting from the southern part of the northern hill of the Buddhist temple of Karatepa

12.2.4 Tavka

Archaeological excavations in the territory of early-medieval Tokharistan have yielded the richest material for characterizing its unique artistic culture. The principal archaeological resource relates to local Buddhist art, but one important archaeological find sheds light on secular artistic practice. This is a wall painting from the fortress of Tavka, which acted as a customs house in the fortification system of Nonidakhon, located to the south of the Iron Gate, close to the exit from the valley of the Sherabad Darya, Sogd. The painting was first discovered by E. Rtveladze in 1987, and subsequent excavations were made at the site from 1989 to 1993 by Sh. Rahmonov. Although news of the wall painting discovery spread at once, its official scientific publication did not occur for another ten years. Eventually, first in an article written by Sh. Rahmonov and M. Reutova, and then in a monograph written by Sh. Rahmonov, the archaeological findings of this interesting monument were published. Some fragments of the wall painting are now on exhibition in the Termez Regional Archaeological Museum, providing important information on the function of the fortress and the nature of its original decoration.

The excavations at Tavka enriched archaeological knowledge of Tokharistan, as well as confronting scholars with some important questions. These firstly relate to the attribution of the monument and its wall painting. Previous identifications of the fortress as a Manichean temple seem not to be correct, as these were made on the basis of very scant evidence. Instead, as a site with natural defensive features, it was an ideal choice for the location of a fortress, in which a director of customs had a residence. The daily needs of a director of customs in both work and life led to a separation of decoration and function in his house. Thus, on the ground floor there were two depots, while on the first floor there were an entrance hall, living room and ceremonial hall. Two schemes of painting covered the walls of the ceremonial hall, which were discovered by the restorer, M. Reutova. The painting techniques used are traditional for Central Asia. The first scheme, of ornamental motifs, was preserved in fragmentary condition. This was subsequently limewashed over and a second scheme painted, showing scenes of hunting and feasting, which appear to be from a wedding ceremony.

The lower tier of painting presents a dynamic scene of galloping horsemen engaged in hunting gazelle (known locally as 'jeiran'). The horsemen are shown with bows and arrows shooting at animals in front and behind; the horses' harnesses are richly decorated with gold. In the upper tier, a scene of feasting complements the hunting scene. Over twenty noble personages, both men and women, participate in the celebration. Unfortunately, the composition only survives in a fragmentary condition. Probably, like the wall painting at Balalyktepa, the figures were arranged in couples positioned opposite each other. They are turned three-quarters towards the viewer, and their attention was most likely focused on an old woman, shown fastening a necklace around the bride's neck. The necklace consists of a strand of beads and cross-shaped pendants. A similar necklace is worn by the old woman herself and by other women at the feast. This important scene occupied the central section of the composition.

A number of figures in the painting wear traditional robes distinguished by triangular lapels on the right of their garments. The women wear silk mantles over their robes, and beneath these, their dresses have close-fitting well-rounded collars. Originally they were probably each shown holding a large white flower on a long stem over their shoulders. The men wear turbans rolled up around their caps. The headgear worn by the female figures is more varied, but it is consistently in the style of a turban. With this information, it can be conjectured that the participants are noble Tokharistans at an Islamic wedding ceremony.

12.2.5 Kazakly-Yatkan

Currently, the oldest example of wall painting from ancient Khorezm is a mural fragment showing a beardless youth drawing a bowstring. It was discovered during the excavation of an observatory-temple in Koi-krylgan-kala, and is dated to the fourth-third centuries BC. A wealth of sculpture and wall painting of the second half of the second and third centuries AD was discovered by scientists during the previously mentioned excavations in Toprak-kala, the residence of Khorezmian shakhs. In subsequent years, they discovered wall painting fragments showing decorative motifs in Kaparas temple in Southern Khorezm, and figurative fragments in the temple complex, Kalalygyr-2, in the Sarykamysh delta of Amudarya, dating to the fourth-second centuries BC.

A real breakthrough in the study of this most ancient period in the art history of Khorezm, and, perhaps, of the entire region of Central Asia, occurred as a result of the work of an international research group of scientists, under the auspices of the Karakalpak-Australian archaeological field team. This group carries out systematic studies of the ancient Tashkyrman oasis, situated in the Beruni District of the Republic of Karakalpakstan. The primary object of the fieldwork is an ancient settlement site discovered in 1956 and known to academics as Kazakly-yatkan. Local people use another, perhaps more ancient name for this monument—Akshahan-kala. Excavations that commenced in 1995 and are still ongoing have enabled scientists to establish that this was the largest settlement site of ancient Khorezm and to make an assumption that it was the capital of Khorezm after its secession from the Akhemenid Empire.

Since 2004 the main effort of the field team has focused on studying a palace-and-temple compound in the so-called "sacred city" of Kazakly-yatkan. The north-western section of the "sacred city" was occupied by a group of buildings which, as the excavation results have shown, used to be the temple-palace compound [8].

The first indications of ancient painting in the temple came with discoveries made as early as 2004–2005. Later on, when the field team was joined by restoration experts, it became possible to begin large-scale clearance, conservation and scientific study of the discovered monuments of ancient art.

Traces of wall painting were found both inside the temple, and on the walls of a gallery that surrounded the temple along its outer perimeter, as well as on the walls of the palace. Up to now a portion of corridor about 60 m in length has been fully uncovered along the western side of the temple, revealing considerable wall painting. The total length of the corridor is about 250 m, which gives us hope for new discoveries. In the southern half of the corridor the already revealed painting shows people on foot, a mounted procession and, probably, horse riders.

12.3 Conservation Training for Preservation of Cultural Heritage

Material cultural heritage is the legacy of physical artifacts that are inherited from past generations, maintained in the present, and bestowed for the benefit of future generations. Understanding is the first step in the process of learning to protect and manage our cultural heritage well. Uzbekistan has an immense cultural heritage in need of proper documentation, recording and assessment. Well-prepared information is required to enable allow site managers and local authorities to make better and informed decisions.

In this context, special mention should be made of the work of experts from the Japan Foundation, the Ikuo Hirayama Foundation, Asia Cooperation on Conservation Science (ACCU), the Nara National Research Institute for Cultural Properties (NABUNKEN), the National Research Institute for Cultural Properties, Tokyo (TOBUNKEN) and other organizations that have carried out training in the study and preservation of cultural heritage in Uzbekistan. One of the most important training projects so far was the "ACCU Workshop on Cultural Heritage Protection in the Asia-Pacific Region 2008". This six-day workshop was organized by ACCU Nara in October 2008, in cooperation with the UNESCO office in Tashkent. Its main theme was documentation, principally measured drawing and imaging of archaeological artifacts.

From 2011 to 2013, The Japan Foundation carried out a key project titled "Human Resources Development and Technical Transfer for the Protection of the Culture Heritage (Uzbekistan)". A workshop held as part of the project provided training in the theory and practical techniques in the conservation and restoration of clay statues, aimed at practitioners in the field of cultural heritage in Uzbekistan. The workshop coordinators were Prof. Aoki Shigeo and Dr. Furusho Hiroaki (Fig. 12.6). The principal aims were: research and restoration of clay objects found during excavation of the ancient settlement of Fayaztep and stored in the State Museum of History of Uzbekistan, as well as training of specialists and technology transfer; study of the materials and techniques for creating clay objects found in the Republic of Uzbekistan and stored in the Museum of History; and improving qualifications of specialists in the field of preservation in Uzbekistan, including technology transfer and exchange of professional knowledge.

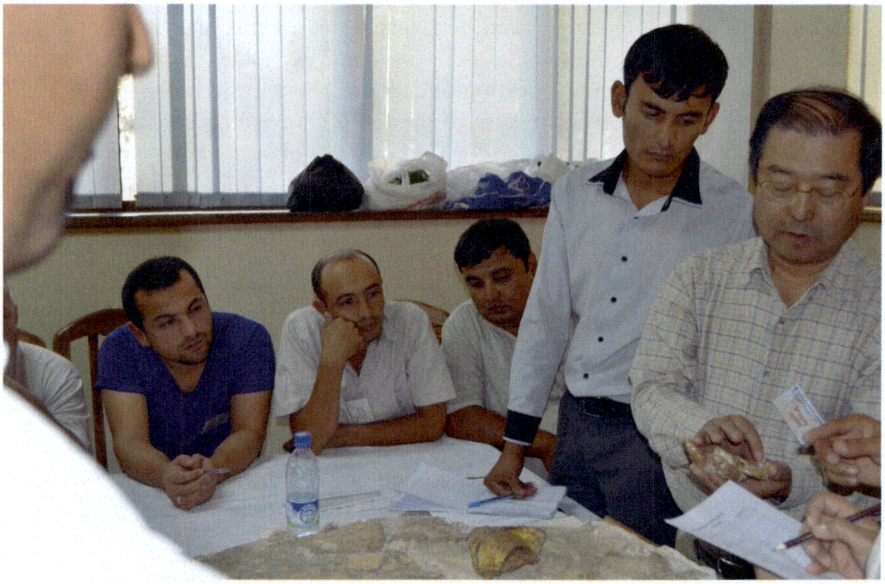

Fig. 12.6 Professor AOKI Shigeo teaching at a workshop in the conservation of clay-based statues at the State Museum of History of Uzbekistan

During this collaboration, Dr. Hayakawa Yasuhiro carried out XRF spectography analysis of three fragments of gilded and colored statues, as well as of fragments of wall painting from the collection of the State Museum of History of Uzbekistan [9]. The three sculpture fragments belong to the body, the left hand and part of the face of what is thought to be a statue from the Buddhist temple of Fayaztepa (second–third century AD, Surkhandarya region). A colored wall painting fragment, possibly excavated from the ruins at Afrasiab (Samarkand, seventh–eighth century AD), was one of those selected for analysis. It shows a head, which is the only surviving part of an almost life-size painting of a warrior, thought to be a stylized depiction of a Sogdian man of the seventh–eighth century AD. His face is conspicuously white, while his finely delineated eyes and mouth are enlivened by subtle gradations of red. His helmet, which is painted in a vivid blue, is also very impressive. Another fragment of wall painting selected for analysis was excavated from the Varahsha ruins (Bukhara, sixth–seventh century AD). This shows a panther-hunting scene, in which the hunter is depicted on the back of a white elephant. It is one of the most famous exhibits in the collection of the State Museum of History of Uzbekistan. The part of the painting that was analyzed shows a procession of partridges. The XRF analysis obtained useful information about the original coloring materials of the statues and wall paintings.

The "Cultural Training Partnership for Artifact Conservation" is a three-year program of capacity building and advanced training for Artifact Conservators' from the national museums of the five Central Asian republics—Kazakhstan, Kyrgyzstan,

Fig. 12.7 Workshop of the 'Cultural Training Partnership for Artifact Conservation' for conservators from the five Central Asian Republics, instructed by Fabio Colombo

Tajikistan, Turkmenistan, and Uzbekistan—funded with the support of the US Department of State through the US Embassy in Tashkent. Starting in 2018, the Oriental Institute (OI) of the University of Chicago, working in partnership with the State Museum for the History of Uzbekistan (SMHU), organized a series of three annual intensive two-week training workshops for 16 conservators from the national museums of the Central Asian republics and provincial museums in Uzbekistan. Teaching is delivered by conservation experts from leading international centers.

The workshop coordinators are Prof. Gil Stein and Mr. Fabio Colombo, who also serves as the head of field conservation for the University of Chicago's Cultural Heritage Preservation projects in Afghanistan. Mr. Colombo was the instructor for the first workshop held in 2018 (Fig. 12.7). The topics covered in each workshop are designed to provide training in internationally recognized standards and practices of treatment for the main classes of artifact types, and with respect to their principal constituent materials, that form the majority of the holdings in the national museums of the Central Asian republics. This is the first systematic program that brings together heritage preservation specialists from the national museums of the Central Asian republics for training, in order to develop a shared set of standardized best practices for conservation of museum objects. It is intended as a first step in building connections among the National Museums of these five countries, to encourage institutional cooperation in cultural heritage management.

The first workshop took place at the State Museum for the History of Uzbekistan from 3rd to 15th September, 2018.

Topics covered included:

a. an overview of theoretical approaches to conservation;

b. a review of the main aspects of applied science that underpin conservation;
c. stabilization of objects;
d. applied practical training in conservation methods, with a focus on wall paintings.

In approaching the conservation of any given object or material, there is almost always a range of potential treatment strategies. For this reason, in addition to instruction in specific techniques, the workshop focused on teaching how to evaluate objects and materials, and how to decide which specific treatments are most appropriate to employ in any given case.

It cannot be emphasized enough that the conservation of cultural heritage artifacts is both an important and complex undertaking. Collaborative conservation programs should be seen as essential steps in a longer-term commitment to capacity building for cultural heritage preservation in the countries of Asia. Certainly, such collaborations are necessary in helping to preserve and safeguard the tangible cultural heritage of Uzbekistan. Of particular importance is the strengthening of site management and conservation practices, to ensure the promotion and transmission of this cultural heritage to the next generation.

References

1. Birstein, V.J.: On the technology of central Asian wall paintings: the problem of binding media. Stud. Conserv. **20**(1), 8–19 (1975)
2. Kostrov, P.I. S.: Restoration of monumental painting on loess plaster using synthetic resins. Stud. Conserv. **6**(2/3), 90–106 (1961)
3. Al'baum L.I.: Balalyk-tepe. Tashkent, pp. 126–162. Fig. 96–109 (1960)
4. Vyatkin V. L.: Ob arkheologicheskikh raskopkah v Samarkande in: "Samarkand" newspaper, No. 29 (8/21.VI) (1904)
5. Shishkin V.A.: K istorii arkheologicheskogo izucheniya Samarkanda i ego okrestnostey. Afrasiab, Tashkent, V. I, pp. 3–121 (1969)
6. Al'baum L. I.: Zhivopis' Afrasiaba, Tashkent, pp. 39–40 (1975)
7. Pidaev, Sh.R.: Nastennaya jivopis' v buddiyskom hrame na Karatepa. Fan va turmush. Tashkent. **3–4**, 45–51, Fig. 1–5 (2016)
8. Kidd, F., Negus-Cleary, M., Yagodin, V.N., Betts, A., Baker-Brite, E.: Ancient Chorasmian mural art. Bull. Asia Inst. **18**, 69–95 (2004)
9. Hayakawa Y., Furusho, H., Aoki, Sh., Aripdjanov, O.: Material Analysis of objects collected in The State Museum of History of Uzbekistan using a handheld x-ray fluorescence spectrometer (早川 泰弘, 古庄 浩明, 青木 繁夫, オタベック アリプトジャノフ. ハンドヘルド蛍光X線分析装置によるウズベキスタン国立歴史博物館所蔵資料の材料調査)in: Science for Conservation (保存科学). Tokyo National Research Institute of Cultural Property, No. 52, pp. 5–69 (2013)

Chapter 13
The Wall Painting Techniques and Materials of Kizil Grottoes

Zhibo Zhou, Ling Shen, and Hui Zhang

Abstract Ancient Kucha (Kuqa, Qiuci), situated at a midpoint on the Silk Road that traverses the Eurasian Continent, was one of the most important centers of trade and Buddhist culture in Central Asia. Kucha played a crucial role in the spread of Buddhism along the Silk Road. A large number of grottoes decorated with wall paintings remain in Kucha, showing us the prevalence of Buddhism in this area. The Kizil grotto site is the largest and most influential of the Kucha grottoes group. The paintings preserved here display an astonishing range of styles and techniques, testifying to the cultural and commercial importance of the site and its vital role in the dissemination of Buddhism along the Silk Road in ancient times. A number of previous studies have been undertaken into the materials and techniques of the Kizil wall paintings. On the basis of these, the authors have conducted further investigations, in the course of which some new discoveries were made. Both the inorganic and organic substances in fragment samples collected from several caves dating from the fifth to the seventh century were analyzed. Morphological studies were performed by optical microscope (OM), cross-section microscope and scanning electron microscope (SEM). Inorganic materials were analyzed by X-ray diffraction (XRD), Raman spectrometry, and elemental maps obtained by SEM with energy dispersive X-ray analyzer (EDX). Organic materials were studied by Fourier transform infrared spectroscopy (FTIR), liquid chromatography coupled to electrospray ionization quadrupole-time-of-flight mass spectrometry (LC–ESI-Q-TOF-MS) and by enzyme-linked immunosorbent assay (ELISA). We identified the white pigment as pyromorphite-mimetite ($Pb_5Cl[(P,As)O_4]_3$), and the yellow colorant as gamboge, which is applied over atacamite, turning it to a delicate shade of green. Stick lac was

Z. Zhou (✉)
Kizil Grottoes Institute, The Kucha Academy of Xinjiang, Baicheng County, Aksu District, 842300 Urumqi, China

L. Shen · H. Zhang
School of Art and Archaeology, Zhejiang University, No. 148 Tianmushan Road, Xixi Campus, Hangzhou 310028, China
e-mail: lingshen@zju.edu.cn

H. Zhang
e-mail: huizhang@zju.edu.cn

S. Aoki et al. (eds.), *Conservation and Painting Techniques of Wall Paintings on the Ancient Silk Road*, Cultural Heritage Science,
https://doi.org/10.1007/978-981-33-4161-6_13

used in both the resinous mordant for gilding and for the red colorant in the wall paintings; and the resinous component (shellac) was also probably applied over tin foil to imitate gold. Plant gum was found to be the binder for a lump of blue pigment discovered in front of Cave 189. These studies throw light upon technical advances, trade intercourse and cultural exchanges along the ancient Silk Road.

Keywords Kizil grottoes · Wall painting · Analysis · Pigment · Binding medium

13.1 The Kucha Grottoes Group

The ancient settlement of Kucha (Kuqa, Qiuci) was located at the northern margins of the Taklamakan Desert, in the southern foothills of the Tianshan Mountains. Kucha was centered on the Kuqa oasis, which during its greatest period of prosperity encompassed the area occupied by the modern-day counties of Luntai, Kuqa, Shayar (Xayar), Baicheng (Bay), Aksu and Xinhe (Toksu) in the Xinjiang Uyghur Autonomous Region of China [1].

Since Zhang Qian's opening of the Silk Road (140 B.C.), Kucha has played a crucial role in the long-term cultural exchange and interaction between Eurasian civilizations. Kucha also played a significant role in the transmission of Buddhism along the Silk Road to China, the Korean Peninsula and Japan.

From the third to the eleventh centuries, officials, monks, and artists taking advantage of the unique natural and cultural environment of the Kucha area, constructed the grottoes of Kizil, Kumtura, Kizilkargha, Simsim, Mazarbaha, Wenbhash, Taitai'er, Tograk-eken and A'ai, which are collectively referred to as the Kucha Grottoes Group [2–6]. The art of the Kucha Buddhist grottoes, produced over a period of more than 800 years, is technically and iconographically rich, and the sheer quantity of surviving painting is staggering.

13.2 The Kizil Grottoes

The Kizil grottoes, the largest and most influential of the Kucha grottoes group, is situated in Kizil Town, Baicheng County, Xinjiang, China. The caves are excavated into the cliff face of Mingwutage Mountain in Weigan Valley (named the Muzart River Valley near the caves), and extend from east to west for 1.7 km, scattered on various levels of the cliff (Fig. 13.1). There exist 349 caves dating from the third century to the ninth century containing 5000 square meters of wall paintings, a small number of sculptures, and a number of architectural remains in front of the caves [7, 8].

There are a great many different types of cave at Kizil, including central pillar caves, Great Buddha caves, square caves, monks' quarter caves, niche caves, special-shaped caves, and combinations of these. The caves dating from the third to the fifth

Fig. 13.1 Aerial image of the Kizil grotto site and details of the wall paintings. (a) aerial image of the Kizil grottoes: the Sugete Valley divides the caves into four natural groups: the western ravine (marked with blue), the inner ravine (marked with yellow), the eastern ravine (marked with green), and the rear hill (marked with red); (b) Flying Apsaras from newly-found Cave 1; (c) Musical Apsaras from Cave 38; (d) Śākyamuni story paintings from Cave 114; (e) Śākyamuni story paintings from Cave 171; and (f) painting from Kizil detached by German exploration team

century are predominantly central pillar caves (Kucha style caves) and Great Buddha caves, and are decorated with wall paintings illustrating themes from traditional (*Hinayana*) Buddhism. These cave types originated at Kizil, and their distinctive western features testify to a fusion of the Indian cave style with Chinese traditions to form a unique Kucha style. These developments at Kizil exerted considerable influence on Buddhist cave art in the Hexi region (Dunhuang Mogao Grottoes), Longyou region, Central China (Longmen Grottoes and early Yungang caves), as well as Central Asia [9].

Abundant Buddhist narrative wall paintings are preserved in the Kizil caves, mainly illustrating Śākyamuni stories, including *Jataka* stories, *Karma* tales and Buddha's life stories. There are more than 100 *Jataka* stories and *Karma* tales, and over 60 Buddha's life stories, reflecting Buddhist worship in the Sarvastivadin school (说一切有部), a branch of Hinayana Buddhism, prevalent in Kucha region (Xuanzang 646). In addition to the religious content of the paintings, which was instrumental in disseminating and publicizing the Buddhist doctrine, the paintings of Kizil are a unique repository of illustrations of the different races living or trading in Kizil, the clothing people wore, the musical instruments played, as well as scenes of work, dancing and other aspects of daily life, providing a precious and vivid insight into otherwise lost aspects of Kucha civilization ([1], Xuanzang 646).

Kizil's claim to be one of the most important cultural sites along the Silk Road is demonstrated by the many artistic styles and painting techniques present in the grotto paintings. The Kucha style incorporated Greek influences via Gandhara in the earlier period, and also absorbed influences from Indian, Persia and the Central Plain [10]. Strong regional characteristics of the style include rhombic-shaped compositions (a significant feature of Kizil paintings), *Qutiepansi'* (a style endemic to the western regions, consisting of powerful outlines full of vitality, and unconstrainted and free script), composite line and halation methods (to enliven and invigorate painting, and to create plump, full images).

13.3 The Painting Techniques and Materials of Kizil

13.3.1 Previous Studies of Kizil Painting Techniques and Materials

The wall painting technology of Kizil is similar to that at other grottoes along the Silk Road, such as Bamiyan, Mogao, and Maijishan. The caves were excavated into a cliff face composed of conglomerate rock, and the walls were then plastered and smoothed over with layers of earthen plaster composed of earth, sand and natural plant fibers (wheat straw). In preparation for the paintings, a ground-layer (gypsum) was applied over the plaster. The plaster of Kizil is different from Mogao in that there is only one layer in the majority of paintings. There are even several caves where the paintings were executed onto a ground-layer applied directly onto the sandstone support without any plaster layers (newly-found Cave 1, Cave 69, Cave 171). The painting layers consist of pigments and colorants mixed with organic binding media.

Previous studies of the technology of the wall paintings of the Kizil Grottoes were carried out by Gettens in 1938 [11], and by Riederer in 1977 [12], who studied painting fragments detached by foreign exploration teams (Germany & Japan). Considerably later, Su Bomin investigated the pigments and binding media of the wall paintings [13, 14]. In 2015, the Tokyo University of the Arts and the Kucha Academy research team conducted further investigations of Kizil's wall painting materials and techniques [15]. The results of these previous studies results are summarized in Table 13.1.

Table 13.1 Previous studies of painting techniques and materials used at Kizil paintings

		Blue	Green	Red	White	Yellow	Black	Brown	Grey	Metal foil	Binder
Su Bo-min	Cave No.	1, 38, 114, 186, 180, 179, 171	1, 38, 114, 186, 180, 179, 171	1, 38, 77, 100, 186, 180, 171	1, 38, 77, 100, 186, 135		1, 38, 100, 186, 179	38, 77, 114	171		
	Methods	XRD PM	XRD PM	XRD PM	XRD PM	XRD PM	XRD PM	XRD PM	XRD PM	XRD PM	HPLC
	Results	Lapis Lazuli	Atacamite, Paratacamite	Red lead, Cinaber, Red Ochre	Gypsum, Anhydrite[a] Calcite		Lead dioxide	Lead dioxide			Animal glue
Kucha Academy & Tokyo University of the arts	Cave No.	69, 224	69, 224, 167	69, 167, 224	69	69	167, 69			69	
	Methods	XRF	XRF	XRF	XRF	XRF	XRF, OM				
	Results	Lapis lazuli	Atacamite	ed lead, ed ochre, Ac(T)	Gypsum	Lithargite	Carbon black			Gold, Tin	
Gettens, R. J(G) Riederer, J(R)	Methods	PM, Chemical analysis (G), PM, cross section, XRD (R)	PM, Chemical analysis (G), PM, cross section, FTIR, XRD (R)	PM, Chemical analysis(G), PM,cross section, XRD (R)	PM, Chemical analysis (G), PM, cross section, FTIR, XRD (R)	PM, Chemical analysis (G), PM, cross section, XRD (R)	PM,cross section, (R) FTIR, XRD (R)	PM, Chemical analysis (G),	PM, Chemical analysis (G),		Wetting PM (G)

(continued)

Table 13.1 (continued)

		Blue	Green	Red	White	Yellow	Black	Brown	Grey	Metal foil	Binder
	Results	Lapis Lazuli (G) (R) Indigo (R)	Chrysocolla (G), Atacamite (R)	Red lead(G) (R) Red ochre (G) (R)	Gypsum, Anhydrite (R)	Yellow Ochre(R) Orpiment (R) Lithargite (R)	Carbon black (R)	Lead dioride (G)	Uncertain, Pb, Ca(G)		Animal glue (G)

[a]Anhydrite: in Xinjiang Management Committee of Cultural Heritage 1997, the XRD spectrum of anhydrite is not given, but the identified results are summarized in a table. In the results summary, the anhydrite was found to be present both in the white paint layer and in combination with the blue pigment lazurite (lapis lazuli) in Caves 38, 171, 179 and 186. Elsewhere, anhydrite has also been detected in white paint layers in Maiji grottoes, and in Mati and Lingbing temples, in Gansu province [27], and in combination with lazurite in Dunhuang [28]. Anhydrite is insoluble in water, has no adhesive properties, and is not easy to use in wall paintings. Given that the anhydrite is detected with gypsum in all reported studies, it can be supposed that anhydrite is present in association with the raw mineral gypsum

13.3.2 Current Investigations of the Painting Techniques and Materials of the Kizil Grottoes

The previous studies discussed above imparted some basic information, but the latest investigations provide a much a deeper understanding of the wall painting techniques and materials employed at the Kizil grottoes. Pyromorphite-mimetite ($Pb_5Cl[(P,As)O_4]_3$) was found to have been used as a white pigment, while gamboge was used to give a yellow glaze over atacamite, to achieve a delicate shade of green. Stick lac, a complex red resinous substance yielding both lac dye and shellac, which is secreted by a number of species of the lac insect, was used both as a mordant for gilding and as a red colorant at Kizil, which is a new finding. Shellac was also probably used on tin foil to imitate gold. Plant gum was detected in the lump of blue pigment found in front of Cave 189.

13.3.2.1 Lead-Containing White Pigment and an Organic Yellow Colorant in Cave 69

Two samples were collected from painting on the vault of the back corridor in Cave 69. The first was a sample of white from the body of a flying apsara, and the second was a sample of green from the lower garment (Fig. 13.2).

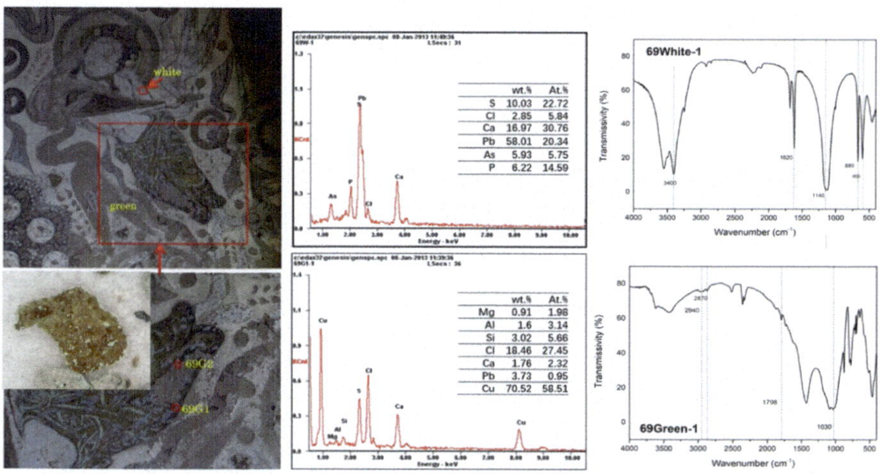

Fig. 13.2 Samples collected from Cave 6; locations shown on the left. 69 W-1 is the white from the flesh of the body of the Flying Apsaras. Samples 69 G1 and 69 G2 are green covered with a yellowish substance; there is a fine grey layer beneath green. There is no plaster layer and the painting has been applied directly onto the sandstone. In the middle and on the right, the EDX images and FTIR spectra of the samples

The white sample was taken because its whitish-yellow hue differed from the white of the ground, presenting the possibility that a different pigment was used in this case. Analysis by Scanning Electron Microscopy (SEM) with Energy Dispersive X-ray analyzer (EDX) indicated that Pb, Ca, As, P, S, Cl were present in the white sample. Results showed that the element Pb predominated at 58.01 wt%, indicating that a lead-based pigment was employed. Lead white ($Pb_3(CO_3)_2(OH)_2$) is the most common lead-based white pigment and has been used in numerous wall paintings. However, in the FTIR spectrum of the white paint layer from Cave 69, the C-O carbonate bonds usually occurring at 1420 cm^{-1} were not detected, suggesting that the pigment is not lead white. Considering the nature of the other elements in this pigment (As, P, Cl, S and Ca), it can be assumed that the white pigment is pyromorphite ($Pb_5Cl(PO_4)_3$) with gypsum. Pyromorphite usually contains a small amount of arsenic because phosphorus (P) and arsenic (As) often coexist in nature and replace each other in the molecular structure, forming pyromorphite-mimetite ($Pb_5Cl[(P,As)O_4]_3$) [16]. Fourier transform infrared spectroscopy (FTIR) spectra with absorption bands at 1200–1100 cm^{-1} can be attributed to (SO_4^{2-}) antisymmetric stretching vibration; peaks at 1620 cm^{-1}–1680 cm^{-1} can be attributed to bands of OH−.

Pyromorphite-mimetite is not a familiar pigment, although it is quite widely found in a number of ancient murals. In its natural state, it is a secondary mineral of lead ore, having a white to yellow appearance. It is used as a white pigment in the Maijishan grottoes [17], for example, and as a yellow pigment in tomb murals in Gansu province [18]. However, its use as a pigment in the Kizil grottoes has rarely been reported.

The green sample was collected from the garments of the flying apsaras, which are yellowish in color. Observation under the microscope showed that there is a brown colorant layer above the bright green pigment. SEM-EDX analysis results indicate that Cu, Cl and Ca are present in both samples. Ca and S come from the ground ($CaSO_4{\cdot}2H_2O$). The presence of Cu and Cl indicates that atacamite ($Cu_2(OH)_3Cl$) is probably the green pigment. In its FTIR spectrum, the absorption bands at 2940 cm^{-1} and 2860 cm^{-1}(C-H asymmetric stretching vibration of CH_2 and C-H symmetric stretching vibration of CH_2) suggest the presence of organic materials. Based on its appearance, a yellow-brown colorant—most likely gamboge—was probably applied over the surface of the green to modify its colour. Among previous material studies in archaeological remains, the use of gamboge has been identified in the Astana tomb murals (680–880 A.D.), in Turfan, near ancient Chotscho, to the east of ancient Kucha [19].

13.3.2.2 The Blue Pigment

A lump of blue pigment was found in front of Cave 189 during remedial conservation in 2013. Its appearance is shown in Fig. 13.3a. Analysis was carried out to understand its properties. X-ray diffraction (XRD), Fourier Transform Infrared Spectrometry (FTIR) and enzyme-linked immunosorbent assay (ELISA) were used for studying inorganic and organic materials. The results show that the blue pigment is lapis lazuli

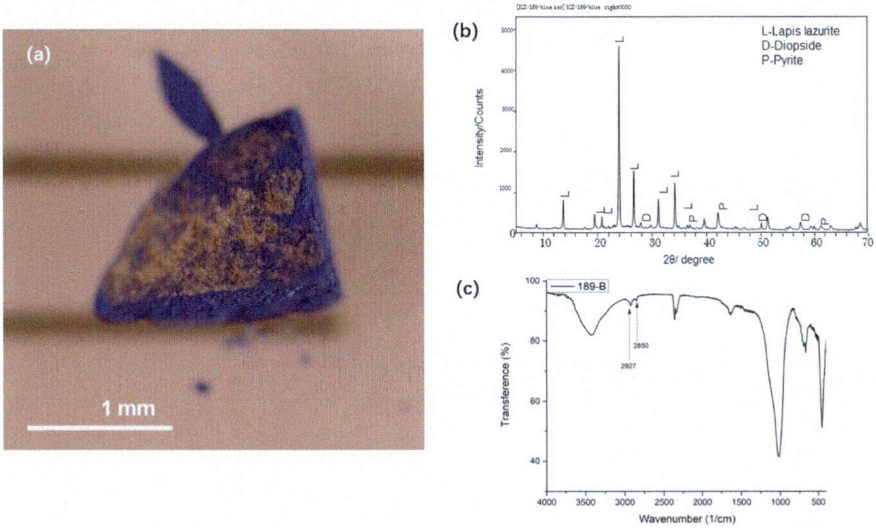

Fig. 13.3 The lump of blue pigment and analysis results: (a) image of the unearthed lump of blue pigment; (b) the indicative peaks on the XRD spectra are marked by L (Lapis Lazuli), D (Diopside) and P (Pyrite); (c) FTIR spectrum

and its binding medium is plant gum. This lump of pigment was presumably left by the ancient painters who created the wall paintings.

In the XRD spectrum, the characteristic peaks of lapis lazuli are very clear (Fig. 13.3b). In addition, the natural impurities diopside and pyrite can be seen, indicating that the blue pigment is made from natural lapis lazuli. The FTIR spectra (Fig. 13.3c) show bands at 1089 cm^{-1} (Si-O-Si Antisymmetric stretching vibration) and 694 cm^{-1} (S-O stretching vibration), which can be attributed to lapis lazuli. Other bands around 3421 cm^{-1} (O-H stretching vibration of carbohydrate) and 2927 cm^{-1} (C-H asymmetric stretching vibration of CH_2) indicate the presence of organic materials. Furthermore, the ELISA results shown in Table 13.2 indicate that the organic binder used with the blue pigment is a plant gum (sample 189B1). In order to confirm if plant gum is also used in the wall painting as a binder, an additional sample was collected and analyzed by ELISA (sample 189R). However, none of the OD absorbance values provide a clear result to identify the kind of binding media used.

While it can be supposed that the plant gum was probably used as a binding medium for the wall paintings, its presence in the lazurite block found in front of the cave may alternatively indicate that plant gum was used in the processing of the mineral for the separation out of fine particles.

Table 13.2 ELISA results [OD absorbance of samples (20 min)]

	bs0813R (casein)				JIM13 (plant gums)				MAC265 (tragacanth)			
	40μL	20μL	20μL	10μL	40μL	20μL	20μL	10μL	40μL	20μL	20μL	10μL
189R	0.091	0.093	0.095	0.096	0.099	0.107	0.118	0.116	0.104	0.058	0.102	0.111
189B1	0.103	0.105	**0.117**	0.099	**0.392**	**0.417**	**0.436**	**0.66**	0.117	0.121	0.084	0.122
ave + 3SD	**0.111**				**0.361**				**0.124**			
	ab19811 (collagenI, exclude goat)				ab34710 (collagenIIexclude rabbit)				ab1225 (ovalbumin)			
	40μL	20μL	20μL	10μL	40μL	20μL	20μL	10μL	40μL	20μL	20μL	10μL
189R	0.098	0.092	0.097	0.1	0.131	0.119	0.124	**0.171**	0.236	0.216	0.224	0.372
189B1	**0.105**	**0.107**	0.1	0.102	0.145	0.146	0.149	0.158	0.355	0.355	0.333	**0.419**
ave + 3SD	**0.103**				**0.163**				**0.392**			

13.3.2.3 Gilding Techniques in the Kizil Grottoes

Six samples were taken from Caves 171 from areas where metal foil and/or resinous coatings had been observed: from inside the Buddha niche; from the sculpture above the niche; from the central pillar and from the main chamber (the sample locations are shown in Table 13.3, together with optical microscope [OM] images and detailed descriptions of the samples). Cross-sections were prepared and morphologically studied by OM and Scanning Electron Microscopy (SEM) observation. Elemental maps were obtained using SEM-EDS (Scanning Electron Microscope equipped with Energy Dispersive Spectrometer).

The cross-section images and elemental maps show that gold (Au) is present in two samples: KZ-C171-S8 and KZ-C171-S3. Tin (Sn) was detected in five samples: KZ-C171-S2, KZ-C171-S5, KZ-C171-S12, KZ-C171-S3 and KZ-C171-S7. Analytical results show that two samples, KZ-C171-S5 and KZ-C171-S3, have two layers of metal foils (Fig. 13.4b).

In BSE (Backscattered Electron) images, also shown in Fig. 13.4, a dark layer could be observed in all the samples just beneath the metallic leaf, suggesting the layer is composed of elements with smaller atomic numbers. Referring to the OM and cross-section images, this layer varies considerably in the samples, from a red-purple to brownish color. All these observations suggest it may be an organic mordant for the tin leaf. Lead (Pb) was detected in the orange-coloured lower layer of the sample, indicating use of minimum (red lead, Pb_3O_4). In all except sample KZ-C171-S8 Fig. 13.4, which was from the figure's dress, gold (Au) was detected above a red ground. Organic-looking mordant layers could be found in all the samples. In KZ-C171-S8, an extremely thin dark layer could be observed just beneath the gold leaf in the Backscattered Electron (BSE) images, meaning there are elements with smaller atomic numbers. Lead (Pb) was also detected in the orange layer beneath the gold leaf in KZ-C171-S12 and KZ-C171-S5, suggesting the use of minium. Moreover, in

Table 13.3 Detailed description of each sample with images of sampling locations and fragments

KZ-C171-S2		Niche wall		Grey-brown layer with red support and white plaster
KZ-C171-S12		Mount Summeru	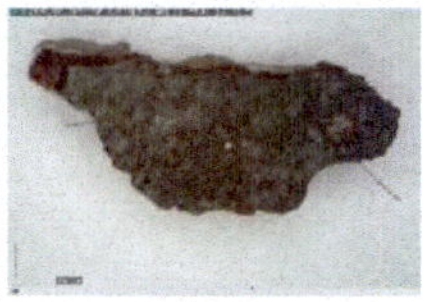	Grey-brown layer applied over the dark-red layer
KZ-C171-S7		Decorative band on the wall of the main chamber	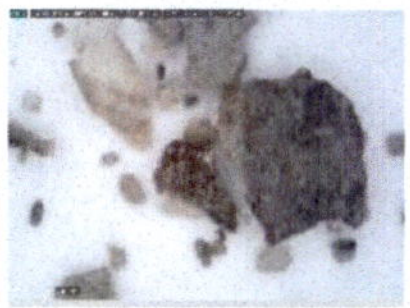	Grey-brown layer applied over the dark-red layer
KZ-C171-S8		Figure's dress on the wall of the main chamber		Gold-coloured layer with a metallic luster, dark-red layer beneath
KZ-C171-S5		Mount Sumeru		Brown resin and purple-red visible at the lower right edge of the sample
KZ-C171-S3		Budda mandorla, niche wall		Gold-coloured layer with brown material beneath

samples KZ-C171-S12, KZ-C171-S5 and KZ-C171-S7 where minium was present in the stratigraphy, a thin dark-red layer, probably made from a red colourant, was applied on top, presumably to adjust the color hue.

13.3.2.4 Organic Materials Used in the Application of Metal Foils

FTIR spectra of samples are shown in Fig. 13.5. The μ-FTIR spectra were obtained

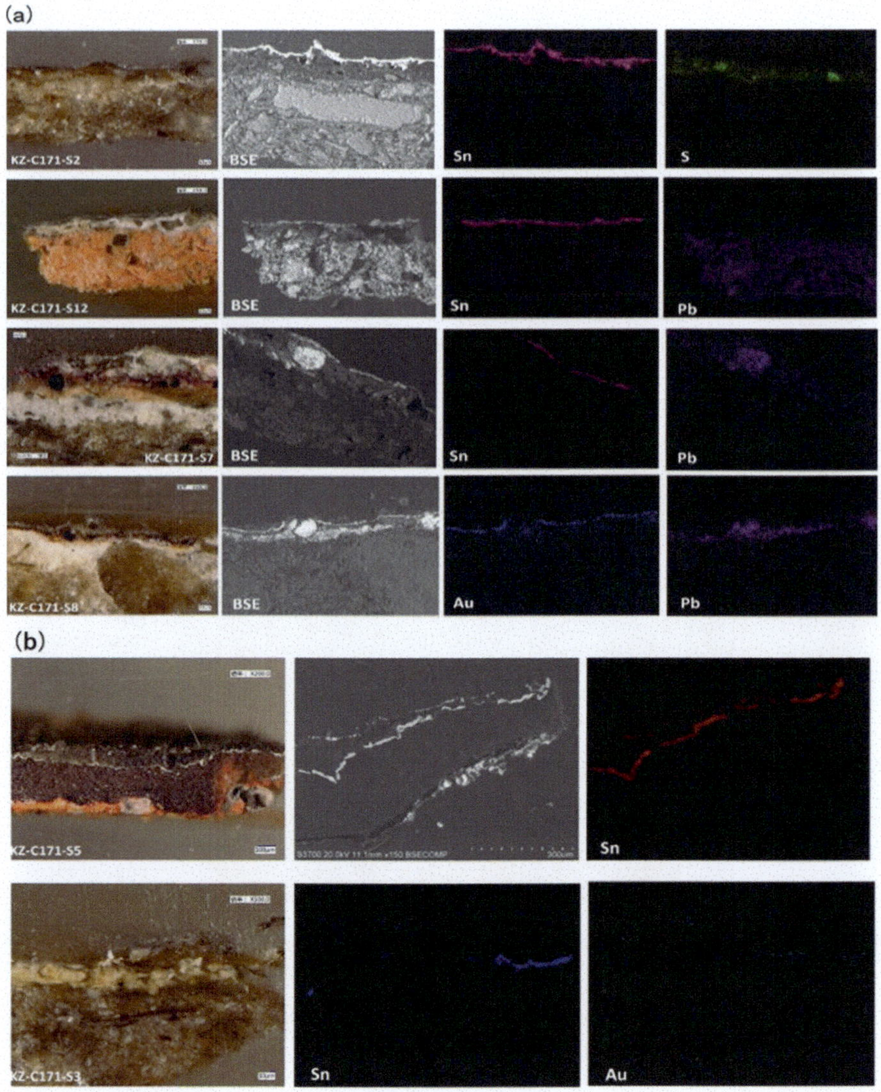

Fig. 13.4 (a) cross-section images, backscattered electron images and elemental mapping of samples with a single layer of metal foil; (b) cross-section images and elemental mapping of samples with a double layer of foil

from the yellow layer and the red layer beneath it in sample KZ-C171-S5 Fig. 13.4a and the yellow layer in KZ-C171-S2. Peaks occur at 1735 cm^{-1} [ν (C = O) ester] and 1710 cm^{-1} [v (C = O) aldehyde, ketones, carboxylic]; shoulder peaks at 1775 cm-1 [ν (C = O) lactone, cyclic esters], 1465 cm^{-1} [δ (CH_2)], 1416 cm^{-1} [ω (CH_2)], 1377 cm^{-1} [δ (CH_3)], 1246/1165/1100 cm^{-1} [ν (C-O)] and 723 cm^{-1} [γ(CH_2)] are

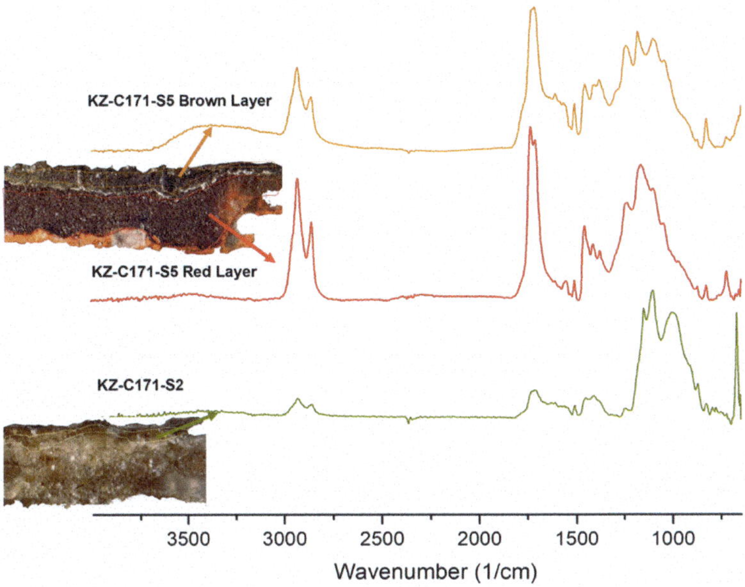

Fig. 13.5 FTIR spectra of organic materials: stick-lac (internet purchase from India); sample KZ-C171-S5 and KZ-C171-2 from wall painting (for the sample location and description, see, Table 13.3)

detected in both the yellow and red layers, suggesting the presence of carboxylic acids.

As shown in the FTIR spectra, the absorption bands at 2920 cm^{-1}, 2842 cm^{-1}, 1730 cm^{-1}, 1711 cm^{-1}, 1460 cm^{-1}, 1422 cm^{-1}, 1375 cm^{-1}, 1240 cm^{-1}, 1170 cm^{-1}, 870 cm^{-1} and 722 cm^{-1}were detected in the yellow-brown layer over the tin leaf, in the red-purple layer beneath the tin leaf of sample KZ-C171-S5, and in the yellow-brown layer of KZ-C171-S2, indicating all these organic compounds contain comparable functional groups, leading to similarities in their vibrational signatures. 1730 cm^{-1} can be attributed to C = O stretching vibration of esters, while 1710 cm^{-1} can be attributed to C = O stretching vibration of aldehyde, carboxylic acids and ketones. The IR bands of organic compounds at 1175 cm^{-1} (C-O stretching), 1240 cm^{-1} (the bending vibration of carboxyl), 1460 cm^{-1} (C-H_2 Bending vibration), 2845 cm^{-1} (C-H symmetric stretching vibration of CH_2), 2920 cm^{-1} (C-H asymmetric stretching vibration of CH_2) and antisymmetric vibration of CH_2 at 1422 cm^{-1} can be clearly seen. The characteristic peaks in the IR spectra coincide well with the typical peaks of carboxylic acids, indicating the possibility that oil was used in these layers.

In previous research, it has been speculated that lac was probably used in the wall painting of Cave 224 [15]. On this basis it was decided to try to establish whether it was also used as a colorant in Cave 171. The FTIR spectrum of the stick lac sample identified its main component as aleuritic acid, which, in common with other carboxylic acids, can indicate the presence of drying-oils. In order to get more information about the organic materials and confirm if the dark-red layer in samples KZ-C171-S5, KZ-C171-S12 and KZ-C171-S8 is red lac, analysis was performed using liquid chromatography coupled to electrospray ionization quadrupole-time-of-flight mass spectrometry (LC–ESI-Q-TOF-MS) in negative ion mode (results are shown in Table 13.4). The proposed compounds of shellac for target screening were obtained through the literature [20] and further identification of the molecule was achieved by comparison of its retention time with that of a prepared reference shellac sample (extracted in ethanol from raw stick lac). Comparative results of sample fragments from the wall paintings and of the raw stick lac are shown in Table 13.4. The results of target screening show the presence of typical natural shellac components including free hydroxy aliphatic acids and cyclic sesquiterpene acid in all the mural samples. Specifically, aleuritic acid ($[M\text{-}H]^-$, m/z 303, RT 13.56 min), a major ingredient in (30–40wt%) [21], is detected in all samples except KZ-C171-S2, in which the dark-red layer is absent, indicating that the red layer is composed of stick lac. Furthermore, the fact that typical components of stick lac were detected in the samples taken from the red background color supports the speculation that red lac was used both as a red colorant for painting by ancient painters and also as a mordant exclusively for gold foil.

13.4 Discussion

Our study of the materials and techniques of the Kizil wall paintings is only a beginning, but the results of analysis and examination have already provided ample evidence for the transmission of materials and techniques along the ancient Silk Road, which is enough to inspire us. As gamboge was not local to Kucha, it might have been transported from the Central Plains. Stick lac, which originated in Persia (Iran), Sindhu (India) and South Asia might have been introduced via trade to Kucha, thence to the Central plains. The technique of applying glazes to tin in imitation of gilding has not previously been found in Chinese wall painting. While the technique of applying safflower seed oil over silver foil and baking it to produce a gold appearance is recorded in the ancient book "Tian Gong Kai Wu" (天工开物), a technological encyclopedia written by Song Yingxing 宋应星 (1600–1660 A.D.) in 1637 ((Ming Dynasty) Song Ying-xing, "Tian Gong Kai Wu"), there is no record of using tin foil in the same manner. Tin foil has also been found in paintings in Afghanistan [22], and in Indian [23] and western paintings [24–26], but the tin foil used in Kizil seems to be the earliest recorded example (i.e., fifth–seventh century A.D.). It is therefore very likely that this unique imitation gilding technique originated in Central Asia, and then spread westward.

It is perhaps inevitable that the limited exploration undertaken so far of the painting materials and techniques used in Kizil has resulted in an incomplete and sometimes incorrect understanding of ancient painting procedures. However, we very much hope that these studies have thrown some new light upon the technical advances, trade

Table 13.4 Extracted ion chromatograms of aleuritic acid from mural samples and reference shellac. EICs obtained by LC–ESI-Q-TOF-MS analysis shows the nearly identical retention time of aleuritic acid with the reference, implying the presence of the compounds of stick lac

Chemical formula	Identified compounds of stick lac	Reference	KZ-C171-S5	KZ-C171-S8	KZ-C171-S7	KZ-C171-S12	KZ-C171-S2
$C_{14}H_{28}O_3$	Butolic acid	+	+	+	+		
$C_{16}H_{32}O_5$	Aleuritic acid	+	+	+	+	+	
$C_{16}H_{30}O_5$	Oxidised aleuritic acid	+		+		+	
$C_{14}H_{28}O_4$	9,10-dihydroxytetradecanoic acid	+	+	+		+	
$C_{14}H_{26}O_3$	6-oxotetradecanoic acid	+	+	+	+	+	
$C_{16}H_{32}O_4$	9,10-dihydroxyhexadecanoic acid	+	+	+	+	+	
$C_{16}H_{30}O_4$	9,10-dihydroxyhexadecenoic acid	+	+	+	+	+	+
$C_{15}H_{20}O_6$	Shellolic acid, Epishellolic acid	+		+	+		
$C_{15}H_{20}O_4$	Laccijalaric acid	+			+	+	
$C_{15}H_{18}O_5$	Oxidised jalaric acid	+		+	+		
$C_{31}H_{50}O_8$	Laccijalaric-aleuritic	+	+			+	
$C_{31}H_{48}O_8$	Laccijalaric- oxidised aleuritic	+	+	+			
$C_{31}H_{50}O_6$	Laccijalaric-(16-hydroxyhexadecanoic)	+	+	+	+		
$C_{30}H_{48}O_9$	Aleuritic-Llak	+	+	+		+	

intercourse and cultural exchanges along the ancient Silk Road. We are honored to have had the opportunity to study from the brief perspective of our own work the thousands of years of human achievement distilled into the art of Kizil.

Acknowledgements We thank Ms. Takashima Miho from the National Museum of Western Art (Tokyo, Japan) for help with ELISA analysis, and Professor Kirino Fumiyoshi and Professor Tsukada Masahiko from the Tokyo University of the Arts for the help with FTIR analysis of samples from Caves 69 and 189. We thank Dr. Chenguo Feng and Dr. Fang Zhang from the Shanghai University of Traditional Chinese Medicine for their help with mass spectroscopy analysis, and Ms. Yang Gao from the Bruker (Beijing) Technology Co. Ltd for her help with the ATR-FTIR analysis.

References

1. Xinjiang Research Institute of Kucha Caves Collected Essays on Kucha Buddhist Culture, Xinjiang Fine Arts Photography Press, 142 (1993). 新疆龟兹石窟研究所. 龟兹佛教文化论文集. 新疆美术摄影出版社, 142 (1993)
2. Su, B.: Types and ages of Kizil grottoes in Bai-cheng, Xinjiang. Researches of Stone Caves in China, Cultural Relics Publishing House, 34–35 (1996[1]). 宿白. 新疆拜城克孜尔石窟部分洞窟的类型与年代. 中国石窟寺研究. 北京:文物出版社, 34–23 (1996[1])
3. Su, B.: "Types and Chronology of Parts of Xinjiang Baicheng Kizil Caves" 宿白.新疆拜城克孜尔石窟部分洞窟的类型与年代 (1996[2])
4. Su, B.: Liangzhou Caves Ruins and 'Liangzhou Style', China Temple Caves Study. Cultural Relics Publishing House (1996[3]). 中国石窟寺研究,文物出版社 (1996[3])
5. The Kucha Academy of Xinjiang: Comprehensive Record of Contents of Kizil Grottoes. Urumqi: Xinjiang Fine Arts Photography Press, 4 (2000). 新疆龟兹石窟研究所. 克孜尔石窟内容总录. 新疆美术摄影出版社, 4 (2000)
6. The Kucha Academy of Xinjiang: Comprehensive Record of Kumtura of Kizil Grottoes. Urumqi: Xinjiang Fine Arts Photography Press (2008). 新疆龟兹石窟研究所. 库木土喇石窟内容总录. 新疆美术摄影出版社 (2008)
7. Su, B.: Buddhist Caves in China, Cultural Relics Publishing House, 11 (2010). 宿白. 中国佛教石窟寺遗迹. 文物出版社, 11 (2010)
8. Vignato, G.: Wooden architecture in front of the Kizil Caves. Heritage **10**, 75–83 (2004) 魏正中 (Giuseppe Vignato). 克孜尔石窟前的木构建筑. 文物, **10**, 75–83 (2004)
9. Department of Archaeology of Peking University and Baicheng Cultural Heritage Institute of Kizil Caves. Report on Excavation of Kizil Caves in Xinjiang, vol. 1, 1st edn. Cultural Relics Press (1997). 北京大学考古系、 克孜尔千佛洞文物保管所. 新疆克孜尔石窟考古报告 (第一卷、第一版). 文物出版社 (1997)
10. Su, B.: Periodization and Chronology of Parts of Kizil Caves: A Preface, in Kizil Caves, vol. 1, Cultural Relics Publishing House (1989). 宿白. 克孜尔部分洞窟阶段划分与年代等问题的初步探索—代序. 克孜尔石窟 (第一卷), 文物出版社, 1989
11. Gettens, R. J.: The materials in the wall paintings from Kizil, Chinese. Tech. Stud. Field Fine Arts **6,** 89 (1938)
12. Riederer, J.: Technik und farbstoffe der frühmittelterlichen Wandmalereien Ostturkistans. In H.H hcrausgegeben (ed.), Beltrage zur Indienforschung Preussischer Verödffentlichungen des Museums für Indische Kunst Berlin. Stiftung Kulturbesitz Staatliche Museen Berlin, 353–423 (1977)
13. Su, B.: Wall painting pigments research of Kizil Grottoes. Dunhuang Res. **1**, 65–75 (2000). 苏伯民, 李最雄. 克孜尔石窟壁画颜料研究[J]. 敦煌研究 **1**, 65–75 (2000)

14. Su, B.: HPLC analysis of binding medium of pigments of wall painting at Kizil Grottoes. Dunhuang Res. **4**, 57–61 (2005). 苏伯民. 克孜尔石窟壁画胶结材料的 HPLC 分析[J]. 敦煌研究 **4**, 57–61 (2005)
15. Tokyo University of the Arts: The Kucha Academy of Xinjiang. The Silk Road. Painting Materials and Techniques of Kizil Grottoes (2015) (in Japanese)
16. Zhou, G.: Lead-containing pigments in ancient murals and polychrome sculptures from the northwest China. Sci. Conserv. Archaeol. **24**(1), 95–101 (2012). 周国信. 中国西北地区古代壁画彩塑中的含铅白色颜料[A]. 文物保护与考古科学, **24**(1), 95–101 (2012)
17. Zhou, G.: XRD analysis of inorganic pigments of murals and polychrome sculptures of Maijishan Grottoes. Archaeology **1**, 744–750 (1991). 周国信. 麦积山石窟壁画、彩塑无机颜料的X射线衍射分析[J]. 考古 **8**, 744–750 (1991)
18. Xue, J., Ma, Q.: Analysis of mural pigment in Jiuquan and Jiayuguan, Gansu Province. Archaeology, **3**, 277–281 (1995). 薛俊彦, 马清林 等. 甘肃酒泉、嘉峪关壁画墓颜料分析[J]. 考古, **3**, 277–281 (1995)
19. Si, Y., Jiang, H.: Raman microspectroscopic analysis of polychrome wooden artifact from the Astana tomb of the Tang dynasty, Xinjiang, China. Spectrosc. Spectral Anal. **33**(10), 2607–2611 (2013). 司艺, 蒋洪恩. 新疆阿斯塔那墓地出土唐代木质彩绘的显微激光拉曼分析[A]. 光谱学与光谱分析. **33**(10), 2607–2611 (2013)
20. Azouka, A. Huggett, R. Harrison, A.: The production of shellac and its general and dental uses: a review. J. Oral Rehabil. **20**, 393–400 (1993)
21. Tamburini, D., Dyer, J. Bonaduce, I.: The characterisation of shellac resin by flow injection and liquid chromatography coupled with electrospray ionisation and mass spectrometry. Scientific Report 7, 14784 (2017)
22. Taniguchi, Y., Marini, C., Checkround, E., Okeda, H: Constituent material analysis of the Bamiyan Buddhist mural paintings (II): a study of pseudo-gold leaf technique discovered at Cave N(a) using synchrotron-based μFTIR (in Japanese). Conserv. Sci. **46**, 181–188 (2007)
23. Bayerová, T.: Buddhist wall paintings at Nako Monastery, north India: changing of the technology throughout centuries. Stud. Conserv. **63**(3), 171–188 (2017)
24. Duran, A., Perez-Rodríguez, J.L., Jimenez de Haro, M.C., Herrera, L.K., Justo, A.: Degradation of gold and false golds used as gildings in the cultural heritage of Andalusia, Spain. J. Cult. Heritage **9**(2), 184–188 (2008)
25. van der Werf, I.D., Gnisci, R., Marano, D., de Benedetto, G.E., Laviano, R., Pellerano, D., Vona, F., Pellegrino, F., Andriani, E., Catalano, I.M., Pellerano, A.F. Sabbatini, L. San Francesco d'Assisi (Apulia, South Italy): Study of a manipulated 13th century panel painting by complementary diagnostic techniques. J. Cult. Heritage **9**(2), 162–171 (2008)
26. Carretti, E., Natali, I., Matarrese, C., Bracco, P., Weiss, R.G., Baglioni, P. Salvini, A., Dei, L.: A new family of high viscosity polymeric dispersions for cleaning easel paintings. J. Cult. Heritage **11**(4), 373–380 (2010)
27. Yu, Z., Zhao, L., Li, F., Li, Z.: Analysis of pigment in wall paintings of Mati temple, Tianti mountain and Lingbing grotto temple. Dunhuang Res. **4**, 67–70 (2005). 于宗仁、赵林毅、李燕飞、李最雄. 马蹄寺、天梯山和炳灵寺石窟壁画颜料分析. 敦煌研究 **4**, 67–70 (2005)
28. Wang, J.: Study of unearthed blue pigment in Dunhuang grottoes. Archaeology **3**, 74–81 (1993) 王进玉. 敦煌莫高窟出土蓝色颜料的研究[J]. 考古 **3**, 74–80 (1993)
29. Xuanzang: (646) (Tang dynasty) Xuan Zang. Da Tang Xi Yu Ji, vol. 1. Da Zheng Zang(chapter 51): 870. (唐) 玄奘, 大唐西域记(卷一). 大正藏(51册): 870 页

Chapter 14
Wall Painting Materials and Techniques of the Mogao Grottoes

Su Bomin

Abstract Among sites on the World Heritage List, the Mogao Grottoes, Dunhuang, China, preserves the largest number of wall paintings, which embody painting techniques and materials used for hundreds of years. Following many years of analysis and research, the Dunhuang Academy, which is responsible for overseeing the caves, has succeeded in establishing the main pigments and organic colorants used in the creation of the paintings, and has developed a set of analytical approaches and techniques for better understanding their technology. As an example, this paper examines in detail the wall painting technology of Cave 98.

Keywords Wall painting · Mogao Grottoes · Pigment · Techniques · Analysis

14.1 Introduction

The Mogao Grottoes are located in Dunhuang at the western end of the Hexi Corridor in northwestern China. Constructed from the 4th to the fourteenth centuries, the Mogao Grottoes are 1680 meters long and preserve 735 caves (including 492 with wall paintings and painted sculptures), with over 2,000 painted sculptures, about 45,000 square meters of wall paintings and 5 original wooden porticos. Over the span of ten dynasties—the Sixteen States, the Northern Wei, the Western Wei, the Northern Zhou, Sui, Tang, the Five Dynasties, Song, the Western Xia and Yuan—the artistic styles of the wall paintings in the Mogao Grottoes developed. The type and content of the paintings cover seven categories: statue paintings, narrative paintings, sutra illustrations, illustrations of Buddhist history, illustrations of Chinese mythologies, portraits of donors, and decorative patterns. The discovery of the Library Cave at the Mogao Grottoes in 1900, together with the tens of thousands of manuscripts and relics it contained, has been acclaimed as the world's greatest discovery of ancient Oriental culture. In 1987, the Dunhuang Mogao Grottoes were inscribed on the World Heritage List by UNESCO.

S. Bomin (✉)
Conservation Insutitute of Dunhuang Academy,
Mogao Grottoes, Dunhuang 736200, Gansu, China

S. Aoki et al. (eds.), *Conservation and Painting Techniques of Wall Paintings on the Ancient Silk Road*, Cultural Heritage Science,
https://doi.org/10.1007/978-981-33-4161-6_14

14.2 Overview of Wall Painting Materials and Techniques of the Mogao Grottoes

The scientific research of the paint materials used in the creation of the Mogao Grottoes began in 1983 [1] and has subsequently been expanded by many scholars at the Dunhuang Academy [1–4]. From 1997 to 2010, the collaborative project of the Dunhuang Academy and the Getty Conservation Institute (Los Angeles, USA) to conserve Cave 85 also resulted in the accumulation of considerable information on original technology [5]. After 2000, the Dunhuang Academy similarly cooperated with the National Research Institute for Cultural Properties, Tokyo, Japan, to investigate wall painting materials and techniques in Cave 285 [2, 3, 6–9]. These are only a few of the many studies that have been carried out at the Mogao Grottoes.

14.2.1 Painting Materials

Based on these studies, a number of broad conclusions can be made about the technology of the wall paintings. The main inorganic pigments used are those listed in Table 14.1. As for organic colorants, only a few have been definitively identified so far, such as lac and indigo, although others are probably present but remain undetected. Studies of binding media used in the paintings have identified collagen-based proteinaceous materials, such as animal glue, in the majority of the caves examined from the Northern Liang (397–439) to the Yuan (1227–1368) dynasties [10]; the presence of amino acids in other analysed samples correlated broadly with fruit gums and mucilage in the form of glycosides, indicating the use of various plant

Table 14.1 The main inorganic pigments and ground materials used in the Mogao Grottoes wall paintings

Colour	Mineral
Red Pigments	Vermilion (α-HgS), Red lead (Pb_3O_4), α-Fe_2O_3 (including red ochre, iron oxide red, calcined red ochre), Realgar (As_4S_4), Litharge (PbO, Tetragonal)
Blue Pigments	Azurite ($Cu_3(CO_3)_2(OH)_2$), Lapis lazuli ($((Na,Ca)_8(AlSiO_4)_6(SO_4,S,Cl)_2)$)
Green Pigments	Malachite ($Cu_2CO_3(OH)_2$), Atacamite ($Cu_2(OH)_3Cl$)
Yellow Pigments	Orpiment (As_2S_3), Goethite (α-FeOOH)
White Pigments and ground materials	Kaolinite ($Al_2Si_2O_5(OH)_4$), Calcite ($CaCO_3$), Muscovite ($KAl_2Si_3AlO_{10}(OH)_2$), Talc ($Mg_3Si_4O_{10}(OH)_2$), Gypsum ($CaSO_4 \cdot 2H_2O$), Anhydrite ($CaSO_4$), White lead ($Pb_3(OH)_4CO_3$)
Black Pigments	Carbon black (C), Iron black (Fe_3O_4)

gums as binding media, too [10]. A collagen-based material such as bone glue or hide glue was identified as the principal binding medium used in Cave 85 [4].

14.2.2 Painting Techniques

The wall paintings of the Mogao Grottoes are composed of multiple layers applied to the conglomerate rock support. These comprise earthen plaster layers, sealants and ground layers, paint layers and glazes, coatings and surface decorations (Figs. 14.1, 14.2, and 14.3). Local earth deposits, and added sand and plant fibers are the three basic components of the plaster layers. Evidence indicates that alluvial deposits from the river in front of the grottoes were the main source of the earth ("Dengban soil"), which principally contains illite and chlorite in its clay fraction, matching the mineral composition of plasters that have also been analyzed. Generally two layers of plaster were applied to the conglomerate rock walls: the lower layer, varying in thickness from approximately 2–3 cm, is usually a coarse mixture of the local earth and a relatively low amount of added sand, with wheat straw, reeds and other coarse plant fibers; the upper finer layer, ranging in thickness from 2–5 mm, appears to have

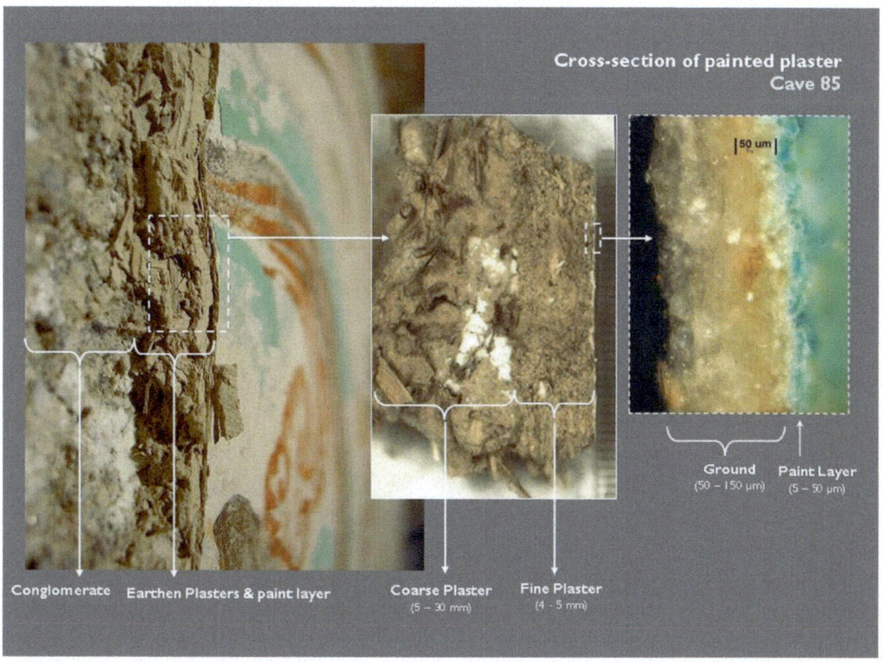

Fig. 14.1 The stratigraphy of the painted plaster in Cave 85 illustrates the basic structure of the wall paintings at the Mogao Grottoes

Fig. 14.2 Detail of the Late Tang dynasty painting on the west wall of Cave 85, Mogao Grottoes (left). Cross-section of a paint sample from Cave 85 showing the ground layer beneath the paint

Fig. 14.3 Analysis indicates that in Cave 85 a sealant layer composed of animal glue mixed with alum (KAl(SO4)2·12H2O) was applied over the upper plaster

slightly more sand added to the earth, and other additives include cotton, beaten hemp or wool, and other fine fibers.

Over the fine plaster, a sealant layer was applied to prepare the surface for the ground and paint layers. Literary sources on Chinese painting techniques and analysis indicate that these sealants were usually made of animal glue mixed with alum ($KAl(SO_4)_2{\cdot}12H_2O$). The ground layer of the wall paintings is generally composed of varying mixtures of white minerals such as talc, gypsum, calcite and muscovite. However, in the earlier wall paintings, such as those from the Northern Wei, no uniform ground layer is present, and instead primary paint layers such as red, pink and white are used as substrates according to desired optical effects.

Both prior to and during painting, a variety of techniques were used for setting-out and transferring designs, including incised and drawn lines, compass incisions, and cartoons. There is evidence too of preliminary drawing in dilute paint. As mentioned above and listed in Table 14.1, a wide range of inorganic and organic paint materials were employed. Gilded extruded work is widely employed as an applied decoration.

14.3 Investigative and Analytical Methods Used at the Mogao Grottoes

Following many years of research and development, the Dunhuang Academy has developed a set of analytical approaches for investigating the wall painting technologies of the Mogao Grottoes, based on a wide range of instrumental techniques and equipment. Although laboratory-based analysis of small samples taken from the paintings is still a component of this approach, in situ non-destructive investigative and analytical methods are now widely used, too. These latter methods include multi-spectral imaging, hyper-spectral imaging, portable microscopy, portable X-ray fluorescence spectroscopy, portable mid-Fourier transform infrared spectroscopy, portable vis-NIR reflectance spectroscopy, portable Raman spectroscopy, portable optical coherence tomography, portable laser-induced breakdown spectroscopy (LIBS). Laboratory analysis includes procedures such as X-ray diffraction, scanning electron microscopy-energy dispersion spectroscopy, cross-section analysis, and so on (Fig. 14.4). The full range of investigative and analytical equipment and procedures used at the Mogao Grottoes are shown in Table 14.2.

The main reasons for carrying out these investigative and analytical procedures are not only to determine the original materials and techniques of the wall paintings, but also to understand their long-term performance and changes over time.

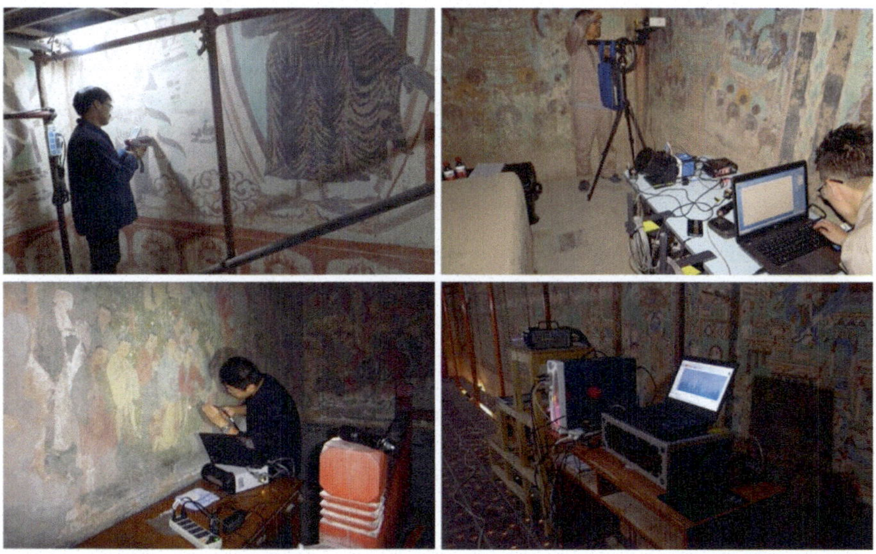

Fig. 14.4 Examples of in situ non-destructive investigative and analytical techniques used by the Dunhuang Academy at the Mogao Grottoes (upper left: XRF; upper right: Raman; lower left: reflectance spectroscopy; lower right: LIBS)

Table 14.2 The investigative and analytical equipment and procedures used at the Mogao Grottoes

Investigative/analytical method	Instrument	Use
Multi-spectral imaging	Modified camera and filters	Characterization of painting techniques
Hyper-spectral imaging	PHILUMINA NUVNIR-350	Mapping of pigments
Microscopy	Keyence VHX-600E	Microscopic observation of morphology
XRF	Thermo Fisher NITON XL3t-800	Elemental analysis
Mid-FTIR	Bruker ALPHA	Molecular identification
Reflectance spectroscopy	ASD LabSpec5000	Identification of pigments
Raman spectroscopy	Horiba Jobin Yvon HE785 and HE532	Molecular identification
OCT	Thorlabs Ganymede II	Observation of stratigraphy
LIBS	LIBSCAN 100	Elemental analysis
XRD	Rigaku Dmax/2500	Molecular identification
SEM-EDS	JEOL JSM-6610LV and Oxford INCA x-act	Microscopic observation of morphology and elemental analysis
cross-section analysis	Leica DMLP	Observation of stratigraphy

14.4 Investigation of the Painting Materials and Techniques Used in Cave 98 at the Mogao Grottoes

14.4.1 History and Description of Cave 98

Cave 98 was constructed in the period of Yijin Cao (AD 914–935) in the Five Dynasties. It is a ground-level cave located just south of the Nine-Storey Pagoda in the southern section of the grotto cliff face. The immense cave interior is decorated with 693 m^2 of wall paintings. Their subject matter includes 11 sutra illustrations on the four walls, "thousand buddha" schemes on the four slopes, and the Four Heavenly Kings on the corners of the ceiling. A pattern of coiled dragons and parrots is painted on the caisson. Important personages are depicted along the lower walls, including the king of Khotan, female donors of the Cao Family, and officials and famous monks. The rich and very extensive repertoire of painting makes Cave 98 the most important example of cave decoration of this period at Mogao (Fig. 14.5).

Fig. 14.5 The location of Cave 98 on the ground tier of the cliff face (left) and the portrait of the king of Khotan in the wall paintings (right)

14.4.2 Scope and Nature of the Technical Investigations and Analysis

Technical investigations and analyses were carried out on the ceiling, the four slopes, and the east wall, south and north walls of the main chamber of Cave 98. Non-destructive in situ methods and procedures included visual examination, hyperspectral imaging, portable microscopy (Fig. 14.6), portable XRF, NIR-Reflectance spectroscopy, and Raman Spectroscopy. Laboratory-based procedures included XRD and cross-section analysis.

Fig. 14.6 Examination of the wall paintings in Cave 98 using a portable digital microscope

14.4.3 Results and Discussion

14.4.3.1 Pigments

The analytical results demonstrate the use of a consistent palette in Cave 98. The red pigments include vermilion, red lead, hematite and litharge; green pigments include atacamite; the blue pigment is lapis lazuli; yellow pigments include orpiment and yellow iron oxide; white pigments and ground materials are comprised of white clay minerals; and the black pigment is carbon black. Gold is also present. In areas of brown-black painting, lead dioxide was detected, testifying to the alteration of lead-based pigments.

A curious discovery was made in an area of green painting on the north wall, where Emerald green was identified (Fig. 14.7). Emerald green (sometimes known as Paris green) was first manufactured in the nineteenth century [11]. This finding indicates that the paintings in Cave 98 were partly restored in their later history.

The original palette must also have included a range of organic colorants. However, the light-induced fading and deterioration of organic materials makes their detection difficult, and further research is necessary in this area.

14.4.3.2 Painting Application Techniques

The painting in Cave 98 is characterised by skilful paint application techniques. In the depiction of the Amitābhasutra on the south wall, for example, following the application of the preparatory white ground layer, the paintwork was built up in a sequence of layers or distinct areas. Red preparatory lines were employed to establish the composition, over which passages of paint were superimposed or applied in adjacent areas, using atacamite green, red lead and hematite red, and lapis lazuli blue, as shown in Fig. 14.8. Final outlines were made in black. Although organic colourants have mostly disappeared, there is remaining evidence to suggest that they were once abundant.

The faces of the donors on the lower walls of the cave provide further evidence of how the paintings were created. On these, the white ground layer provides the main background for the skin colour, over which features such as the eyes and eyebrows, cheek patches and jewellery were established by a variety of pigments, applied singly, or in combination, or in overlapping layers (Fig. 14.9).

The "thousand buddha" schemes on the slopes of the ceiling demonstrate division of artistic labour. There are 703 well-preserved Buddha figures, which are painted the same size (about 48 cm high and 28 cm wide), each identified by an inscription. The figures are depicted as alternating pairs, which systematically duplicate colour conventions and paint sequences (Fig. 14.10). These types of paint application procedures indicate how painting practice and designated tasks were carefully divided among the original artists.

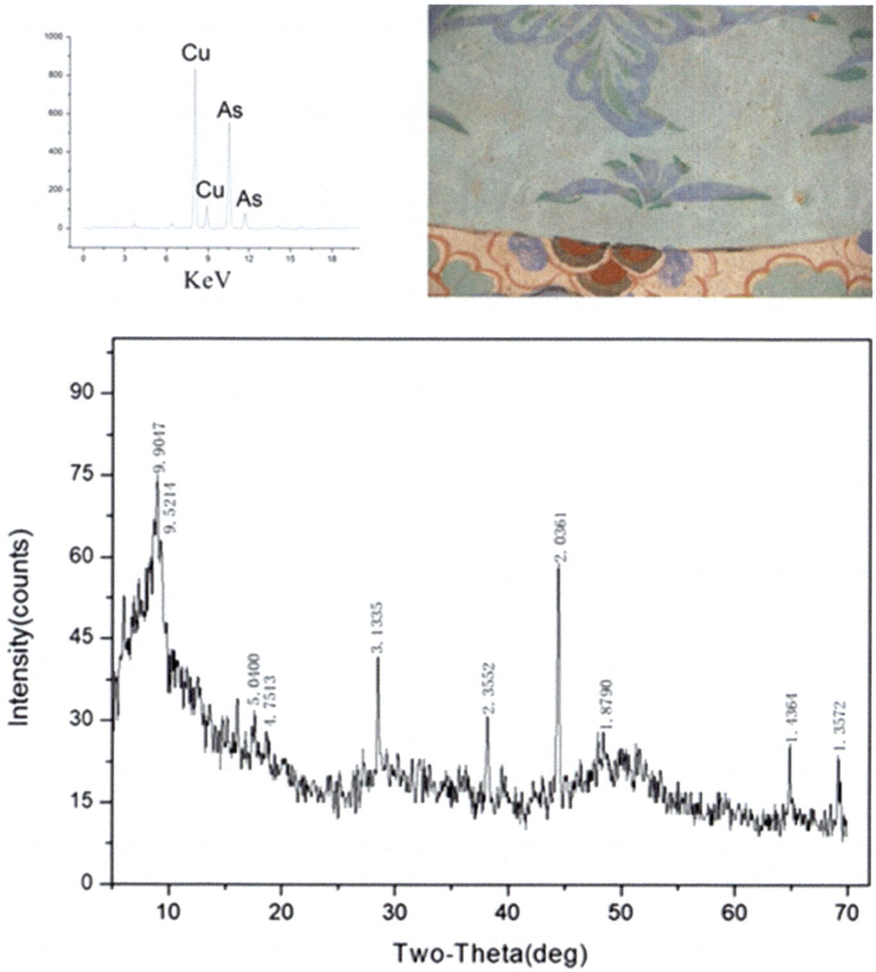

Fig. 14.7 Use and identification of Emerald green (sometimes known as Paris green) in the partial restoration of the wall paintings in Cave 98

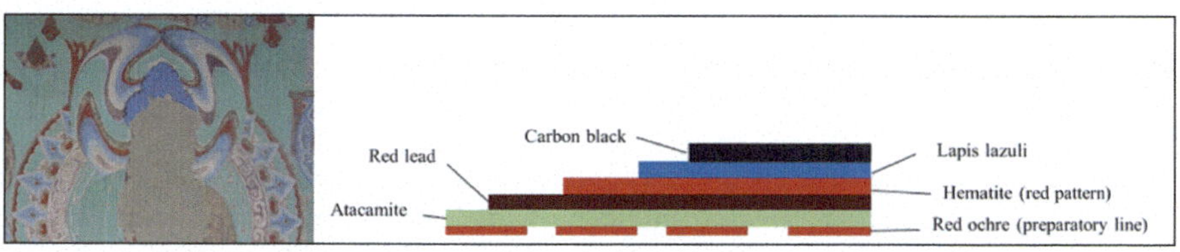

Fig. 14.8 Layering of paint in the depiction of the Amitābhasutra on the south wall of Cave 98

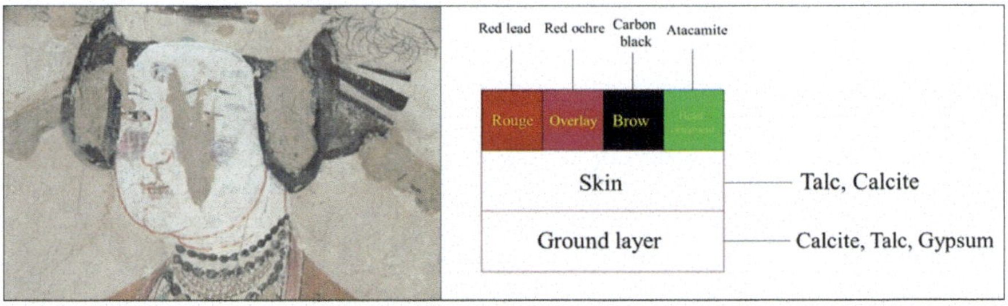

Fig. 14.9 Superimposition of paint layers on the face of one of the donors in Cave 98

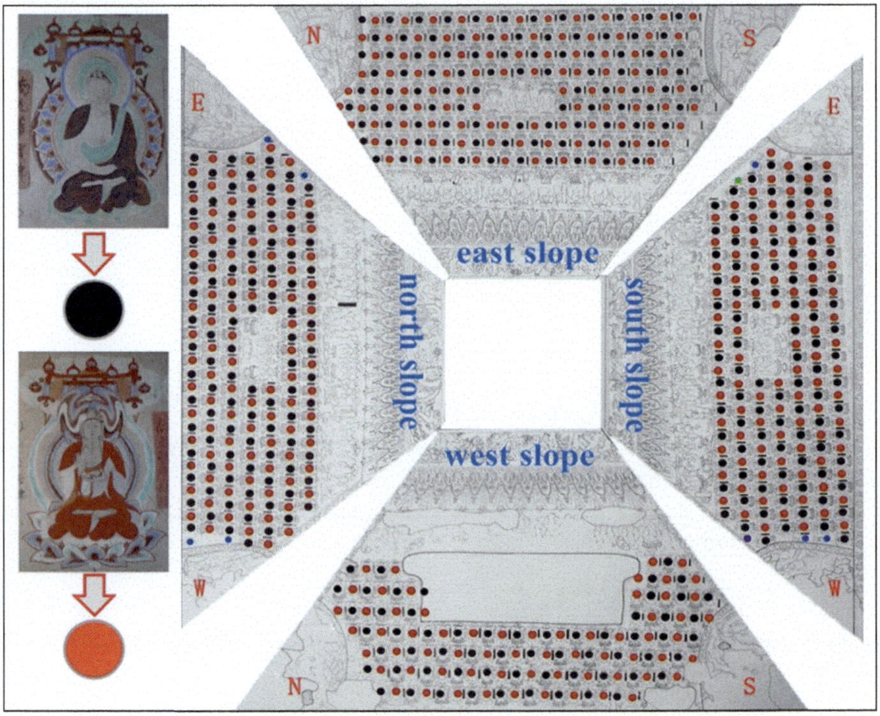

Fig. 14.10 The "thousand buddhas" in Cave 98

Among the 11 sutra illustrations in Cave 98, the Gratitude Sutra and the Lotus Sutra are distinguished by a large number of inscriptions painted on rectangular cartouches. The background colours of the cartouches are predominantly grey and red, but some are also pink, orange or green. Although some of this variation is simply due to the use of different colours, preferential organic coatings are also visible on some of the cartouches. Alteration and preferential exfoliation of these coatings also contributes to the colour differences.

14.5 Conclusion

The Buddhist wall paintings of the Mogao Grottoes reflect the changing nature of painting materials and techniques over a 1,000-year period. They are a unique record of the interchange and integration of painting technologies along the Silk Road. Based on years of research and analysis, the Dunhuang Academy has developed a considerable understanding of the wall painting materials and techniques used at the Mogao Grottoes. Nevertheless, this remains an incomplete task. The complexity of both the paintings and their deterioration stand in the way of their complete understanding. Questions regarding the sourcing of paint materials remain little researched. The presence and use of organic components in the paintings continue to challenge analytical detection. Much work remains to be done, which will be continued by the Dunhuang Academy.

References

1. Xu, W., et al.: X-ray analysis of Mogao Grottoes wall paintings and painted sculptures inorganic pigments. Dunhuang Res. (inaugural issue), 187–197 (1983) (in Chinese)
2. Fan, Y., et al.: Multi-spectral imaging and non-invasive investigation on the south wall of the Cave 285 at Mogao Grottoes(I). Dunhuang Res. **5**, 49–53 (2007) (in Chinese)
3. Inuzuka, M., et al.: Blue materials and techniques for the east wall in Mogao Cave 285. Sci. Conserv. **52**, 71–79 (2013) (in Japanese)
4. Schilling, M.R., et al.: Analytical research in Cave 85. In: Conservation of Ancient Sites on the Silk Road, pp. 438–449. The Getty Conservation Institute, Los Angeles (2010)
5. Wong, L., Agnew, N.: The Conservation of Cave 85 at the Mogao Grottoes, Dunhuang, pp. 155–189. Getty Conservation Institute, Los Angeles, CA (2013)
6. Research on the scientific recovery of Dunhuang art. Tokyo: Tokyo National Research Institute for Cultural Properties (2014) (in Japanese)
7. Takabayashi, H., et al.: Non-contact analysis of color materials used in the mural of Mogao Cave 285. Sci. Conserv. **47**, 89–101 (2008) (in Japanese)
8. Sato, K., et al.: Colouring technique on the mural of north wall in Mogao Cave 285. Sci. Conserv. **48**, 75–84 (2009) (in Japanese)
9. Takabayashi, H., et al.: Colouring materials and technique on the south wall in Mogao Cave 285. Sci. Conserv. **48**, 85–98 (2009) (in Japanese)
10. Li, S.: Quantitative analysis of binding media in Dunhuang wall paintings. Dunhuang Res. **3**, 29–46 (1995) (in Chinese)
11. Eastaugh, N., et al.: Pigment Compendium A Dictionary and Optical Microscopy of Historical Pigments, pp. 155–156. Butterworth-Heinmann (2008)

Chapter 15
Analysis and Diagnosis of the Buddhist Wall Paintings in the Josadang Shrine, Buseoksa Temple, Korea

Hwa Soo Lee, You Na Song, Gyu Seong Han, and Kyeong Soon Han

Abstract Korea's National Treasure No. 46, the Josadang Shrine of the Buseoksa Temple, was produced in the 3th year of King Woowang's rein in the Goryeo Dynasty. It is of critical research importance, as it preserves the earliest Buddhist wall painting in Korea. Investigations of the mural identified notable physical damages, such as cracks, peeling and exfoliation of the paint layers, as well as excessively glossy residues and stains left by previous fixing attempts. The wall paintings are mainly affected by traverse cracking, but cracks, peeling and other forms of damage are also present in an area where a reinforcing mortar was used. Since salt deterioration was also associated with the reinforced area, correction measures are required. It is difficult to understand fully the nature of the wall painting stratigraphy because it is concealed by a wooden frame applied during the Japanese colonial period. The structure of the wall paintings was therefore determined by carrying out ground penetrating radar (GPR). At least two earth-based layers are present. Ultrasonic testing revealed that the wall paintings generally have good physical properties. However, ultrasonic velocities were very low around cracks on the mortar-reinforced parts, implying poor conditions in these areas. In addition, thermal imaging identified

H. S. Lee (✉)
Department of Formative Arts, Konkuk University,
268 Chungwondae-ro, Chungju-si, Chungcheongbuk-do 27478, Republic of Korea
e-mail: comp@kku.ac.kr

Y. N. Song
CON-TECH, Korea National University of Cultural Heritage,
367 Baekjemun-ro, Buyeo-gun, Chungcheongnam-do 33115, Republic of Korea
e-mail: yuna3517@naver.com

G. S. Han
Department of Wood and Paper Science, Chungbuk National University, 1 Chungdae-ro, Seowon-gu, Cheongju-si, Chungcheongbuk-do 28644, Republic of Korea
e-mail: wood@chungbuk.ac.kr

K. S. Han
Institute of Conservation of Paintings, Konkuk University,
268 Chungwondae-ro, Chungju-si, Chungcheongbuk-do 27478, Republic of Korea
e-mail: conservation@kku.ac.kr

S. Aoki et al. (eds.), *Conservation and Painting Techniques of Wall Paintings on the Ancient Silk Road*, Cultural Heritage Science,
https://doi.org/10.1007/978-981-33-4161-6_15

areas of internal plaster separation. Surface moisture measurements varied on the wall paintings, partly according to when these were made. On the other hand, surface temperature readings of the wall paintings tended to decrease consistently in sequence from left to right, regardless of the measurement period.

Keywords Buddhist Wall Painting · Josadang Shrine of Buseoksa Temple · Analysis · Diagnosis · Conservation

15.1 Introduction

A comprehensive understanding of wall painting conservation has only recently gained acceptance and importance in Korea. Previous treatment efforts were marked by a number of trials and errors, as their focus was on remedial interventions that were made in the absence of an agreed philosophy of conservation, which ultimately led to the damage of cultural assets. Most Buddhist mural paintings in Korea are earth-based. Therefore, once damaged, they are difficult to preserve. For the same reason, a number of works are still exposed without repair, while others were lost during the restoration of the buildings that contained them [1].

The history of conservation of Buddhist wall paintings in Korea started in 1916 with the treatment of the wall paintings of the Josadang Shrine, Buseoksa Temple, in Yeongju city. Since then, gradual improvements have been made in conservation theory and practice. However, the earliest conservation efforts merely copied the methods and materials used in Europe at this time, and little research was done to assess their properties and outcomes. In particular, the use of acrylic materials and lime-based treatments between 1980 and 1990 resulted in various adverse side-effects. Since 2000, wall painting conservation approaches have been developed on the basis of more advanced technologies and better scientific studies, to compensate for past mistakes [2].

National Treasure No. 46, Wall Painting of the Josadang Shrine, in Buseoksa Temple, consists of six Buddhist wall paintings on the inner wall of the Josadang Shrine in the precinct of Buseoksa Temple. Based on ink-written letters discovered during repair-works, they are deemed to have been produced in the third year (1377) of King U, in the Goryeo Dynasty. The wall paintings are therefore reported as the oldest existing scheme in a Korean temple complex, and have important research value for the history of wall paintings in the country. They were applied on earthen walls using ‘dry’ painting techniques. Regarding their subject matter, the image of Indra is on the left side of the exhibition room, and that of Brahma Deva on the right side; four Devas are located at the center.

The six wall paintings were removed from their original built context in 1916, when the Josadang Shrine and Muryangsujeon Hall were disassembled and repaired (Fig. 15.1). In 1925, they were conserved and presented in wooden box-frames by engineers from the Japanese Ministry of Culture in Tokyo, and then transferred to the Muryangsujeon Hall in Buseoksa Temple, though it seems they may have been

Fig. 15.1 The Buddhist wall paintings of the Josadang Shrine before their detachment in 1916. *Photo source* Baek [3, pp. 73–75]

Fig. 15.2 Cleaning of the painting in 1985. *Photo source* Baek [3, p. 71]

kept in storage there and only put on display after 1985. Their exact physical history remains uncertain (Fig. 15.2).

In 1985, the wall paintings housed in Muryangsujeon Hall were transferred to storage at Bojanggak and underwent conservation treatment under the auspices of what is now the Cultural Heritage Administration. The conservation treatment comprised salt-reduction procedures, cleaning (Fig. 15.3), and consolidation to reinforce the painting layers. For removing salts from the surface of the wall paintings, poulticing techniques developed in Florence, Italy were tried. For consolidation, the acrylic co-polymer, Paraloid 72®, was applied in concentrations varying from 1.5 ~ 25% [3]. Following their treatment in 1985, maintenance of the wall paintings was almost non-existent, resulting in damage to some of the paintings, and the accumulation of dust and contaminants on their surfaces. To address this situation, their display environment was improved in 2002 by changing the lighting conditions and introducing temperature/humidity controls. Some remedial treatment of delaminated areas of painting was also carried out. The wall paintings were then transferred to a newly built museum at the temple complex in 2010, and displayed in their current exhibition cases (Fig. 15.4). A recent survey identified serious physical damage to the wall paintings [4]. A principal problem is transverse cracking in the plasterwork, but more particularly damage is also concentrated in areas where gypsum plaster was

Fig. 15.3 Treatment of the painting in 2002

Fig. 15.4 Buddhist wall paintings of the Josadang Shrine, Buseoksa Temple (National Treasure No. 46)

used for reinforcement. It was therefore necessary to establish plans to conserve the wall paintings based on appropriate diagnosis and environmental assessment. This study presents the results of the first non-destructive investigation of the conservation status of the wall paintings, which was carried out to provide useful information for their preservation.

15.2 Method

Scientific investigations and analysis were performed to understand the characteristics of the materials and structure of the wall paintings. For the plaster and painting layers, as well as the reinforcing plaster applied in the previous treatment, a field emission-type scanning electron microscope (Ultra-plus, Zeiss, Germany) with energy dispersive x-ray light splitter (NS7, Thermo Fisher Scientific, USA) (SEM-EDX) was used to study the morphology and chemical components of the particles of the materials. To identify the mineral composition and particle-size distribution of the earthen plaster, x-ray diffractometry (XRD) analysis (D/MAX-2500/PC, Rigaku, Japan), particle-size analysis (analytical sieves, JIS Z8801, Japan), and laser particle-size analysis (Mastersizer 2000, Malvern, England) were carried out. In addition, fibre species were identified by determining the optical characteristics of their cell structures (such as type, shape, size, and arrangement), using a microscope (ECLIPSE 80i, Nikon, Japan).

The condition of each wall painting was carefully recorded, using high-resolution imaging, Adobe Illustrator® and Auto CAD®. Types and extent of damage were precisely documented. UV-examination was carried out to reveal and confirm areas of retouching and other non-original repair materials. Examination and imaging in the infrared range were performed to identify features such as underdrawing. Thermal imaging was used to determine areas of separation and attachment within the stratigraphy of the wall paintings by comparing differential temperature readings. In order to understand other differences in material properties, ultrasonic measurements were taken (Ultracon-170, MKC-korea, Korea). A spot thermometer (HT-11, Minolta, Japan) was used to measure the surface temperature of the wall paintings. The moisture content of the murals was measured using a moisture meter (Testo 616, Testo, Germany). Ground penetrating radar (GPR) was conducted in three-dimensional mode to identify the location and shape of internal features of the wall paintings, using a Structure Scan Mini HR, with a penetration depth of about 11.8 inches (Geophysical Survey Systems, Inc. USA).

15.3 Results

15.3.1 Earthen Materials Analysis

X-ray diffraction (XRD) was used to examine the crystal structure of the particles composing the finishing layer of the support. Quartz, the main component of sand, was identified as a major mineral, and felspars (such as anorthite), a product of stone weathering, were also found (Fig. 15.5). SEM-EDX analysis of both the finishing plaster and the layer below identified sandy particles and clay micro-particles, forming an aggregated mixture. The principal chemical components were silica (Si) and aluminum (Al), confirmed the presence of a clay-soil mixture (Fig. 15.6).

Sieve-size analysis of the sandy fraction of the finishing plaster revealed that particles larger than medium sands represented 37.11%, and those smaller than fine sands, 62.89%. This suggests that the ratio of these components in the plaster was about 3.7:6.3 (Fig. 15.7). In laser particle size analysis, fine sand accounted for 41.52%, silt accounted for 57.96%, and clay (particles smaller than 2 μm), 1.98% (Fig. 15.8).

Meanwhile, fibres mixed in the finishing layer were identified through optical examination as hemp (*Cannabis sativa*). Observations of dissociated fibres showed

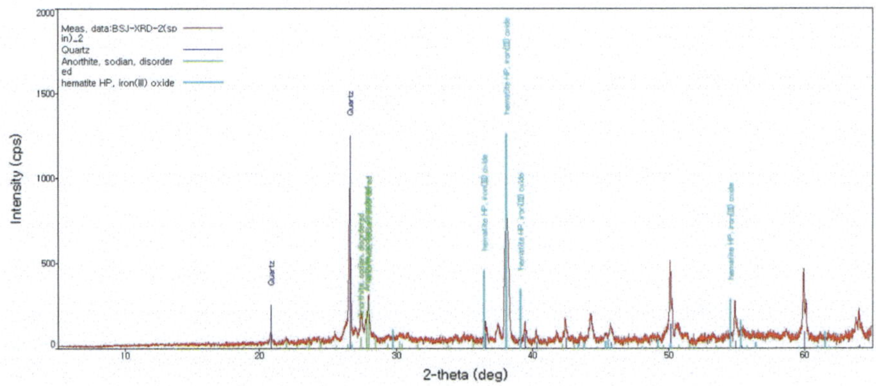

Fig. 15.5 XRD analysis of the finishing plaster layer (the Brahma Deva)

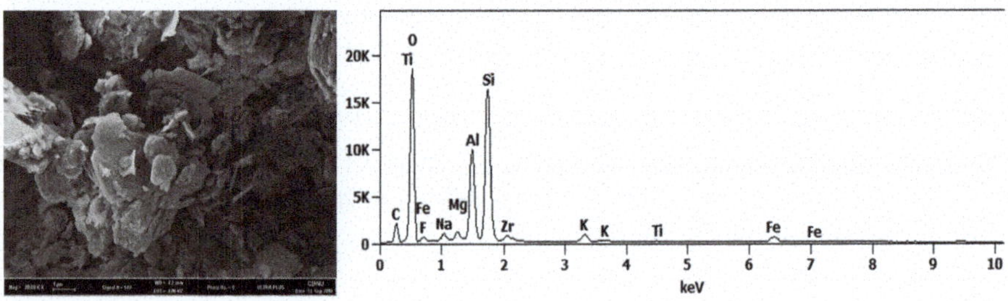

Fig. 15.6 Results of EDX analysis of the support layer (the Four Devas, South)

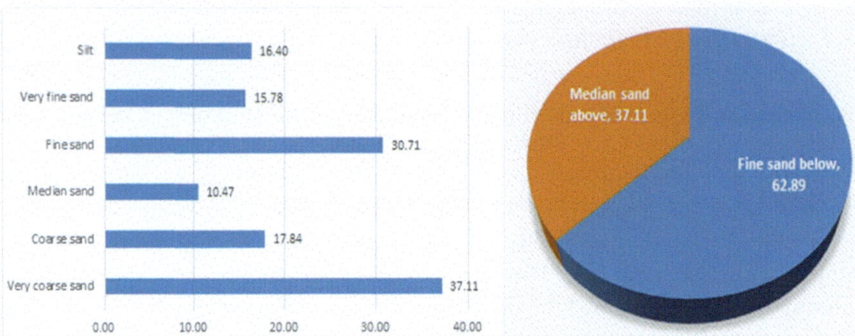

Fig. 15.7 Particle size distribution in the finishing plaster layer (the Brahma Deva)

no epidermal cells but fibre cells, which classified them primarily as bast fibres. Longitudinal streaks, nodes and fibre distortions were also observed, classifying the fibres as hemp. Fibre widths ranged from 12.8–46.0 μm, with an average of 26.6 μm, which is also consistent with hemp identification (Fig. 15.9). Although

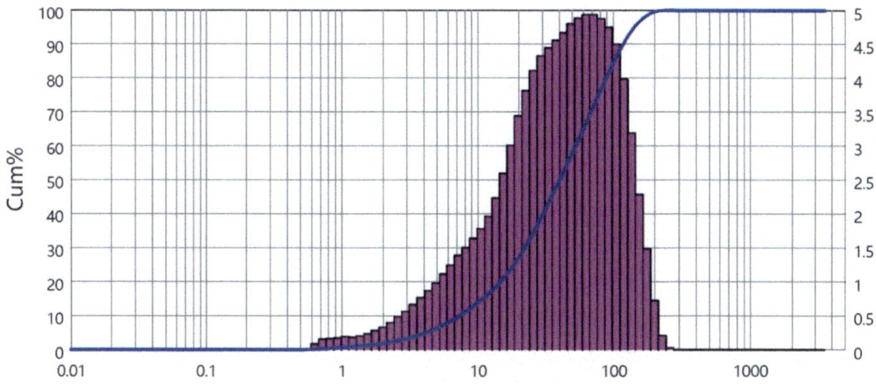

Fig. 15.8 Laser Particle size distribution in the finishing plaster layer (the Brahma Deva)

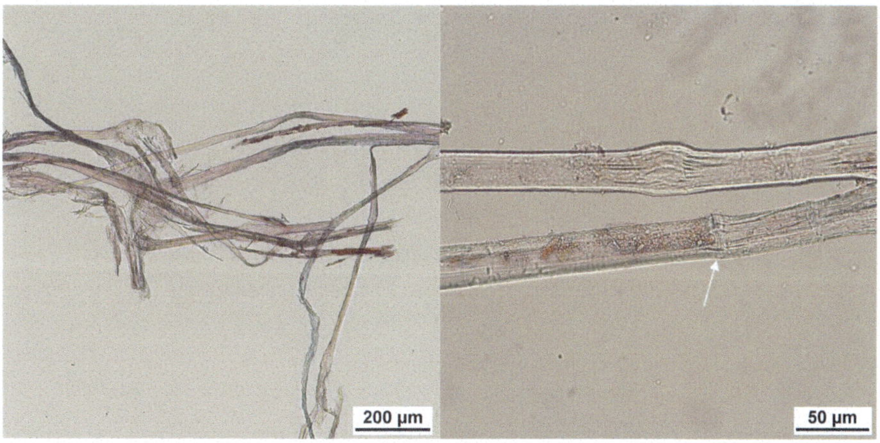

Fig. 15.9 Microscopic examination of fibres from the finishing plaster layer (left: dissociated fibre, right: fibre node)

flax has a similar width-range, this plant was reportedly only introduced to Korea in the 1900s, via Japan.

In summary, the two layers that constitute the secondary supports of the wall paintings are composed of sandy earthen plasters, with hemp included in the upper layer as an additive. This study also analyzed the wood species of the interwoven branches that are incorporated into the structure of the wall paintings, identifying it as originating from a pine tree (Fig. 15.10). The wood used for the wall painting frames was identified as belonging to the Cupressaceae family in the Chamaecyparis genus (Fig. 15.11). Analysis of the reinforcing plaster applied in the previous treatment of the murals identified gypsum ($CaSO_4$). On samples examined by SEM, typical gypsum particles of angular columnar shape were observed (Fig. 15.12). EDX analysis identified the main elements as calcium (Ca) and sulfur (S) (Fig. 15.13). Raman analysis identified gypsum. The combined results confirmed this identification (Fig. 15.14).

Two pigment samples were examined and analyzed, a red and a green. The paint is applied as a solid, dense layer on the upper plaster, having a thickness of approximately 2 μm ~ 4 μm (Figs. 15.15 and 15.16). Iron (Fe) was the major component identified in the red pigment layer (Fig. 15.17). Iron (Fe), potassium (K), alumina (Al), silica (Si) and magnesium (Mg) were the components in the green pigment layer (Fig. 15.18). These findings indicate that the red is an iron oxide, and the green is celadonite ($K(Mg,Fe^{2+})(Fe^{3+},Al)[Si_4O_{10}](OH_2)$). In both paint samples, alumina (Al) and silica (Si) were also present, originating from the earthen plaster below.

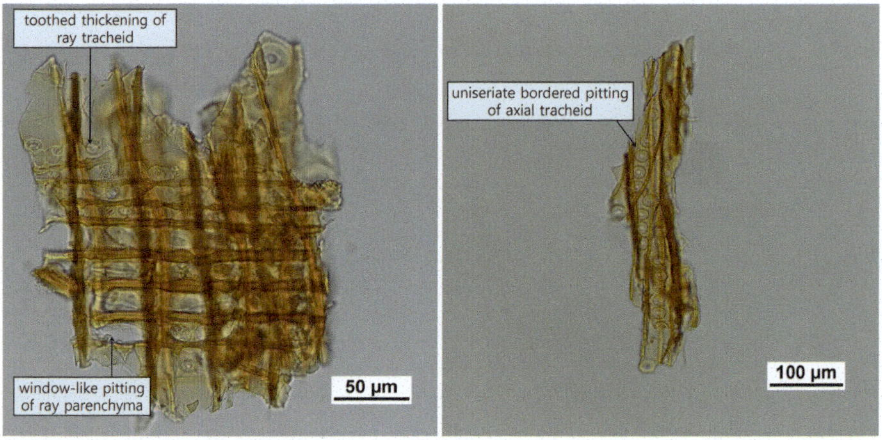

Fig. 15.10 Analysis of the wood species used for the frame

Fig. 15.11 Analysis of the wood species of the interwoven branches in conported into the of the wall painting

Fig. 15.12 SEM image of the repair plaster showing the presence of characteristic gypsum crystals

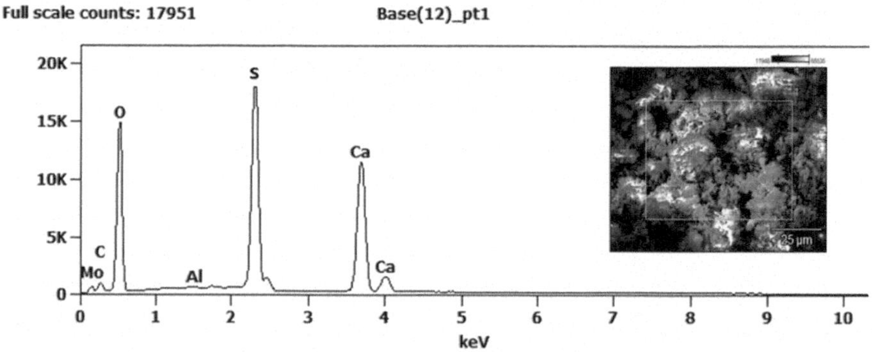

Fig. 15.13 EDX analysis identified the main elements of the repair plaster as calcium (Ca) and sulfur (S)

15.3.2 Diagnosis and Investigation

15.3.2.1 Investigation of the State of Conservation

Some of the most serious problems affecting the wall paintings were present in the paint layers rather than in the plasters, manifesting as exfoliation, delamination and contamination. Deterioration was especially concentrated in the ground-layer, where plate-like exfoliation occurred. Much of the painting along the lower edges of the paintings had already flaked and powdered off, and this zone also showed evidence of physical abrasion. Cracking and separation in the original plaster layers were

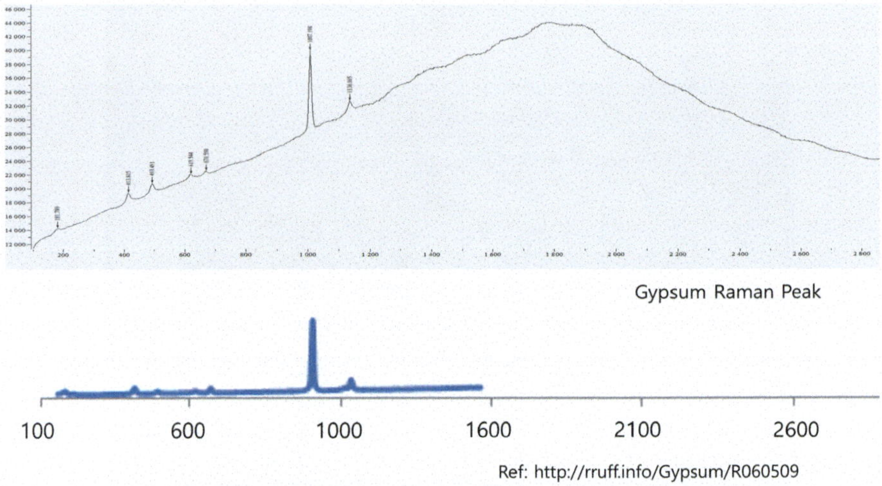

Fig. 15.14 Results of Raman analysis, confirming the repair plaster as gypsum

Fig. 15.15 Cross section of green pigment layer (×80)

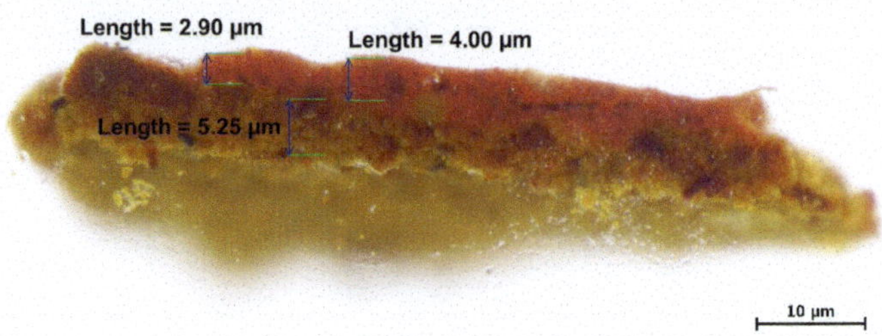

Fig. 15.16 16 Cross section of red pigment layer (×100)

also destabilizing the gypsum repairs. Much of this damage and deterioration can be attributed to past events and previous treatments: for example, the wall paintings have a saturated appearance due to the presence on them of applied consolidation

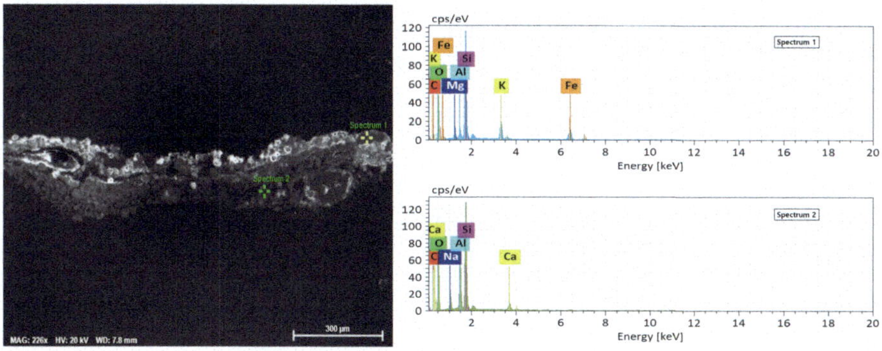

Fig. 15.17 SEM image and EDX elemental peaks for the green pigment

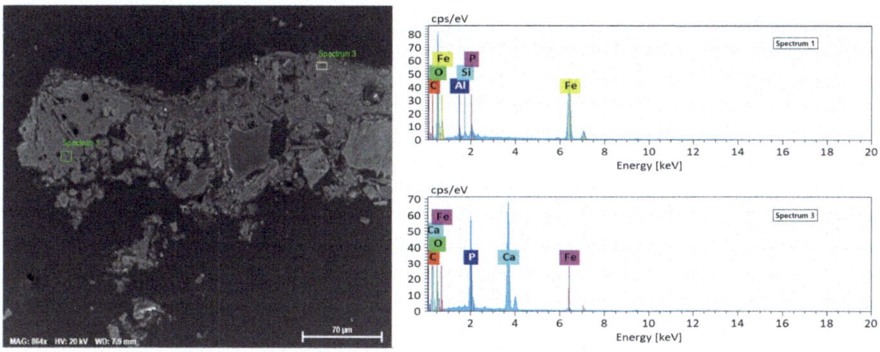

Fig. 15.18 SEM image and EDX elemental peaks for the red pigment

materials. Unfortunately, some of the most important areas of painting have been reduced to a very poor condition (Fig. 15.19).

Visual investigation revealed that a white surface contamination was very severe on the paintings of Indra and Brahma Deva, and on the murals of the East, West, North and South Guardian Kings (Fig. 15.20).

15.3.2.2 Optical Investigation

Ultraviolet investigation of the wall paintings revealed a number of surface contaminants, such as accumulated dust, that could be partly distinguished by their differing fluorescent properties. The UV-fluorescence also helped identify the presence of the reinforcing gypsum repairs, which was not always easily possible in normal light. On some of the murals, fluorescence differences revealed remaining fixatives on the surface. Infrared examination confirmed ink-lines that had mostly not been previously observed. On the Indra and Brahma Deva murals, outlines of ornaments and

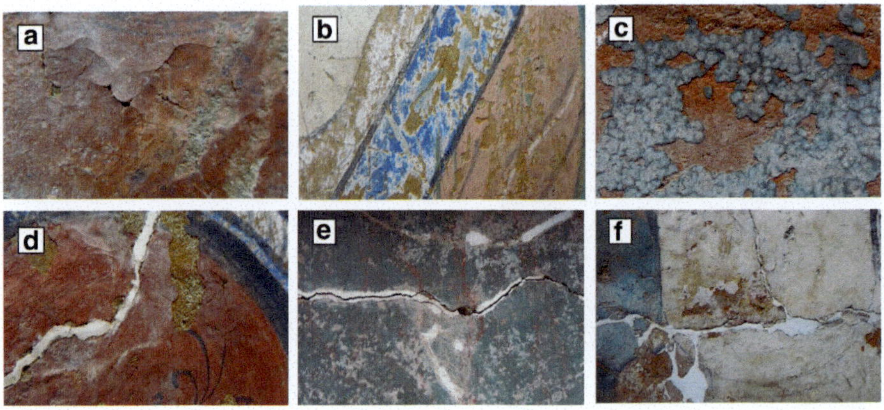

Fig. 15.19 Types of damage A: paint cracking and exfoliation; B: loss of paint layers; C: damage from fixing and consolidation attempts; D: damage from reinforcing materials; E: plaster cracking; F: exfoliation

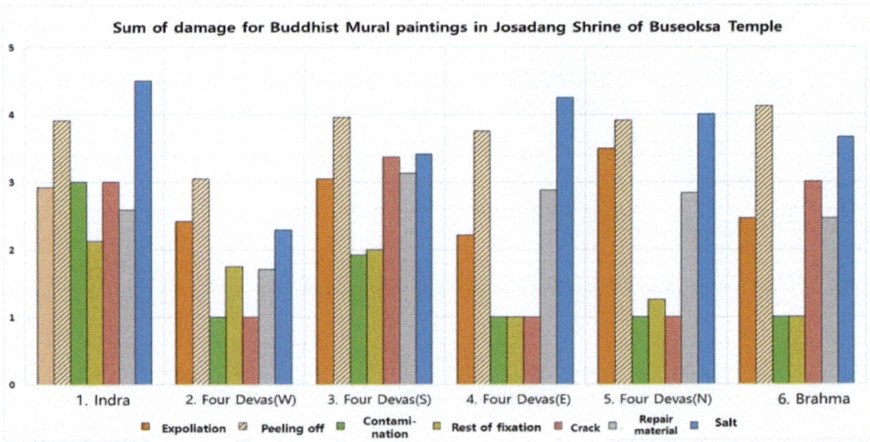

Fig. 15.20 Bar charts showing the proportions of different types of damage affecting the wall paintings

clothing were detected, which cannot be seen under normal light conditions; and on the West Guardian King, animal patterns on both his shoulders were clearly revealed. In the painting of the South Guardian King, an inscription reading "South Guardian King" was found in the top left of the wall painting. Although the infrared investigations carried out so far have been very effective, more are necessary to record other concealed technical features of the paintings (Fig. 15.21).

Multispectral imaging in the 900–1700 nm wavelength range revealed clear ink-lines hidden under areas of green and blue painting in the wall paintings of Indra and Brahma Deva, such as on the halos around the heads and in the costumes. Ink-lines

Fig. 15.21 The optical investigation using UV fluorescence helped identify the presence of the reinforcing gypsum repairs

concealed by green paint in the Brahma Deva wall paintings were clearly done to suggest the shape of the costume painting. Delicate ink-lines detailing adornments on costumes and other compositional features were also found in the Four Guardian Kings including, most notably, sketches of the spiritual beasts decorating the armor. To reveal and record some previously unnoticed features, such as details of lettering in the Brahma Deva mural, careful adjustments had to be made in the infrared wavelengths (Fig. 15.22). A synthesis of the data obtained using various wavelengths provided a comprehensive mapping image of condition and technical features that had not previously been visualized (Fig. 15.23).

Fig. 15.22 Hyperspectral imaging at 1600 nm to reveal concealed ink-lines in the Indra wall painting

Fig. 15.23 Composite image of data revealed by hyperspectral imaging at various wavelengths (part of the Indra wall painting)

15.3.2.3 Infrared Thermal Imaging Investigation

Thermal imaging was a useful tool for detecting areas of potentially endangering separation within the stratigraphy of the wall paintings, by comparing differential temperature readings with those of attached areas. Additionally, temperature differences were recorded in different passages of painting, which stems from the varying heat absorption and reflection rates of paint materials and surfaces. The presence of gilding in some areas was almost impossible to detect under normal conditions, due to its almost complete loss. But since gold has a high heat reflection rate compared with other pigments, even trace remains of gilding were distinguished by thermal imaging. The reinforcing gypsum repairs showed a higher temperature distribution than the earthen plaster surfaces, probably reflecting a difference in density between these original and added materials (Fig. 15.24).

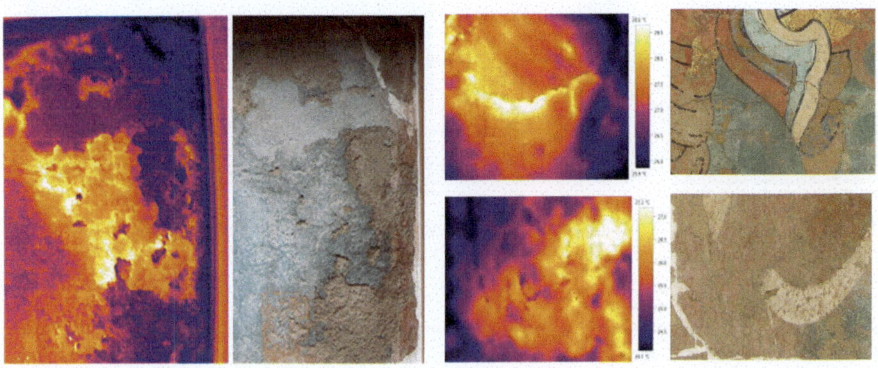

Fig. 15.24 Examples of results of the investigation of the wall paintings using infrared thermal imaging

15.3.2.4 Ultrasonic Testing

Ultrasonic data were used to produce a contour map of the relative velocity distribution of the wall paintings' surfaces. The investigations were carried out according to an indirect measurement method, whereby small ultrasonic probes (54 kHz) positioned at a distance of 150 mm from each other were operated at 1200 V.

The average ultrasonic speed of the wall painting surface was 493.3 m/s. The minimum value was 70.3 m/s, recorded on the East Guardian King, while the maximum value was 1288.3 m/s, on the West Guardian King. The order of the average values was North Guardian King, West Guardian King, Indra, Brahma Deva, South Guardian King, and East Guardian King.

Ultrasonic measurements were made along the plaster cracks to understand their properties. Their average ultrasonic speed was 168.6 m/s: the minimum value was 83.0 m/s for Brahma Deva, and the maximum value was 611.3 m/s for Indra. The order of the average values for cracks was Indra, East Guardian King, West/South Guardian King, Brahma Deva, and North Guardian King. Mapping these results confirmed that the shape or status of some of the interwoven branches that form the support of the wall paintings (see discussion below) affected ultrasonic speeds. In addition, the ultrasonic speeds of cracks were all low. It can be assumed that degradation of some of the embedded branches is responsible for undermining the stability of the wall paintings. Generally, ultrasonic speeds decreased at the bottom of the wall paintings, too. This may be related to events or circumstances that occurred when the wall paintings were in storage, but these factors need to be further investigated (Fig. 15.25).

The differences in ultrasonic values were very varied, even within a single wall painting. The overall nature, condition and conservation status of each mural influenced these results, including their complex stratigraphies and the composition of their constituent materials, and other factors such as the density and thickness of the layers, and the presence of cracks, and of attached and separated areas. As such, the

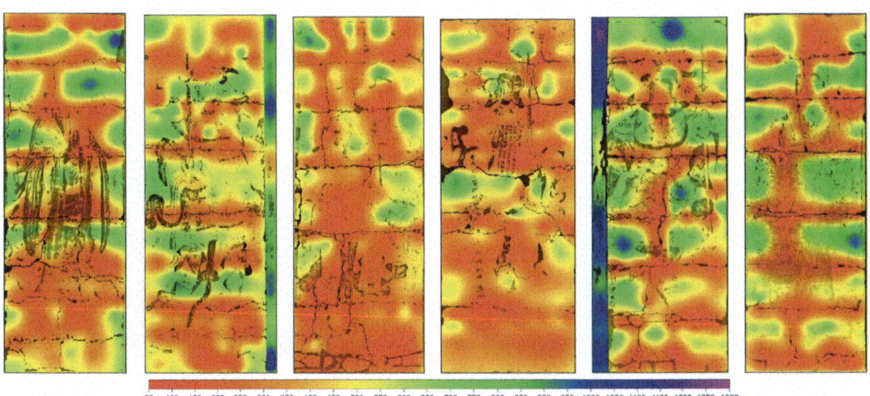

Fig. 15.25 Combined results of the ultrasonic imaging investigation

ultrasonic inspection may not be the most reliable tool for understanding the physical properties of the wall paintings. However, since similar ultrasonic recording techniques have been used for diagnosing artifacts under similar conditions, this comparative material provides a body of data for evaluating the value ranges found on the wall paintings.

In this study it was considered that the ultrasonic inspection provided objective data about certain areas that required conservation treatment, such as areas of cracks and internal separation [5].

15.3.2.5 GPR Exploration

The GPR (Ground Penetrating Radar) investigation revealed an interesting technical feature shared by all six murals: at a depth of about 30 mm within the mural structure are embedded densely woven branches, spaced at regular intervals of about 30 mm. These provide the primary support for the earthen plaster. It had not previously been possible to determine their presence, because the wooden frames conceal stratigraphic information (Fig. 15.26). Although some signals were misleading due to multiple reflected waves, the results generally confirmed the vertical and horizontal arrangement of the branches. This new technical information demonstrated that the support structure of the wall paintings in the Josadang Shrine of the Buseoksa Temple is similar to that of other temple wall paintings in the Joseon era.

Fig. 15.26 The GPR investigation revealed that woven branches provide the primary support of the wall paintings (paired images left: Brahma, roght: the Four devas, North)

15.3.2.6 Measurement of Surface Temperature

Infrared surface temperature measurements of the six wall paintings were taken in September and November 2018 (Figs. 15.27 and 15.28). In both months, maximum surface temperatures were detected on the inner side of the exhibition room (Indra), deceasing gradually towards the opposite side (Brahma Deva). Average and minimum values, except those recorded on the North Guardian King in September, were similar overall. On each of the murals, temperatures fell along a trajectory from left top to bottom right.

The different locations of the wall paintings in the temple are likely to affect their varying surface temperature distributions. In front of the Brahma Deva is the exhibition room, making it more influenced by outside air. In the temperature distribution

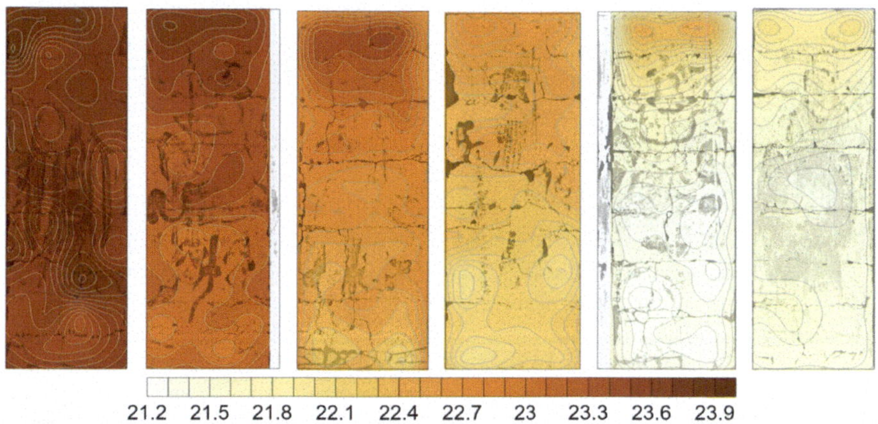

Fig. 15.27 Mapping of the surface temperatures (°C) of the wall paintings (September 2018)

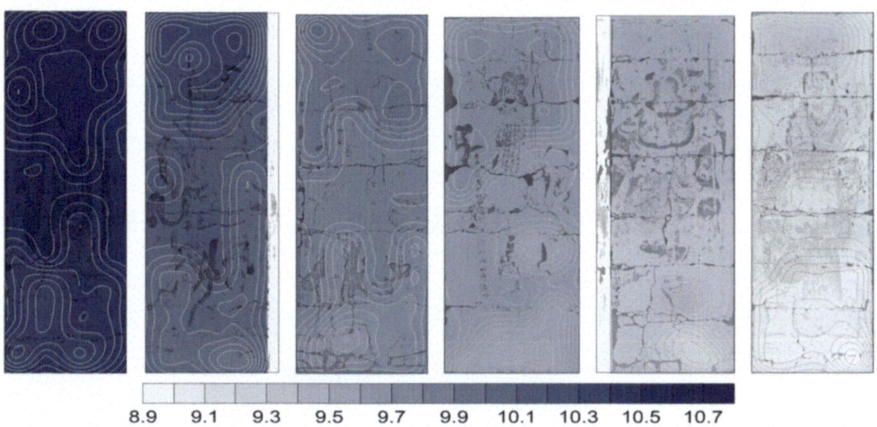

Fig. 15.28 Mapping of the surface temperatures (°C) of the wall paintings (November 2018)

of each wall painting, there is only a minor difference of around 1 °C on average, which is not considered to be a potential problem for the safety of the wall paintings. It is not clear if the lights installed at the top of the exhibition cases could become a problem, but as they are positioned very close to the paintings, this should be investigated and any required improvements made.

15.3.2.7 Measurement of Moisture Content

Moisture content measurements using a surface meter were taken of the six wall paintings in September and November 2018 (Figs. 15.29 and 15.30). The average

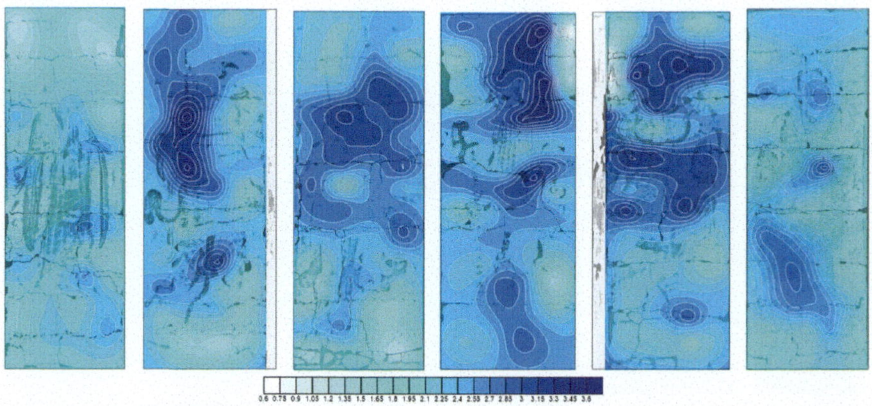

Fig. 15.29 Mapping of the moisture content of the wall paintings using a surface meter (September 2018)

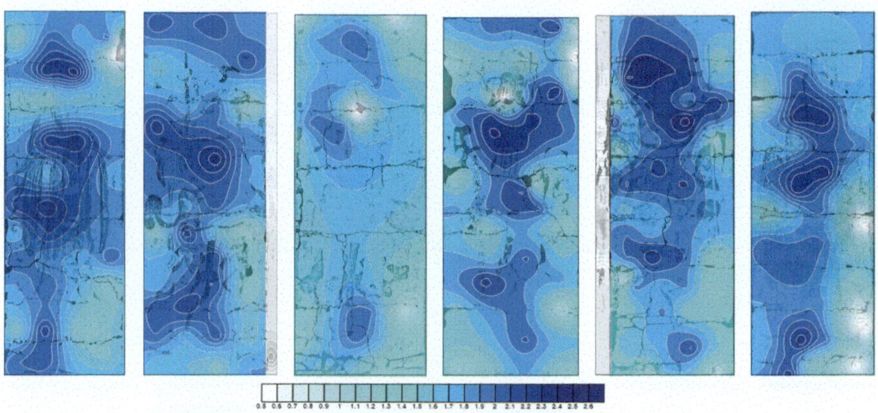

Fig. 15.30 Mapping of the moisture content of the wall paintings using a surface meter (November 2018)

moisture reading in September 2018 were 2.35, while those in November 2018 were 1.73. In September, the average indoor temperature was 21 °C and the average relative humidity 66%, indicating late summer conditions. In November, the average indoor temperature was 12.1 °C and the average relative humidity, 46%, indicating typical winter patterns. Although the difference are small, the results suggest that the moisture content of the wall paintings in September was higher than in November. A general comparison of the recorded moisture readings revealed that they were low in the vicinity of repairs and in areas where excessive synthetic resin was present. It seems that these non-original materials reduce the porosity of the wall paintings, thereby impeding moisture transfer in these localized areas. These results are also evident in the moisture content images. The location of each wall painting is not likely to impact moisture content. Nevertheless, given that the results collected in September were high on the wall paintings of the West, South, and East Guardian Kings, which are positioned at the center of the room, it would be useful to compare this data with other locations.

15.4 Discussion

15.4.1 The Making of the Murals

The scientific investigation and analysis allowed insights into how the wall paintings were made. The main findings can be summarized as follows:

(1) Although the wall painting stratigraphy is difficult to determine because of its concealment by the wooden frames, at least two plaster layers are present, applied over a support structure of woven branches, as revealed by the GPR investigation. A study of other temple wall paintings of the Joseon Dynasty found that they too are applied over supports of woven branches.

(2) The current frames are around 230 mm deep, and the wall paintings' thickness slightly less than this. It is known from conservation records that when the wall paintings were separated from their building and then framed, the support was reinforced with plaster. Therefore, it is difficult to know how much of the structure of the wall painting is original, and how much is added plaster. Based on past literature and data, however, a conjectural diagram of the support structure is shown in (Fig. 15.31).

(3) The lower plaster layer is earth-based, containing clays, silt and coarse sand. The analyzed characteristics and ratios are believed to be particular to the murals of the Josadang Shrine, and the plaster is considered to be an especially durable type compared with other temple murals in the Joseon era. Since this layer is almost completely obscured from view, it is difficult to determine its thickness and nature, except through some losses and damages in the upper plaster. The upper finishing plaster was made of weathered earth and sand, mixed with hemp fibre. Its thickness is about 11 ~ 12 mm, which is marginally thicker than other

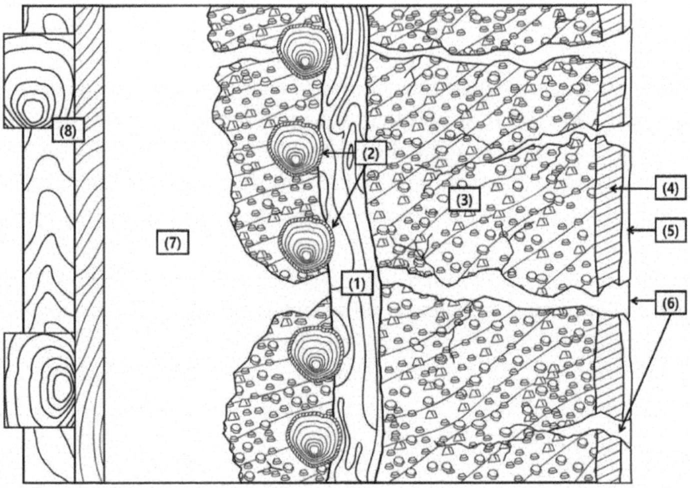

Fig. 15.31 Cross section of the wall painting structure: (1) vertical lath, (2) horizontal lath, (3) lower plaster layer (earth-based), (4) finishing plaster layer (earth-based), (5) painting layer, (6) gypsum plaster in crack, (7) gypsum backing reinforcement, (8) protection frame on the rear (board and square bar)

similar plasters in the Joseon Dynasty. The two plaster layers are characterized by earthen materials of different particle-size distributions.

(4) On the upper plaster, a pigmented preparatory layer was applied to receive the painting. The colours used are natural and artificial pigments. These features are comparable with other temple murals in the Joseon Dynasty.

(5) Results of the laser particle-size analysis indicate that the main particle-range in the plasters is in the silt and clay fractions, with silt particles being in the greater proportion. The presence of such a large proportion of fine particulate material may make the plasters particularly susceptible to cycles of moisture absorption and drying, which may eventually result in physical damage such as cracking and exfoliation due to internal stresses [5, 6].

15.4.2 Conservation Status of the Wall Paintings

The investigations determined that the main forms of damage and deterioration to the murals included paint-layer exfoliation, areas of internal delamination, and surface contamination. Various contaminants are present, including from remaining fixatives and other repair substances. Cracking and delamination are associated with the failure of the previous gypsum repairs, which have also contributed to salt deterioration; salt-related deterioration is also present at the bottom of the murals. As a priority, it is necessary to treat the problems of instability created by the cracking, exfoliation and internal delamination; and to address the excessive glossiness and staining left by the application of previous fixatives. Results of the ultrasonic testing and thermal imaging provided objective data about the instability problems that will help to formulate a conservation plan.

Since the hyperspectral imaging of the murals revealed delicate ink-line drawing and other detailing that had not been previously identified, this raised the prospect that similar painting could remain concealed beneath later crude restorations. This is, in fact, the case. We therefore produced a comprehensive mapping image showing the various layers of non-original and original paint based on a synthesis of the data obtained using various wavelengths.

The results of the moisture investigations indicated that ambient environmental conditions are not constant and that the murals are very sensitive to temperature/humidity fluctuations. The variable moisture readings recorded on the surface of the paintings are believed to be incurred by micro-air currents in the exhibition room, but in-depth and long-term assessment of the micro-environment is needed to confirm these conditions and their impacts.

Changes in the moisture content of the murals could bring about expansion and contraction in the support, and such changes may lead to physical damage. In particular, differential moisture absorption and release between the repair plasters and the original earthen materials may give rise to stresses that lead to cracking and failure. The presence of salts in the murals is a risk that could be worsened in the event of new moisture damage.

15.5 Summary Conclusions

This study presents the results of the first non-destructive investigation of the materials and techniques of the wall paintings of the Josadang Shrine of the Buseoksa Temple, and of their current condition, to provide useful information for their future conservation treatment. While a number of physical conditions have been identified that require remedial attention—such as paint-layer exfoliation, areas of internal delamination, and surface contamination—a key finding is that these measures can only be effectively implemented within a framework of appropriate diagnosis and evaluation. In particular, it is essential that suitable environmental conditions are established, monitored and maintained to ensure the future preservation of the murals.

References

1. Han, K.S.: Restoration and Conservation of the Buddhist Mural Paintings of Korea. Ph.D. thesis, University of Herzen State Pedagogical, Saint Petersburg (2003)
2. Lee, H.S: Consideration on current state of restoration of murals in temples in Korea. The Art History Research Institute of Korea (ed.), Art Hist. J. **38**, 351–369 (2012)
3. Baek, C.G.: Conservation of Murals of Josadang Shrine of Buseoksa Temple in Yeongpung-gun. The National Institute of Cultural Heritage (ed.), Conserv. Sci. Stud. **6**, 58–76 (1985)

4. Yeongju-si and Haisung Conservation Science for Cultural Heritage Co. Ltd.: Research on the scientific preservation status of mural paintings and report on the production of photographic edition (a display replacement) in Buseoksa, pp. 124–303 (2018)
5. Lee, H.S., Han, K.S., Kim, S.H.: Diagnosis and Evaluation of Conservation State of Mural Paintings in Payathonzu Temple in Bagan Heritage Site in Myanmar. The Korean Society of Conservation Science for Cultural Heritage (ed.), J. Conserv. Sci. **35**(5), 494–507 (2019).
6. Cho, H.J. Lee, T.J. Lee, H.S. Chung, Y.J.: Study on Physical Change in the Earthen Finish Layer of Tomb Murals Due to Drying. The National Institute of Cultural Heritage (ed.), MUNHWAJAE Korean J. Cult. Heritage Stud. **50**(4), 148–165 (2017)

Chapter 16
Conservation of Wall Paintings on Plaster in the Tumuli of Japan

Noriko Hayakawa

Abstract In Japan, cultural sites with wall paintings are not found as commonly as in Europe or other Asian countries. In this paper, two rare cases of the restoration of wall paintings on plaster are considered. These are the paintings from the tumuli at Takamatsuzuka and Kitora constructed and painted in the seventh–eighth century, which were discovered in the second half of the twentieth century. The wall paintings of Takamatsuzuka were found in 1972 and depict the four divine creatures, four groups of male and female figures, and constellations. Having decided to preserve the tumulus in situ, it was conserved on site for 40 years. Despite the many efforts to control biodeterioration, outbreaks of mold were not completely avoided. Following severe outbreaks of mold, it was eventually decided in 2001 to dismantle the site and conserve the wall paintings ex situ. After dismantling, the painted stones were maintained in a conservation facility. The objectives of the conservation programme were threefold: (a) to record the condition of the wall paintings; (b) to stabilize the weakened plaster; (c) to treat the problem of biodeterioration. In the case of the Kitora tumulus, the existence of wall paintings came to light in 1983, but their full extent was not confirmed until 2001; the tumulus was excavated in 2004. In its burial chamber were also depicted the four divine creatures and constellations, and drawings of the 12 signs of the zodiac. It was decided that the paintings should be removed from the burial chamber for ex situ conservation treatment, because some areas of plaster were already dangerously separated from the stone. The conservation measures were divided into three main areas, as follows: (a) maintenance of the wall paintings inside the burial chamber; (b) detaching and removing the wall paintings from the burial chamber; (c) reassembling the plaster fragments.

Keywords Takamatsuzuka · Kitora · Microorganisms · Biodeterioration · Conservation · Restoration · Detachment · Reassembly · Enzyme cleaning · UV irradiation

N. Hayakawa (✉)
Center for Conservation Science, Tokyo National Research Institute for Cultural Properties, 13-43 Ueno-Park, Taito, Tokyo 110-8713, Japan
e-mail: niriko@tobunken.go.jp

S. Aoki et al. (eds.), *Conservation and Painting Techniques of Wall Paintings on the Ancient Silk Road*, Cultural Heritage Science,
https://doi.org/10.1007/978-981-33-4161-6_16

16.1 Introduction

In Japan, cultural sites with wall paintings are not found as commonly as in Europe or other Asian countries. Indeed, wall paintings with multi-layered stratigraphies and varied material components can only be seen in the architecture of temples in the Asuka period, from the seventh century, and they are hardly seen at all until the Edo period, with the exception of the wall paintings in the Amida-do hall at Houkaiji. In contrast, there are many cases of wooden architecture, on which a plaster or kaolin ground-layer provides a substrate for polychromy. On the other hand, when it comes to tumuli rather than architecture, there are many sites remaining from the fourth and seventh centuries that preserve painted wall surfaces. However, almost all of these were directly painted onto stone. Paintings with additional preparatory layers have only been found at the Takamatsuzuka and Kitora tumuli, whose wall paintings are on plaster, and the Torazuka tumulus, where a kaolin-based ground-layer is used [1]. The Torazuka painting shows a geometric pattern, whereas, for the purpose of the present discussion, the figurative paintings at Takamatsuzuka and Kitora may be considered to preserve a more 'developed' type of wall painting. Other wall paintings in temple buildings of the Asuka period, such as at Houryuuji, Kamiyodohaiji, and Wakakusagaran, could also be considered within the scope of this conference, but, of these, only the small walls of Houryuuji have endured without significant injuries, such as damage by fire. Hence, this paper will focus on the conservation of tumulus wall paintings painted onto plaster.

16.2 Wall Paintings in Takamatsuzuka Tumulus

The Takamatsuzuka tumulus wall paintings were found in one of the tumuli in Japan from the end of the Kofun period, known as 'terminal-stage' tumuli, which are estimated to have been built between the late-seventh and early-eighth centuries. It is a two-tiered circular tumulus located in Asukamura, Takaichi-gun, Nara prefecture. Excavation was carried out in 1972, when the wall paintings were discovered. Until then, no figurative tumulus wall paintings executed on plaster had been found, so this exciting discovery was widely reported and attracted a high profile. The entire tumulus was designated a special historic site in 1973 and, following discussions about conservation approaches, it was decided to conserve the paintings in situ. Subsequently, in 1974, objects excavated from the tumulus were designated as important cultural property, and the wall paintings as national treasures. Thus, in recognition of its extremely important status, Takamatsuzuka received two kinds of designation, as a 'special historic site' and as a 'national treasure'. Following this, the conservation of the Takamatsuzuka tumulus was planned, which was implemented from 1976 onwards. Under Japan's laws on cultural property protection, the site was designated as tangible cultural property (fine arts and craft), and the restoration works carried out on this basis. Details of the conservation programme are described below.

16.2.1 The Wall Paintings

The Takamatsuzuka tumulus wall paintings depict the four divine creatures that are said to rule over their four respective directions (the portion corresponding to the Vermilion Bird has disappeared with the loss of the plaster). On the east and west walls, too, are painted four groups of male and female figures on each wall, each group having four figures (Fig. 16.1). These 16 figures wear continental-style clothing.

Constellations are painted on the ceiling, and, separately, images of the sun and moon are painted on the east and west walls, together with representations of mountains and clouds. The colors used include red, blue, green, yellow, gold, and silver. The plaster is several millimeters thick, which was applied very smoothly, although very little of this original surface still survives, due to deterioration and disturbance. Some parts of the paintings were also covered in mud at the time of their discovery.

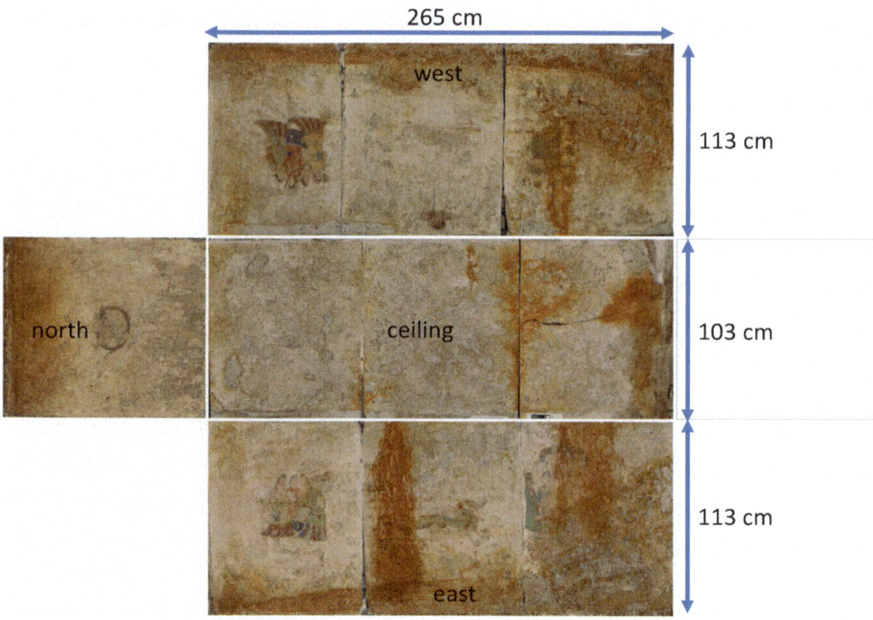

Fig. 16.1 Elevation and ceiling images of the Takamatsuzuka tumulus

16.2.2 Conservation Facilities at the Takamatsuzuka Tumulus

Having decided to preserve the tumulus in situ, a custom-made facility was constructed for the conservation process (Fig. 16.2). The principal objective was to suppress changes to the internal atmosphere and environment during the entry and exit of workers to and from the site. As such, the aim was not to regulate the atmosphere and environment within the stone burial chamber directly, but to divide the entrance into sealable partitions, so as to take in as little air from outside the stone burial chamber as possible, and avoid creating damaging environmental changes on the interior. Before workers entered, they were fumigated with paraformaldehyde; they then put on dustproof work-wear and sterilized themselves again. Although complete environmental control was not possible at the juncture between the stone burial chamber and the partitioned entry facility (known as 'shuffle sections'), the earth and sand surrounding the burial chamber remained as they were at the time of excavation. This may be considered a crucial conservation consideration.

The internal environment of the stone burial chamber differed considerably from the exterior. Changes in internal temperatures were delayed by about 3–4 months compared with those of the air outside. The internal humidity was constantly close to 100% RH, dropping to about 90% when workers entered the chamber. These parameters were maintained even after the conservation facilities were constructed and the influence of the exterior air was carefully excluded.

Fig. 16.2 Cross-section model of the conservation facility of Takanatsuzuka tumulus

16.2.3 In Situ Conservation Treatments

The main aims of the work inside the stone burial chamber were to carry out a condition assessment and to treat the wall surfaces. Conditions investigated included microbiological deterioration, and aspects of failure in the plaster and pigment materials. Where mold growth was found, this was removed with tools such as brushes and then sprayed with alcohol. Weak plaster was consolidated using Paraloid B-72® (Rohm & Haas). Paraloid B-72 is a copolymer of methyl acrylate and ethyl methacrylate with a molecular weight of 60,000–70,000, and has a few decades of usage in conservation practice [2]. It was recommended for the treatment of the Takamatsuzuka tumulus wall paintings when they were inspected by Professor Paolo Mora, of the Instituto Centrale del Restauro, in Italy. It was applied dissolved in trichloroethylene. Trichloroethylene has a relatively high density compared with other organic solvents, being 1.5 times higher than water. Since the plaster in the stone burial chamber contained a lot of water, these properties facilitated a higher penetration of the acrylic resin than if an alternative solvent had been used. Due of its adverse effects on the human body, including being a carcinogen, use of trichloroethylene has since been substituted by other solvents. But at the time, it was widely used, including in industry. The acrylic resin was applied to address problems such as separating and cracking plaster, using tools such as syringes and brushes. Reportedly, the bond strength was sometimes inadequate and the process of application had to be repeated.

16.2.4 Microbial Damages and Dismantling of the Stone Burial Chamber

Despite the efforts to control the influx of exterior air as much as possible, outbreaks of mold were not completely avoided. At the Takamatsuzuka tumulus, two severe outbreaks of mold occurred, in around 1978–1981 and after 2001. The aforementioned fumigation with paraformaldehyde was carried out after 1981, following the first outbreak in 1978–1981. Until then, although natural vaporization of paraformaldehyde was carried out on the inside of the stone burial chamber, there are records that vaporization stopped occurring due to increased condensation inside the chamber. Thereafter, fumigation treatments started to be made. In addition to the application of alcohol, the treatments taken against mold at this time made use of other substances, such as trichloroethylene and a solution of formalin and ethanol. However, mold outbreaks continued. The area around the White Tiger was particularly prone to mold outbreaks, and, while treatments were ongoing, it was observed that painted lines delineating the tiger seemed to have become less distinct. After the severe outbreak of 1978–1981, subsequent long-term treatment resulted in the suppression of the mold growth, reducing it to only slight effects. These positive results were confirmed by annual inspections over a thirty-year period.

However, in 2001, the second severe outbreak of mold was confirmed. The reasons for this were that collapses of sediment and sand had become intense in the "shuffle sections" mentioned above, and, while efforts were made to prevent this, soil was brought into the stone burial chamber that had not undergone treatments to suppress microorganisms. That year, inspection inside the stone burial chamber was postponed because mold was discovered in the shuffle sections. Half a year later, following checks to confirm that there was no further mold activity in the shuffle sections, an inspection was carried out inside the burial chamber. However, a mold outbreak had already occurred there. Despite treatment efforts, the mold could not be suppressed. Discussions were held about the constraints of achieving effective conservation at a site with high humidity and in which working conditions were extremely difficult, owing to the small size of the burial chamber, and it was eventually decided to dismantle the site and conserve the wall paintings ex situ.

16.2.5 Dismantling of the Stone Burial Chamber

The stone burial chamber comprised a total of 16 stone blocks: four on the ceiling, four on the floor, three on each of the east and west walls, and one each on the north and south walls. To dismantle these, the jointing mortar between each of the stones first had to be removed. Then, the surfaces of the paintings were protected with Rayon® facings, adhered with hydroxypropyl cellulose (HPC), which can be easily re-dissolved with water or ethanol. The facing was not applied to entire surfaces, but only to those parts preserving painting. For those areas where surface conditions were especially poor, protection was achieved by spraying with methylcellulose (MC). Following these measures, each stone was sandwiched into a jig, lifted with a crane, and removed.

After being dismantled, the stone surfaces were rubbed with alcohol as a first precaution. They were then transported to a conservation facility for their full treatment. The protective materials applied to the surface of the stones were immediately removed after their arrival at the facility, and it was confirmed that there had been no damage to the painting.

16.2.6 Conservation After Dismantling

The dismantled stones were all treated at the national Takamatsuzuka tumulus conservation facility. The stones were mounted horizontally, with the paintings face up. Conservators were assigned their own decicated stones to work on. Considering that mold generally breaks out at relative humidities greater than 60%, the environment of the treatment facility was maintained at a temperature of 21 °C and a relative humidity of 55% for the duration of the conservation programme (Figs. 16.3 and 16.4).

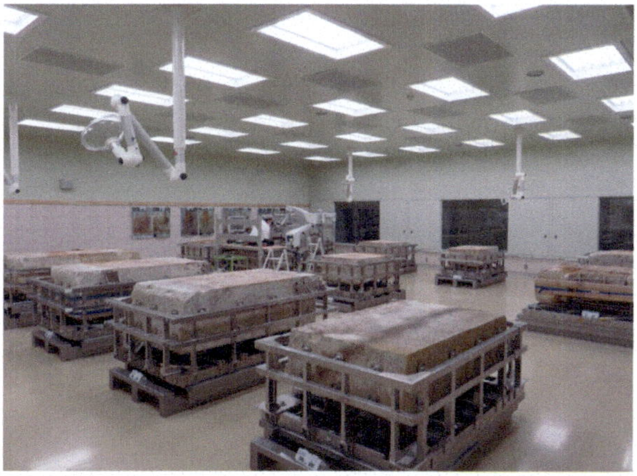

Fig. 16.3 Current conservation facility of Takamatsuzuka tumulus painings

Fig. 16.4 Conservation treatment

Although the principal reason for dismantling the tumulus was the occurrence of biodeteriortaion on the paintings, there had also been weakening of the plaster, which was identified as another major focus of treatment. The objectives of the conservation programme were therefore threefold: (a) to record the condition of the wall paintings after dismantling; (b) to stabilize the weakened plaster; and (c), to treat the problem of biodeterioration.

16.2.6.1 Condition Recording

A condition record of the paintings was achieved by visual examination and manual palpation. Conditions recorded included separation, cracking, and 'chalking' (weakening of the plaster). Locations of applied treatment materials and their effects were also recorded, as were areas of mold activity. Condition categories were organized as layered graphic documentation and archived.

16.2.6.2 Stabilization of the Plaster

The thin layer of plaster applied to the walls of the Takamatsuzuka tumulus was about 1.5 mm thick, but this was weakened and undermined in many areas. Separation of the plaster from the stone had occurred in many areas, and before the dismantling process, consolidation with the acrylic co-polymer, Paraloid B-72®, was carried out to address this problem. As a consequence of the constrained working conditions in the burial chamber, the resin was unevenly applied. In the treatment facility, excess resin was removed using a solvent, and areas where the plaster still existed in a weak condition were further consolidated. Initially, funori was used as a consolidation material, but because this inhibited subsequent treatments with enzymes, treatment reverted to the use of Paraloid B-72®. To avoid the health risks associated with the prior use of trichloroethylene in situ, a solution with organic solvents such as acetone or ethyl acetate was used instead. Even after re-treatment, some areas of the plaster remain inadequately consolidated, and this will be a focus of future work.

16.2.6.3 Removal of Microbial Stains

Staining by microorganisms was mainly caused by a gelatinous biofilm, which is made up of substances such as microbial remains and metabolized saccharides. Many parts of the wall paintings were colored by this biofilm. In addition to inhibiting appreciation of the paintings, it had also penetrated considerably into the plaster, contributing to its deterioration due to effects of dehydration and dimensional change. Thus, it was a priority to remove the biofilm in order to stabilize the plaster.

In the early stages of the project, trials were made using sodium hypochlorite. Although this method partially bleached the biofilm, making it colorless, it also made it difficult to disintegrate or remove. The colorless biofilm would contract after drying, increasing its bond with the plaster surface. It also permeated the upper layer of the plaster and paint, so trying to remove it physically using tools such as brushes caused damage. Consequently, the following two methods were developed to remove the biofilm, and these were employed in conjunction with the sodium hypochlorite treatment [3]:

Fig. 16.5 UV irradiation system

(1) Cleaning using ultraviolet radiation

A method for removing microbial staining was developed, based on the selective use of ultraviolet irradiation. The materials that make up biofilms are organic polymers. It was considered that ultraviolet radiation would partially disintegrate these, and that de-polymerized materials could then be safely removed without damaging the plaster. To avoid the risk of disintegrating original paint materials, this was only judged safe to use on unpainted areas of plaster. Although prior testing on pigmented areas in the wavelength 360–365 nm did not show any visible alterations, it was considered that undetected changes could still be a risk. Colorimetric tests carried out on the jointing mortar between the stones before and after irradiation with this wavelength did not show any changes, however.

Therefore, it was decided to use a wavelength of 365 nm for cleaning the unpainted plaster, utilizing two types of light source. One was capable of wide-beam irradiation of φ10 mm,; the other was a pen-light model capable of irradiation with an alterable beam of between φ3 mm and 0.3 mm. These were used for different applications (Fig. 16.5). Where the surface of the plaster was uneven, the pen-light model was able to reach into hollows and irregularities that could not be reached by the wider beam irradiation. Using both irradiation beams in combination allowed the biofilm to be removed efficiently.

(2) Cleaning using enzymes

Since ultraviolet radiation could not be used on the pigmented areas, cleaning methods using bacteriolytic enzymes were considered as a safe alternative. Enzymes are substances secreted by living organisms in order to metabolize materials that are sources of nutrients. They are a type of protein that functions according to specific chemical bonding, and they are therefore not prone to unexpected problems such as adverse side-reactions. For this reason, if an enzyme is chosen that only acts on a specifically intended chemical bond, it can be considered safe to use. If enough purification is carried out, the enzyme is capable of suppressing all extraneous reactions. To achieve this level of purification, an enzyme with high heat-resisting properties was chosen and used. The enzyme selected in this case was a mixture of β-1,

Fig. 16.6 Detail of the painting before (left) and after (right) enzyme treatment

3 glucanase and chitinase (Enzyme Mixture CTB-1 [EMCTB-01], Thermostable Enzyme Laboratory Co., Ltd.) [4].

Before application of the enzyme, the stability of the pigments was checked, to confirm that no colour alterations occurred; it was also confirmed that there was no effect on any of the applied conservation materials. The pigments that were checked were based on those that had been identified as having been used in the Takamatsuzuka tumulus wall paintings, as well as more generally in other Japanese paintings. The conservation materials examined were cellulose ether, paper, funori, starch and casein, as types of protein; and oil decomposition products such as palmitic acid. Having confirmed that the enzyme caused no adverse effects in these cases, it was decided that it would be used.

The enzyme was applied by dropping it onto the painting from a brush at an optimal temperature. The catabolites produced by the enzyme reaction were removed with a melamine sponge or the tip of a brush, and it was thereby possible to remove the biofilm in delicate parts of the painting (Fig. 16.6). Additional enzymes were similarly used at the same time. These were achromopeptidase (014-09661, Wako Pure Chemical Industries, Ltd.) and lysozyme (L6876, Sigma-Aldrich).

16.2.7 Present Condition of the Takamatsuzuka Tumulus Wall Paintings

At present, a large-scale condition assessment and further cleaning are being carried out on the Takamatsuzuka tumulus wall paintings, and it can be reported that the situation regarding microbial staining is much improved (Fig. 16.7). However, some parts of the plaster remain alarmingly undermined, and, in areas where the biofilm is still present, a complex combination of treatments is required: the weakened plaster needs to be consolidated, while allowing the biofilm to be swelled with water as a precursor to applying the enzyme treatment. Since it would be impossible to consolidate the plaster with the acrylic resin, Paraloid B-72®, which is hydrophobic, the

Fig. 16.7 Condition of the painting of the female figures from the Takamatsuzuka tumulus in 2007 (left) and in 2018 (right)

delicate task of using a water-soluble adhesive and an enzyme was reconsidered and subsequently implemented. This treatment ended in March 2020.

The conservation project responded to the highly urgent circumstances that accompanying the dismantling of the Takamatsuzuka tumulus. The principal aims were to stabilize the plaster and remove the microbial staining from the surface of the paintings. Those aims have almost been achieved. For the future, however, it is still necessary to decide on exhibition and storage options for the dismantled stone blocks, and, following this, to determine final treatment decisions. For wall paintings in other parts of the world, in situ conservation has become standard practice, whereby treatment decisions are directed at preserving the integrity of original materials in their context. However, the Takamatsuzuka tumulus wall paintings have been removed from their original location, and, for the time being, it has been decided that they will be conserved in their present artificial environment. The basic principle of this is to minimize the need for and the scope of any further interventions. For example, since the dismantled stones are stored horizontally rather than vertically, the need for additional consolidation treatment is minimized. Taking all these factors into account, including any retreatment that may be required in the future, the underlying principle is one of doing as much as necessary and as little as possible, in order to maintain the status quo of the last ten years.

16.3 Kitora Tumulus Wall Paintings

The Kitora wall paintings were found inside a tumulus in Asukamura, Takaichi-gun, in Nara prefecture. Like the Takamatsuzuka tumulus, it is a two-tiered construction from the 'terminal-stage' tumulus period, but its date is estimated to be a little earlier than that of the Takamatsuzuka tumulus. A survey carried out in 1983 using a fiberscope revealed the existence of a Black Tortoise on the north wall, and an additional survey in 1998 revealed an Azure Dragon on the east wall, a White Tiger on the west wall, and astronomical drawings on the ceiling. It was designated as a 'special historic site' in 2000. In 2001, another survey confirmed the presence of a Vermilion Bird on the south wall, and the 12 signs of the zodiac dispersed over all the walls. These scenes and subjects have no other parallels in Japan, with the exception of the Takamatsuzuka tumulus wall paintings, which also preserve astronomical drawings.

16.3.1 The Wall Paintings

As in the Takamatsuzuka tumulus, the paintings in the Kitora tumulus were executed on a thin layer of plaster, which varied in thickness from as much as 7–8 mm to less than 1 mm; the stone support was tuff. The paintings of the Black Tortoise, Azure Dragon, Vermilion Bird, and White Tiger, the four gods said to rule over the four directions, occupied their designated positions on each of the walls. Of the 12 signs of the zodiac, only the Rat, Ox, Tiger, Horse, Dog, and Boar could be initially confirmed. The presence of the Dragon, Serpent, and Monkey could not be confirmed because the plaster was covered with dirt; and the Hare, Ram, and Rooster have unfortunately been lost with the plaster in their respective locations. The astronomical chart painted on the ceiling has stars applied in gold leaf, while the constellations linking them together are in red, and their trajectories are painted with red lines.

16.3.2 Conservation and Restoration of Kitora Tumulus Wall Paintings

The excavation of the Kitora tumulus, which was begun in February 2004, clarified the extent and nature of the surviving wall painting, confirming the presence of large areas of separated plaster that were at risk of collapse (Fig. 16.8).

Since this situation demanded immediate attention, it was decided that the already separated plaster on which the paintings were executed would be removed from the burial chamber for ex situ conservation treatment. The conservation measures were divided into three main areas, as follows:

(a) Maintenance of the wall paintings inside the burial chamber;
(b) Detaching and removing the wall paintings from the burial chamber;

Fig. 16.8 The condition of the west wall in the Kitora tumulus in Kitora tumulus in 2004

(c) Reassembling the detached wall paintings [3].

These phases of work have been completed, and, in September 2016, the Center for the Presentation of the Kitora Tumulus Mural Paintings, "Shijinno Yakata", in Asukamura, Takaichi-gun in Nara prefecture, put on public display the west wall painting of the White Tiger, the south wall painting of the Vermilion Bird, and the astronomical ceiling. Since then, the paintings have been exhibited to the public four times a year, although the center is open year-round, displaying materials such as a full-scale model, details of the discoveries of the tumulus and the wall paintings, and high-definition images of the walls on monitors.

16.3.3 Maintenance of the Wall Paintings Inside the Burial Chamber

The decision to detach and transfer the wall paintings was made in September 2004. Subsequently, working in stages, it took over six years to completely remove them. Over this lengthy time-span, various provisions were made and treatments carried out to prevent the wall paintings from further deteriorating.

The inside of the stone burial chamber had a microclimate of close to 100% relative humidity. If this were to drop suddenly, the environmental shock to the wall paintings would lead to drying, shrinkage and separation. Thus, a conservation facility was constructed to maintain the RH at a high level. Continuous monitoring of temperature and humidity was carried out inside the stone burial chamber to ensure this.

Since there was also a concern that microorganisms could generate under these temperature and humidity conditions, careful precautions were taken against bringing in microorganisms, such as making it compulsory for people entering the room to pass through air showers, wear dustproof work-wear and masks, and disinfect themselves.

Fig. 16.9 Use of germicidal lamps in the Kitora tumulus in 2004 after removal of the plaster

These precautionary measures against microorganisms were carried out diligently, and were accompanied by surface inspections of the wall paintings.

Since only areas of plaster preserving painting were selectively removed, a considerable amount of unpainted plaster was left inside the stone burial chamber. As there was no need to consider effects on paint materials, biodeterioration risks were managed through periodic irradiation with germicidal lamps, to suppress the outbreak of microorganisms. Regular inspections were made, which demonstrated that this method was effective (Fig. 16.9).

16.3.4 Removing the Wall Paintings

The first areas of separated painted plaster that were removed at the start of the project were relatively thick and easy to cut, and a variety of spatulas and tools were used to achieve this. Later, however, the plaster was found to be in a variable condition, with some areas being very thin and brittle. A diamond wire-saw was developed to respond to these conditions, so that the plaster could be cut in a controlled and uninterrupted way, even through bonded areas, while also minimizing losses at the edges. A further

Fig. 16.10 Removing the painting of the Vermilion Bird from the Kitora tumulus with a diamond-wire saw

aim was to remove the plaster and its immediate substrate at a uniform thickness. The surface of the plaster was uneven, as was the stone support. Various trials were carried out, including on replica wall paintings. Eventually it became possible to remove the Vermilion Bird, which was painted onto a relatively thin plaster layer of about 3 mm thickness (Fig. 16.10).

The removal process also required facing materials to be applied to the paintings, to prevent damage during the operation. The choice of facing materials was governed by the needs of the location, such as those imposed by the curvature of the ceiling; by the condition and thickness of the plaster; and by the requirement that applied materials should be easily removable afterwards.

For the first facing layer, a rayon fiber sheet containing hydroxypropyl cellulose (HPC) was custom-made and applied by the conservators. The rayon fiber sheet was selected for its resistant to microorganisms. The sheet was impregnated with the adhesive HPC, dissolved in water or ethanol. After drying, it was attached in situ by swelling the embedded adhesive with an ethanol-aqueous solution. The facing materials were selected for their properties of strength and transparency, and for their ability to conform to the shape and condition of the paintings. The same materials and techniques were later put to use during the dismantling and recovery of the Takamatsuzuka tumulus wall paintings.

For the astronomical ceiling, a facing solution was required that could conform to the curved shape of the ceiling, and provide immediate strength and support. Following research of applications in the field of emergency public works, fiberglass reinforced plastic (FRP) was selected. FRP includes a photocoagulation type of epoxy resin, so after taking the shape of the surface to which it is applied, it is possible to stiffen the resin in position by irradiating it with light. A number of protective intervention layers were placed between the resin and the painting.

16.3.5 Reassembling the Detached Wall Paintings

The wall paintings were detached in 1143 pieces, which then had to be reassembled and reconstructed. After compensating for the vulnerability of the individual plaster pieces, they were reassembled, and the gaps between them were filled. The undersides of the pieces were consolidated for added protection, and they were reassembled onto strong support systems (Fig. 16.11). The consolidation and adhesion materials were selected after many trials. The materials and procedures used were as follows:

(a) Plaster consolidation: methylcellulose followed by Paraloid B-72®;
(b) Underside support: polyacrylate fiber sheet including Paraloid B-72® (layer 2);
(c) Height regulation: Rohacell® board (polymethacrylimide form), applied to compensate for the lack of uniformity in the plaster thickness;
(d) Space fillers: mixture of paper pulp and HPC;
(e) Supporting framework: honeycomb board and carbon board.

For the consolidation of the plaster, materials were used that could be easily reversed in future, if necessary. To support the underside of the plaster, a very strong fiber called polyarylate,—which is used in the production of bulletproof vests—was cast into a sheet with Paraloid B-72®. To fill in the spaces between the plaster pieces, materials were chosen that could be easily removed again, if necessary. These were also mixed in a ratio that would not shrink after drying. Rohacell® board is a polymethacrylimide-based foam material with high strength and lightweight properties. The supporting framework of carbon-fiber board combined with a metal honeycomb board was also chosen for its light-weight properties, and because it is able to keep its shape [5].

In 2009, the stated policy for preserving the Kitora tumulus wall paintings was conserving and exhibiting to the public at an appropriate institution outside the stone burial chamber for the time being. This approach opens the possibility for different

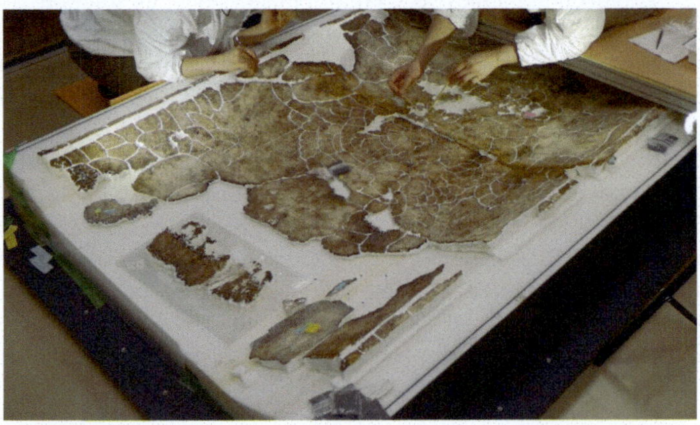

Fig. 16.11 Reassembling the detached ceiling fragments from the Kitora tumulus

circumstances in the future. It is therefore necessary that the wall paintings can be readily dismantled and reconstructed again, if necessary. This means that the materials used in their preservation need both to support and strengthen the paintings appropriately, and maintain the option for their reversal in the case of changed circumstances. These considerations informed the development and implementation of the subsequent conservation programme.

16.3.6 Kitora Tumulus Wall Paintings and Technology for the Restoration of Japanese Paintings

Current plans being considered for the retreatment of the Kitora tumulus wall paintings include the possibility to remove and stabilize the upper painter plaster for display. The tradition of Japanese paintings known as "honshi" ('original') and the soukou style of mounting may be cited as an influence for this approach. Works of painting or calligraphy in East Asia have been periodically remounted all the way back to antiquity, and re-restoration is always integral to the preservation process. An original painting on paper or silk is known as a "honshi". To conserve such items, appropriate mounting must be carried out, which allows for repeated remounting, when processes of deterioration make this necessary. This is known as soukou style.

Although specialist conservators work on wall paintings in other countries, this type of specialism is not common in Japan. Most of the conservation of the Kitora tumulus wall paintings was carried out by technicians from the Association for Conservation of National Treasures. This organization, which operates under the Japanese Agency for Cultural Affairs, is mainly concerned with the preservation of traditional artifacts, principally focusing on paintings and calligraphy that have been designated as special cultural property. In contrast to conservation practice elsewhere in the world, which emphasizes the need to preserve wall paintings in situ, approaches in Japan are characterized by the assumption that dismantling and restoration will be carried out repeatedly. For example, when mounting objects such as hanging scrolls, although they may appear to be made from a single flat sheet, in reality they are divided into many parts, which are designed to be easily dismantled and re-restored. This system of utilizing parts that can be deconstructed and reconstructed is adaptable to the treatment of wall paintings. In our approaches to conserving fragile Japanese wall paintings, methods and materials are selected—such as reversible adhesives and intervention layers—that are intended to provide the possibility of reversing treatments and allowing for re-restoration.

The conservation approaches adopted for the Kitora tumulus wall paintings were formulated on a basis of sound scientific knowledge, involving the collaboration of various experts. At the same time, the emphasis on allowing for the possibility of re-restoration reflects specific cultural approaches that are prevalent in Japan. These approaches, which favour the reconstruction of moveable parts, are distinct from conservation practices in the West, which privilege instead conservation in situ.

References

1. Sano, C., Hayakawa, Y., Miura, S.: Review of the scientific analysis on the painting materials of Takamatsuzuka Tumulus. Sci. Conserv. **48**, 119–131, NRICP (2009) (in Japanese)
2. Tsuga, S., Sonoda, N.: Comparative analysis of old and new "Paraloid B-72". In: Syntetic Materials and the Museum Object, pp. 183–194 (2003) (in Japanese)
3. Hayakawa, N., Kawanobe, W., Ogasawara, T., Yamamoto, N., Tsujimoto, T., Udagawa, S., Tateishi, T.: Conservation of the mural paintings in Takamatsuzuka tumulus as national treasure. In: The proceedings of the Japan Society for the Conservation of Cultural Property 2018 Congress, pp. 32–33 (2018) (in Japanese)
4. Sato, Y., Kigawa, R., Kida, K., Kawanobe, W., Hayakawa, N.: Selection and characterization of enzymes for removal of microbial contamination on the mural paintings of the Takamatsuzuka and Kitora Tumuli. Sci. Conserv. **57**, 11, NRICP (2018) (in Japanese)
5. Hayakawa, N., Kawanobe, W., Tsujimoto, T., Yamamoto, N., Kamei, R., Udagawa, S., Tateishi, T.: Conservation of the mural paintings in Kitora tumulus. In: The proceedings of the Japan Society for the Conservation of Cultural Property 2017 Congress, pp. 178–179 (2017) (in Japanse)

Chapter 17
The Protection and Utilization of Wall Paintings of Ancient Tumuli in Japan

Shigemasa Udagawa

Abstract This paper outlines events from the discovery of the wall paintings of the Takamatsuzuka and Kitora tumuli to their current conservation and management. It also describes the prospects for their conservation and utilization in the future. The brilliantly colored wall paintings of the Takamatsuzuka Tumulus were discovered in 1972. The discovery spurred a booming interest in ancient history and archaeology that continues to this day. During the process of conservation and management that followed the discovery, the wall paintings were found to be damaged. Therefore, the stone chamber was disassembled to allow for fundamental conservation work to be carried out and the wall paintings are now being preserved in a conservation facility. The Kitora Tumulus entered the spotlight in 1983 when wall paintings were discovered during an investigation of the stone chamber using fiberscopes. The wall paintings were removed from the stone chamber and, after 12 years of conservation work, they were put on display in 2016. The paper also describes the state of damage to decorated tumuli caused by large-scale disasters such as the Great East Japan Earthquake of 2011 or the Kumamoto earthquakes of 2018.

Keywords Takamatsuzuka Tumulus · Kitora Tumulus · Wall painting · Plaster · Conservation and utilization

17.1 Preface

From the middle of the third century to the end of the sixth century, many mounded tumuli were created in Japan. This period is called the "tumulus era" in Japan. Some of the tumuli representative of this era are those of the "Mozu-Furuichi Kofun Group: Mounded Tombs of Ancient Japan [1], which collectively were designated as a World Heritage Site on July 6, 2019 at the 43rd World Heritage Committee held in Baku, the capital of Azerbaijan.

S. Udagawa (✉)
Cultural Resources Utilization Division, Agency for Cultural Affairs,
3-2-2, Kasumigaseki, Chiyoda-Ku, Tokyo 100-8959, Japan
e-mail: udagawa@mext.go.jp

S. Aoki et al. (eds.), *Conservation and Painting Techniques of Wall Paintings on the Ancient Silk Road*, Cultural Heritage Science,
https://doi.org/10.1007/978-981-33-4161-6_17

Around 700 tumuli in Japan are decorated with reliefs, petroglyphs, coloring, or other forms of decoration. Those created in the fifth century were decorated with petroglyphs or reliefs, and their coloring was done with only red pigments. By about the sixth century, additional colours were used to paint the reliefs; and by the seventh century, reliefs were replaced by wall paintings. The Takamatsuzuka Tumulus and Kitora Tumulus were created between the end of the seventh century and the beginning of the eighth century. These tumuli are adorned with multicolored figures, including the Four Guardians, painted on plaster. They are completely different from what are considered the traditional kind of decorated tumuli, and they are termed *shuumatsuki-kofun* ('end-of-era' tumuli).

This paper describes how the wall paintings discovered in the Takamatsuzuka and Kitora tumuli have been conserved. It provides a general outline of the events from the discovery of the wall paintings to their preservation, and describes in particular how they have been protected and managed against the threats presented by mold growth. In addition, it includes a discussion of how the wall paintings will be utilized in the future.

17.2 The Takamatsuzuka Tumulus

17.2.1 Description and History

The Takamatsuzuka Tumulus is a small tumulus located in Takamatsu, Hirata, O-aza Asuka Village, Takaichi-gun, Nara Prefecture. The round tumulus consists of two tiers: the circle that forms the lower tier is 23 m in diameter, and the one that forms the upper tier is 18 m in diameter. There is a circular groove two meters wide running around the tumulus. It is believed that the tumulus was built some time between the end of the seventh century and the beginning of the eighth century. As mentioned above, the tumuli built during this period are called *shuumatsuki-kofun* ('end-of-era' tumuli). Many other tumuli survive around Takamatsuzuka Tumulus.

The first time the name "Takamatsuzuka Tumulus" appears in the literature is in *Records of Emperor and Empress Tombs* (1697). In this, the Takamatsuzuka Tumulus is listed as "Mount Takamatsu." It is described as having one pine tree atop a small mound, and was formerly assumed to be the Mausoleum of Emperor Monmu, based on historical records. Currently, a neighboring tumulus is considered to be the Mausoleum of Emperor Monmu, who reigned from the year 683 to 707. In the years following this account, the Takamatsuzuka Tumulus was gradually forgotten.

The Takamatsuzuka Tumulus regained attention in March 1972, when the Archaeological Institute of Kashihara and the Asukamura Board of Education conducted a study to excavate tumuli as part of a project to coincide with the publication of a volume of *Asuka Village Records* [2, 3]. A stone chamber with a horizontal exit made of tuff was discovered. Furthermore, by observing the interior through a hole dug by grave robbers some time in the Middle Ages (between approximately 1185 and 1568), it was found that the stone chamber was plastered and painted with vividly colored murals, the likes of which had not been seen previously in Japan. The discovery was hailed as a "find of the century". The tumulus was visited daily by countless people, spurring an interest in ancient history and archaeology that continues to this day.

17.2.2 The Murals and Artifacts

The fame of the Takamatsuzuka Tumulus derives from its brightly colored wall paintings, most notably the Asuka Bijin ('beautiful women') wall painting. Plaster was applied to the walls and ceiling of the stone chamber, which were then painted. On the walls were figures of men and women and the Four Taoist gods said to reign over the four points of the compass (the Azure Dragon of the east, the White Tiger of the west, and the Black Tortoise of the north, and, formerly, the Vermillion Bird of the south, which was stolen by grave robbers and is now missing). An asterism was depicted on the ceiling. The only other known example of an 'end-of-era' tumulus with wall paintings is the Kitora Tumulus, which, however, lacks paintings of groups of people. The Takamatsuzuka Tumulus wall paintings therefore have unparalleled value in the history of art, and were consequently designated as National Treasures in April 1974 (Fig. 17.1).

In the excavation of 1972, several original artifacts and pieces of a wooden coffin left behind by grave robbers were discovered both within the stone chamber and inside a hole dug by them. These artifacts were designated as 'Important Cultural Properties' in April 1974.

Fig. 17.1 Takamatsuzuka Tomb: view of the stone chamber from the south wall opening

17.2.3 *Past and Current Protection and Management of the Takamatsuzuka Tumulus*

The excavation carried out by the Archaeological Institute of Kashihara and the Asukamura Board of Education was completed in April 1972, and the study and management of the wall paintings were designated to the Agency for Cultural Affairs in view of the wall paintings' importance. The national government established the 'Research Group on Emergency Measures to Protect Takamatsuzuka Tumulus', comprising specialists in conservation science, archaeology, art history, civil engineering, and other fields. The Research Group conducted an investigation of the site and held its first discussions on what should be done to protect it. During this process, specialists from France (invited in 1972) and Italy (invited in 1973) were consulted [4], as they were then regarded as leaders in the field of wall painting conservation (Fig. 17.2). The Research Group then deliberated comprehensively on the following matters and issues:

- the historical and artistic significance of the wall paintings;
- the effects that a changed environment in the tumulus would have on the wall paintings;
- the technical difficulties involved in removing the wall paintings from the walls and ceiling; and
- the effect of microorganisms such as mold on the wall paintings.

It was then decided that the best course of action would be to conserve the wall paintings in situ (October, 1973). Since then, the Agency for Cultural Affairs has proceeded with protective measures based on this guiding principle. In 1976, a facility was established to provide a safe environment for conservation work, which was first directed at securing detaching plaster on the southern side of the stone chamber (the side with the grave passage). During this period, five conservators in charge

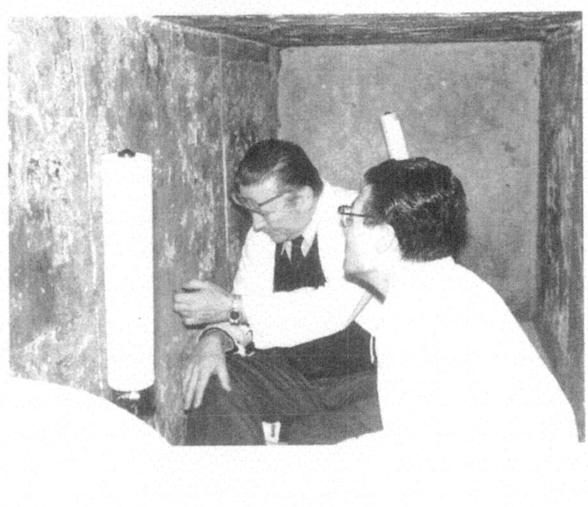

Fig. 17.2 Investigation being made by Professor Paolo Mora (Italy)

of the restoration process were sent to the International Centre for the Study of the Preservation and Restoration of Cultural Property (ICCROM) and the Instituto Centrale Per il Restauro in Italy to learn about wall painting conservation techniques.

From 1975 to 1989, with the advice of Paolo Mora of ICCROM, treatment was carried out to strengthen the wall paintings using the acrylic resin, Paraloid B72® dissolved in trichlorethylene (3–20%). This was introduced into weakened and detached plaster by specialists who had undergone wall painting conservation training in Italy. After completion of this treatment, the interior of the stone chamber was fumigated and disinfected by heating paraformaldehyde. This procedure was carried out twice, in May and November, when the temperature underground (at a depth of about 3 m, the level of the stone chamber) was fairly similar to the outside temperature. Each fumigation campaign was about three weeks in duration. Monitoring of the interior was kept to a minimum, once annually, in order to limit the number of personnel entering the chamber. During this period, mold growth was controlled and stable conditions seemed to have been established.

The connecting space between the conservation facility and the stone chamber was called the *toriaibu* (the 'joint-part'). In this section, the excavated earth of the tumulus mound was exposed. An earth collapse from the ceiling was discovered in around 1975, and from about 1989 more serious collapses occurred. In 2001, repairwork was carried out to remedy this situation. Partly as a result of this intervention, the delicate environmental balance that had helped preserve the wall paintings was changed, which encouraged the rapid growth of mold and other microorganisms, covering the murals with black mold. Another possible reason for the microbiological growth was a rise in the underground temperature. In about 1972, the highest underground temperature was around 16 °C, and the lowest, about 11 °C. From around the second half of 1980, however, the outside temperature began an upward trend, and, perhaps influenced by this, the temperature underground began to reside at around 20 °C or higher from 2000 onwards. Since the humidity inside the stone chamber was very close to 100% RH all the time, it can be said that the temperature rise created a warm and wet environment ideal for the growth of microorganisms.

With the damage due to microorganisms worsening, the Agency for Cultural Affairs established in 2003 the Investigative Commission for Emergency Measures to Protect the Murals of the National Treasure Takamatsuzuka Tumulus, comprised of multidisciplinary experts from various relevant fields. Studies were focused of the types and origins of the microorganisms, and measures were taken to reduce the bioactivity by disinfecting with ethanol. In the following year, the Agency established the Investigative Commission for Lasting Measures to Protect the Murals of the National Treasure Takamatsuzuka Tumulus [5] to consider comprehensive and more permanent measures as a follow-up to the previous Commission. At the 4th session of this second Commission, held in June 2005, it was decided that the wall paintings would be restored permanently by removing them with the entire stone chamber from the tumulus, through a process of dismantlement and reassembly. Although every effort had been made in situ to address the recurring damage due to mold and other microorganisms, the constraints imposed by the small space of the burial chamber and an internal humidity of almost 100% RH meant that other options had to be considered. Insects such as centipedes were able to enter and exit the stone chamber, bringing in mold; dead insects also became nutrition for new mold. A food chain centered on the mold had become established in the stone chamber and adjacent areas. The protective measures that were already in place had to be comprehensively reconsidered. It was then decided that it would be difficult to stop the deterioration of the murals in their subterranean environment, and that the best option would be to remove the stone chamber entirely from the tumulus.

Preparations by the Agency for Cultural Affairs for the removal of the stone chamber began in 2005. To prevent the spread of microorganisms and to keep the temperature inside the stone chamber at around 10 °C, a cooling pipe was placed

in the tumulus portion from September of that year. A building was also constructed on the tumulus to cover the pipe and to prevent rainwater or direct sunlight from entering. To prevent damage to the murals from vibrations or unexpected accidents, they were faced with thin non-woven rayon fabric and hydroxypropyl cellulose (HPC) or methylcellulose (MC), diluted to about 2% concentration in an aqueous solution. An exact replica of the stone chamber and tumulus was created, so that trials could be carried out to determine the best ways of cutting, removing and transporting the stones and paintings, and which tools and equipment to use.

Excavations were made in the tumulus to remove the stone chamber efficiently and safely, adopting the latest techniques and procedures (Fig. 17.3). Archaeological information revealed during this process, including evidence of historic damage, was carefully recorded. For example, numerous signs of earthquake damage were detected in the rammed earth tumulus. Construction evidence in the form of compressed layers of earth and the impressions left by ramming poles was also discovered [6].

In April 2007, work to remove the stone chamber began (Fig. 17.4). The chamber was taken apart stone by stone to minimize adverse impacts on the paintings. They

Fig. 17.3 The excavated stone chamber

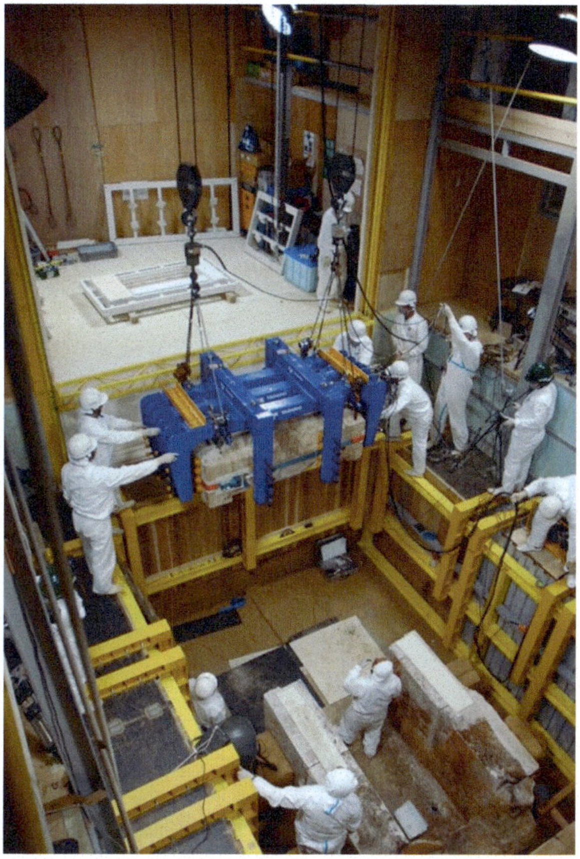

Fig. 17.4 Work to remove the ceiling stones

were then placed on a low-vibration vehicle and transported to a temporary conservation facility, set up within the Asuka Historical National Government Park nearby the tumulus. This was designed to museum specifications, with a temperature of 21 °C and a relative humidity of 55%.

Regarding the current conservation status of the wall paintings and stonework, discussions and investigations are ongoing to decide the best methods and materials to use, based on careful scientific assessment and a consideration of traditional restoration techniques. The scientific investigations aim to collect data on the current state of the site, including its causes of deterioration; and to increase understanding of the original materials and techniques used in the execution of the wall paintings. Detailed observation and non-destructive investigative techniques such as X-ray fluorescence analysis and spectroscopic analysis have helped to clarify the paint materials used [7–9]. Seven elements were detected: calcium (Ca), iron (Fe), copper (Cu), mercury (Hg), lead (Pb), gold (Au), and silver (Ag). Lead was detected in all parts of the underlying plaster, but a greater lead presence was detected within the stratigraphy of the painted areas. Therefore, it was deduced that a thin layer of lead-based paint was applied on top of the underlying plaster, to create the base on which the paintings were executed. The analysis also indicated that the blue pigment, natural ultramarine, is present, as well as the blue-green pigment, verdigris, the latter indicated by the presence of copper. Lead was detected in association with the green or blue pigments, but rather than being mixed with them, it is likely that this was from the underlying layer. In the scene showing a group of women on the west wall of the chamber, mercury was found in their *obi* (sashes) and on their lips. It was also found on the claws of the White Tiger and on the neck of the Azure Dragon, among other places. Calcium was detected in almost all the locations that were analyzed, indicating that this is a principal component of the plaster. Very little iron was detected, and this is likely to be from the soil or bedrock. Gold and silver were found by visual observation.

17.2.4 Considerations Regarding the Current Conservation of the Takamatsuzuka Tumulus Murals

With the long-term conservation of the Takamatsuzuka Tumulus wall paintings in mind, the Investigative Commission for the Protection and Utilization of Tumulus Murals [10] held multiple discussions on the following topics:

1. Conservation of the wall paintings
    (a) Damage due to microorganisms such as mold
    (b) Deterioration of the plaster
2. Conservation of stonework and the stone chamber
    (a) The durability of the stonework
    (b) The durability of the stone chamber in its entirety
3. The conservation of the tumulus
    (a) Maintenance
    (b) Conservation criteria should the stone chamber and the wall paintings be reinstated inside the tumulus
4. The display and utilization of the wall paintings
5. The physical relationship between the wall paintings and the tumulus.

In the discussions about returning the wall paintings to the tumulus, the main issues that arose were:

- the difficulties of creating an environment that would be resistant to mold growth;
- the risk that the original plaster would continue to disintegrate;
- the problems presented by the stonework, which had many cracks and was not very durable;
- the difficulties of trying to reconstruct the stone chamber with weak stone;

- potential problems associated with setting up a conservation facility on the tumulus (e.g., causing damage to the tumulus; adversely affecting the exterior appearance of the tumulus).

These issues were considered at the 15th session of the Investigative Commission for the Protection and Utilization of Tumulus Murals held in 2014, where it was decided that: "it would be preferable to return the wall paintings and stone chamber to the tumulus. With current scientific knowledge and technical standards, however, it will be difficult to create a safe environment for their return. Therefore, to conserve the wall paintings for the future, after they have been further treated, they will not be returned to the tumulus but protected and displayed as they are now." It was also emphasized that "efforts and discussions should continue to find a means of returning the wall paintings and stone chamber to the tumulus in the future."

17.2.5 The Future of the Takamatsuzuka Tumulus Murals

When the vivid wall paintings were discovered in March 1972, media reaction was one of surprise and excitement. Popular demand grew for the wall paintings to be viewed by the public. However, opening the small stone chamber (about 2.6 m in length, 1 m in width, and 1.10 m in height), with its very humid environment, would have created a microbiological problem, such as rapid mold growth. Since the plaster was already separated from the tuff support or existed in a powdery state, there was also the risk of further plaster damage and disintegration occurring. Therefore, to preserve the wall paintings, the chamber was sealed immediately.

To address the popular demand for the wall paintings to be seen, it was decided that they would be copied. Seven Japanese artists carried out this task under the direction of Seison Maeda, a famous traditional painter. They entered the stone chamber with color photographs of the wall paintings and color reference cards, to facilitate accurate colour matching. Then, they painted the colors onto a sheet of handmade *echizen-washi* (a type of Japanese paper) on which the full-sized wall paintings had been printed with monochrome collotype. This method of copying is a method unique to Japan. It was developed when the wall paintings of the Kondo (the Golden Hall) of Horyu-ji Temple were copied in the 1940s. To reproduce the nature of the plaster surface on the printed *washi* paper, a base layer of *gofun* (a white powder made from pulverized oyster or clam shells) was applied and the colors painted on top using a 'pointillism' technique. The copying process took about seven months to compete, from the fall of 1973 to about March 1974. The copies were then displayed at the exhibition "The Takamatsuzuka Tumulus Murals Replicated", which started in the Nara National Museum before travelling to six other museums in Japan. At the same time, Shunichi Terada, a Japanese painter trained in the Western tradition, was working on recreating the murals on top of a layer of plaster. The Takamatsuzuka Mural Hall that was established near to the site shows the history of Takamatsuzuka in its entirety, including copies of the wall paintings made when they were first discovered, as well as the restored replicas, a replica of the stone chamber, and the artifacts found.

In 1963, the prehistoric caves at Lascaux, in France, were closed due to mold growth and other microorganisms that were causing damage to their paintings. In 1983, 20 years later, a replica of the caves was created to address the viewing demands of visitors. Although the Takamatsuzuka Tumulus differs from the caves of Lascaux in its size and display potential, this approach should be considered as one way of utilizing the site and making it available to the public.

Currently, the wall paintings are being restored in a temporary facility located in Asuka Historical National Government Park. To allow the public to understand the work that is being done, the operation room is displayed to the public about four times a year (Fig. 17.5). The conservation process has taken about 12 years and is scheduled for completion in the spring of 2020. The area around the Takamatsuzuka Tumulus has been designated and maintained as national park, and the tumulus itself is still maintained after the removal of the stone chamber. When considering the long-term preservation of the wall paintings, returning them to the tumulus would now be extremely difficult. Current plans are to utilize them and make them available to the public in a new facility.

Fig. 17.5 Restoration laboratory open to the public

17.3 The Kitora Tumulus

17.3.1 Description and History

Kitora Tumulus is a small tumulus located in Aza Ueyama, Oaza Abeyama, Asuka Village, Takaichi-gun, Nara Prefecture. This round tumulus consists of two tiers: the circle that forms the lower tier is 14 m in diameter, the one that forms the upper tier is 9.5 m in diameter. The tumulus was made using rammed earth, just like at the Takamatsuzuka Tumulus. Based on the unearthed artifacts and the structure of the stone chamber, the tumulus is thought to have been made between the end of the seventh century and the beginning of the eighth century, around the same time as the Takamatsuzuka Tumulus.

The wall paintings of the Kitora Tumulus were discovered in November 1983, when the Society to Appreciate Asuka, the Ancient Imperial Capital of Japan, enlisted the support of the Japan Broadcasting Corporation (NHK), and fiberscopes were used to capture videos of the inside of the stone chamber. It was then discovered that the interior had a roof-shaped ceiling, that the entire inner surface was covered with plaster, and that there was a mural of what appeared to be the Black Tortoise on the north wall. This investigation made the Kitora Tumulus only the second tumulus, after Takamatsuzuka, to be found with figurative wall paintings [11, 12]. After the excavation and investigation of the interior of the stone chamber, the Kitora Tumulus was designated as a 'Special Historic Site' in November 2000.

17.3.2 The Murals and Artifacts

The inner surface of the stone chamber was covered with plaster, and paintings survived on the four walls and the ceiling. They depicted, from the east wall counter-clockwise, the Four Guardians: the Azure Dragon, the Vermillion Bird (Fig. 17.6), the White Tiger, and the Black Tortoise. Under each of these Four Guardians were paintings of three of the 12 Earthly Branches. Each of the Earthly Branches was depicted with the head of its representative animal and the body of a human, wearing a wide-collared kimono wrapped to show the right side on top, and holding weapons and other objects in their right hand. The wall paintings were arranged clockwise around the room, with the Rat in the middle of the three animals on the north wall. The compositions were transferred onto the walls using a method involving *nenshi* (paper similar to carbon paper). The transferred designs show evidence of both incised lines and red outline drawing. A constellation of stars was depicted on the ceiling, and the east and west slopes showed the sun and moon. This subject matter is based on the theory of the cosmic dual forces and the five elements, and in the stone chamber the scenes were deliberately arranged in a logical way to emphasize their relationships with each other.

Fig. 17.6 Kitora tumulus: south wall painting showing the red phoenix

The interior had been ransacked by grave robbers during the Kamakura period (latter half of the twelfth century to the first half of the fourteenth century) and slivers of the lacquered wooden sarcophagus were piled on the floor. The excavation of the stone chamber revealed a number of items, including a metal piece with gold and silver bands that was used in Japan to keep a sword at one's side, silverwork, amber beads, and other artifacts. The gold-and-silver-banded piece had a steel base with silver-plating on the inside of the ring, and gold damascening on the outside. This and the decorative metal parts of the silver scabbard are academically important as examples of the very few swords left from the end of the Asuka period (sixth century to the beginning of the eighth century). The murals and artifacts were therefore deemed to be important items from the end of the Asuka period and were designated as 'Important Cultural Properties' in October 2018.

17.3.3 The Past and Current Protection and Management of the Kitora Tumulus

An investigation of the interior of the stone chamber carried out in 2004 showed that the plaster, which is only about 3–5 mm thick, was separated by about 3 cm from its stone support in the areas of the Azure Dragon on the east wall and the White Tiger on the west wall, creating a very dangerous situation (Fig. 17.7). Furthermore, tree-roots had grown behind the separated plaster, and many pieces of plaster had already fallen down. It was decided that re-attaching the plaster to the stone and conserving the murals in situ was impossible. This led the 'Investigative Research Commission for the Protection and Utilization of the Special Historic Site Kitora Tumulus [10] to decide at its 7th session in September 2004 that all of the wall paintings would be removed.

Fig. 17.7 Kitora tumulus: east wall painting of the blue dragon

Before removing the wall paintings, their condition was studied and recorded. Then the surface of the paintings was protected from unforeseen accidents or vibrations in a process known as *omoteuchi*. For this, a thin rayon sheet and hydroxypropyl cellulose (HPC) or methylcellulose (MC) diluted to a concentration of about 2% in an aqueous solution was used. Then, special tools were employed to carefully remove the murals (Fig. 17.8). For the painting of the Vermillion Bird on the south wall, and some other wall paintings where the thin plaster was well bonded to the stone, a diamond wire-saw was used for the removal process (Fig. 17.9). Among these

Fig. 17.8 Kitora tumulus: the astronomical ceiling during its detachment

Fig. 17.9 Kitora tumulus: removal of wall painting using a diamond wire-saw

was the wall painting of the Horse from the 12 Earthly Branches, and other scenes where the pigment layer had been transferred to dried mud on the surface of the wall painting, making them extremely difficult to protect and manage.

The wall paintings of the Four Guardians, the 12 Earthly Branches, the constellations, and other scenes were removed by November 2008. Work to remove areas surrounding the wall paintings then proceeded, and the entire scheme was completely removed by November 2010. Since the wall paintings were removed in 1143 pieces, work to put these back together again then began. There were concerns about the durability of the plaster fragments during this process, and so methylcellulose (MC) in an aqueous solution was dripped onto their reverse sides, and cast sheets made using polyacrylate fiber in Paraloid B-72® solution were used to strengthen and protect them. The original locations of these fragments were identified based on detailed diagrams or photographs of the interior of the stone chamber taken before any work was done. Gaps between the plaster fragments were filled with powdered filter paper mixed with hydroxypropyl cellulose (HPC) in aqueous solution, and the fragments were fixed onto a carbon-fibre honeycomb board using silicone. The reconstruction of the four walls and the ceiling was completed in 2016 [13].

Alongside treatment of the wall paintings, a scientific investigation of the materials and techniques of their execution was carried out. Investigations were based on non-destructive methods, such as X-ray fluorescence analysis and spectroscopic analysis. The analysis detected mercury on the tongue of the Azure Dragon on the east wall, indicating that cinnabar was used as a pigment here. Copper was identified in green and blue areas of painting. In a marked difference from the wall paintings in the Takamatsuzuka Tumulus, lead was not found on the underlying plaster or in any areas of painting.

In 2008, the ‘Investigative Commission on the Protection and Utilization of Tumulus Murals’ was established and discussions were held on the status of the Kitora Tumulus stone chamber and the work to remove its wall paintings; on the temporary protective measures carried out at the Takamatsuzuka Tumulus; on the investigations of the causes of deterioration of the Takamatsuzuka wall paintings; and on other topics. The Investigative Commission also deliberated on how to conserve the wall paintings in the future as one of the pressing issues for the utilization of the Kitora Tumulus. At the 6th session of the Investigative Commission held in August 2009, it was decided that the wall paintings would "be protected, managed, and displayed to the public in an adequate facility outside the stone chamber for the time being." At the 2nd session of another Investigative Commission for the Protection and Utilization of Tumulus Murals held in May 2010, it was decided that the Kitora wall paintings would be eventually protected and managed in a facility located in the Asuka Historical National Government Park Kitora Tumulus Area, and that the detached wall paintings would be conserved for the time being in the Kitora Tumulus Mural Experiential Museum (Shijin no Yakata), located within the Park.

At the 9th session of the Investigative Commission on the Protection and Utilization of Tumulus Murals held in March 2012, 'Basic Guidelines on the Maintenance of Kitora Tumulus' were decided. Regarding the tumulus itself, the original structure of the mound would be conserved, including archaeological evidence revealed during excavation [10]. The conservation of the wall paintings was considered an utmost priority, and the facility to protect and manage them would be one that could display and utilize them as much as possible. When the Asuka Historical National Government Park Kitora Tumulus Area was opened to the public in September 2016, the custom-made conservation facilities were also opened (Fig. 17.10). The facilities have numerous rooms: a wall painting protection room, where the Kitora Tumulus wall paintings are protected and managed; an artifact protection room, where burial accessories unearthed from the tumulus are protected and managed; and a display room with an observation window, which enables visitors to view the wall paintings during the process of their conservation.

Fig. 17.10 Kitora tumulus: storage room for the detached wall paintings

17.3.4 The Protection and Utilization of the Kitora Tumulus

The wall paintings were removed from the stone chamber over a period of six years beginning in 2004. Some 1143 pieces were taken to the temporary conservation facility for the wall paintings of the Takamatsuzuka Tumulus. There, the fragments were put back together and restored over the next six years, until 2016. Even during this process, some of the wall paintings were displayed at the Asuka Historical Museum and other facilities. The wall paintings are currently on display on the first floor of the Kitora Tumulus Wall Painting Experiential Museum (Shijin no Yakata) four times a year for a period of about a month each time. During these occasions, visitors are able to enter the display room and observe the wall paintings inside the wall painting protection room through a glass window.

In the basement is a full-scale model of the stone chamber and other items, as well as a facility to view high-definition videos of the excavations and the restoration works that were carried out at the Kitora Tumulus. A signpost notifies visitors that it is a 'Special Historic Site'. There is also an explanatory display board as well as a topographical model of the area surrounding the tumulus mound. Nearby, there is a metal plate with the wall paintings of the Four Guardians, the 12 Earthly Branches, and the Constellations embossed on it. This is for visitors who wish to create a rubbing of the wall paintings. With the reconstruction of the terrain surrounding the Kitora Tumulus now completed, a major step has been taken in preparing an appropriate environment for the long-term protection and utilization of the site, its paintings and artifacts, and in presenting these components as a single entity.

17.4 The Protection and Utilization of Other Painted Tumuli

Japan has around 700 tumuli, probably created from the fourth century to the first half of the seventh century, in which the stone chamber is decorated with incised designs, petroglyphs, coloring, or some other technique. These tumuli are usually found in two geographic locations: in the northern part of the Kyushu region, and in the area between the northern part of the Kanto region and the southern part of the Tohoku region. They are different in nature from the tumulus at Takamatsuzuka or Kitora, which were created between the latter half of the seventh century and the first half of the eighth century, and which also have fine figurative paintings, including of the Four Guardians. Nevertheless, all the tumuli contain subterranean stone chambers with decoration, and they share problems associated with the conservation and display of both the sites and their artifacts. Defining the best environmental parameters to control mold growth and other biodeterioration are particularly pressing issues.

The experience of having struggled to control the environmental conditions of the Takamatsuzuka Tumulus to prevent the growth of mold and other microorganisms greatly influenced the design and construction of new conservation facilities at other sites. In many tumuli, a conservation facility was constructed with an antechamber and viewing room abutting the stone chamber, making it possible to view the wall paintings twice a year, in spring and fall, when the temperature in the ground and the outside temperature are fairly equal. In this regard, the conservation endeavours represented by the ongoing works at the Takamatsuzuka and Kitora tumuli have been highly significant.

Japan is a country prone to natural disasters such as earthquakes and floods. In the Great East Japan Earthquake that occurred on March 11, 2001, many cultural properties including decorated tumuli were damaged. Faced with such a situation, the Agency for Cultural Affairs decided to conduct a comprehensive survey of all sites throughout the country designated as 'Historic Sites' by the national government—and not only those in the disaster-struck regions—to better understand how decorated tumuli in all parts of Japan were being protected and managed. Furthermore, to consider the survey's findings, the 'Working Group on Decorated Tumuli' was created under the 'Investigative Commission for the Protection and Utilization of Tumulus Murals' in July 2012. The Working Group also visited sites and held hearings to collect facts from those in charge of managing the decorated tumuli. The Working Group's findings were compiled in a report in March 2016. In this, the focus of deliberation was on managing the environment of the decorated tumuli.

In Yamamoto Town in Miyagi Prefecture, a preliminary survey carried out as part of a project to promote group activities related to disaster mitigation after the tsunami, revealed the existence of a decorated passage-grave containing petroglyph wall paintings in the ruins of a site at Kassenhara. For more than six months, the Working Group held discussions on the ethical and technical aspects of protecting the wall paintings. In 2016, they were successfully removed and subsequently restored. In November 2018, they were put on display and utilized at a local exhibition hall.

In the Kumamoto earthquakes that occurred on 14 and 16 April 2016 (Magnitude 6.5 and 7.0 on the JMA scale, respectively), the media reported extensive damage to cultural properties. In particular, there were daily reports on sites such as Kumamoto Castle or the Aso Shrine, attracting public attention both from within Japan and abroad.

In Kumamoto Prefecture, many of the decorated tumuli were damaged by the earthquakes. Reported damages included collapses, intrusions or shifts in the stones of the stone chamber, and soil flowing into the stone chamber. At Idera Tumulus (Kashima Town), the already sloping side wall of the passageway was worsened, stones were dislodged and some fell in the burial chamber, and soil-runs occurred. At Kamao Tumulus (Kumamoto City), the stones in the ceiling of the passageway collapsed, the side wall bulged outwards and the stones there also collapsed, allowing soil into the passageway; in the burial chamber, the stacked stones bulged or collapsed, and cracks widened. For these two tumuli, committees to repair and maintain them have been organized and deliberations have begun on how best to investigate and determine their state of damage. At other decorated tumuli, the damage was so great that work to repair them has not even begun. When considering how to secure the safety of these decorated tumuli, we are faced with a mountain of issues. Takamatsuzuka Tumulus also has numerous cracks in the rammed earth that are thought to have been a result of past earthquakes.

As I have shown in this paper, wide-ranging knowledge has been accumulated through the long-term projects dedicated to the conservation and utilization of the Takamatsuzuka and Kitora tumuli. It is necessary to re-process and make use of this knowledge in future endeavours to conserve and utilize other decorated tumuli in Japan.

References

1. United Nations Educational, Scientific and Cultural Organization: Convention concerning the protection of the world cultural and Natural Heritage World Heritage Committee, Forty-third session, Paris, pp. 249–251, 23 July 2019
2. Archaeological Institute of Kashihara, Nara Prefecture (ed.): Hekiga-kofun: Takamatsuzuka Hekiga-kofun Chuukan Houkokusho (Tumuli with Murals: The interim report on Takamatsuzuka Tumulus, a Tumulus with Murals) (1972) (in Japanese)
3. Takamatsuzuka Tumulus Murals Academic Research Committee (ed.): Takamatsuzuka-kofun Hekiga (Takamatsuzuka Tumulus Murals) (1973) (in Japanese)
4. The Agency for Cultural Affairs (ed.): Kokuhou Takamatsuzuka-kofun Hekiga –Hozon to Shuuri—(The murals of the national treasure Takamatsuzuka Tumulus—preservation and restoration—) Dai-ichi Hoki Co., Ltd., p. 43 (1987) (in Japanese)
5. The Agency for Cultural Affairs Website: Investigative commission for lasting measures to protect the Murals of the National Treasure Takamatsuzuka Tumulus. http://www.bunka.go.jp/seisaku/bunkashingikai/kondankaito/takamatsu_kitora/takamatsukento/index.html. Accessed 10 Dec 2019
6. The Agency for Cultural Affairs (ed.): Tokubetsu-shiseki Takamatsuzuka-kofun Hakkutsuchousa Houkoku: Takamatsuzuka-kofun Sekishitsu Kaitai-jigyou ni Tomonau Hakkutsuchousa (Report on the excavations of the special historic site Takamatsuzuka Tumulus: excavations at the occasion of the disassembly of the stone chamber of Takamatsuzuka Tumulus). Douseisha Publishing Co., Ltd. (September 2017) (in Japanese)
7. Hayakawa, Y., Sano, C., Miura, S.: Handi-keikou Ekkususen-bunsekisouchi ni yoru Takamatsuzuka-kofun Hekiga no Ganryo-chousa (Analysis of the pigments used in the murals of Takamatsuzuka Tumulus by a Handheld x-ray fluorescence spectrometer). Hozonkagaku (Science for Conservation) **43**, 63–77 (2004) (in Japanese)
8. Sano, C., Hayakawa, Y., Miura, S.: Kokuhou Takamatsuzuka-kofun Hekiga no Zairyo no Hensen (Review of the scientific analysis on the painting materials of Takamatsuzuka Tumulus. Hozonkagaku (Sci. Conserv.) **48**, 119–131 (2009) (in Japanese)
9. Hayakawa, Y., Kozuma, Y.: Bunkazai-bunseki: Bunseki-kagaku Jitsugi-shiriizu Ouyou-bunsekihen 7 (Cultural Property Analysis: Analytical Chemistry Practical Series Application Analysis 7), pp. 48–58 (2018) (in Japanese)
10. The Agency for Cultural Affairs Website: Investigative commission for the protection and utilization of tumulus murals. http://www.bunka.go.jp/seisaku/bunkashingikai/kondankaito/takamatsu_kitora/hekigahozon_kentokai/01/index.html. Accessed 10 Dec 2019
11. Academic Research Group for Kitora Tumulus (ed.): Kitora-kofun Gakujutsu Chousa Houkokusho (Report on the academic study of Kitora Tumulus). Asukamura Board of Education (1999) (in Japanese)
12. Nara National Research Institute for Cultural Properties (ed.): Tokubetsu-shiseki Kitora-kofun Hakkutsu-chousa Houkokusho (Report on the excavations of the special historic site Kitora

Tumulus). The Agency for Cultural Affairs, Nara National Research Institute for Cultural Properties, Archaeological Institute of Kashihara, Nara Prefecture, and Asukamura Board of Education (2008) (in Japanese)
13. Hayakawa, N.: Kitora-kofun Hekiga no Hozon Shuuri Houkoku (Report on the protection and restoration of the murals of Kitora Tumulus). Gekkan Bunkazai (Cult. Prop. Mon.) **649**, 7–10, Dai-ichi Hoki Co., Ltd. (2017) (in Japanese)

The manufacturer's authorised representative in the EU is Springer Nature Customer Service Centre GmbH, Europaplatz 3, 69115 Heidelberg, Germany. If you have any concerns regarding our products, please contact ProductSafety@springernature.com

Printed and bound by CPI Group (UK) Ltd, Croydon, CR0 4YY
30/09/2026
02225240-0007